CELL BIOLOGY

CELL BIOLOGY
A Short Course

FOURTH EDITION

Stephen Bolsover
Andrea Townsend-Nicholson
Greg FitzHarris
Elizabeth Shephard
Jeremy Hyams
Sandip Patel

WILEY Blackwell

This fourth edition first published 2022
© 2022 John Wiley & Sons Ltd

Edition History
John Wiley & Sons, Inc. (From Genes to Cells, 1e, 1997) (2e, 2004) (3e, 2011)

Registered Offices
John Wiley & Sons, Inc., 111 River Street, Hoboken, NJ 07030, USA

John Wiley & Sons Ltd, The Atrium, Southern Gate, Chichester, West Sussex, PO19 8SQ, UK

Editorial Office
9600 Garsington Road, Oxford, OX4 2DQ, UK

For details of our global editorial offices, customer services, and more information about Wiley products visit us at www.wiley.com.

Wiley also publishes its books in a variety of electronic formats and by print-on-demand. Some content that appears in standard print versions of this book may not be available in other formats.

Library of Congress Cataloging-in-Publication Data applied for

Names: Bolsover, Stephen, 1954- author.
Title: Cell biology: a short course / Stephen Bolsover, Andrea
   Townsend-Nicholson, Greg FitzHarris, Elizabeth Shephard, Jeremy Hyams,
   Sandip Patel.
Description: Fourth edition. | Hoboken, NJ: Wiley-Blackwell, 2022. |
   Includes bibliographical references and index.
Identifiers: LCCN 2021049318 (print) | LCCN 2021049319 (ebook) | ISBN
   9781119757764 (hardback) | ISBN 9781119757771 (adobe pdf) | ISBN
   9781119757788 (epub)
Subjects: LCSH: Cytology.
Classification: LCC QH581.2 .B65 2022 (print) | LCC QH581.2 (ebook) | DDC
   571.6–dc23/eng/20211007
LC record available at https://lccn.loc.gov/2021049318
LC ebook record available at https://lccn.loc.gov/2021049319

Cover Design: Wiley
Cover Image: © Hybrid Medical Animation/Science Source

Set in 10/12pts TimesLTStd by Straive, Pondicherry, India

SKYF39029D0-723A-4FE8-919B-31E1EE84D449_020722

CONTENTS IN BRIEF

CONTENTS

PREFACE

Our aim in previous editions of *Cell Biology, A Short Course* was to cover a wide area of cell biology in a form especially suitable for first-year undergraduates, keeping the book to a manageable size so that neither the content, the cost, nor the weight was too daunting for the student. For the fourth edition we have thoroughly revised the content to concentrate on the core of cell biology: the structure of the cell, how it is made, and how cells and the parts of cells interact and move.

We begin (Section 1, Chapters 1 and 2) by describing the components of the cell as seen in the microscope. We then (Section 2, Chapters 3–8) turn to the central dogma of molecular biology and describe how DNA is used to make RNA which in turn is used to make protein. We end this section with chapters on protein structure and on recombinant DNA technology and genetic engineering. Section 3 (Chapters 9–11) describes how cells use electricity, ion concentrations, and chemical messengers to interact with and respond to external cues. Section 4 (Chapters 12–14) describes the physically moving cell: intracellular trafficking, the cytoskeleton and molecular motors, and cell division and death. Finally, in Section 5 (Chapter 15) we use cystic fibrosis as a case study that illustrates how all the areas of knowledge covered in this book contribute to an understanding of this disease and have allowed the development of novel clinical treatments.

Boxed material throughout the book is divided into EXAMPLES that illustrate the topics covered in the main text, explanations of the MEDICAL RELEVANCE of the material, and IN-DEPTH sections that extend the coverage beyond the content of the main text. We include BRAINBOXES highlighting a number of the scientists who have made significant contributions to cell biology, either by making critical discoveries, or by creating the tools that made those discoveries possible. REVIEW QUESTIONS in extended matching-set format at the end of each chapter help the reader assess how well they have assimilated and understood the material, while each chapter also poses a "THOUGHT QUESTION" that tests concepts rather than facts.

A comprehensive website accompanies the book at www.wiley.com/go/bolsover/cellbiology4. This includes additional examples, in-depth explanations, and medical discussions for which there was no room in the printed book. Students who would like to test their understanding of the subject will find additional review questions, while teachers can find suggestions of essay titles. The website gives links to other internet resources together with references to primary research publications to allow readers to trace the origin of statements in the text. Lastly, the website allows download of all the figures from the book as slides for use by teachers.

ACKNOWLEDGMENTS

We are very grateful to Professors Jean-Claude Labbé (IRIC, Université de Montréal) and Stephanie Schorge (University College London) for critical reading of the manuscript.

ABOUT THE COMPANION WEBSITE

This book is accompanied by a companion website.

www.wiley.com/go/bolsover/cellbiology4

This website includes:

- Figures from the book in PowerPoint
- Additional examples and revision resources

SECTION 1

THE STRUCTURE OF THE CELL

The cell is the fundamental unit of life. A cell comprises a complex and ordered mass of protein, nucleic acid, and many biochemical species separated from the world outside by a limiting membrane. Cells expend energy to maintain a highly ordered state, and this expenditure of energy and the ability to repair themselves distinguishes living cells from lifeless packets of biological material such as viruses. In the first two chapters we will describe the basic structure of cells and how they can be observed with a microscope. We will describe how, in animals, cells containing the same DNA database assume very different shapes and functions and organize themselves into tissues.

1

A LOOK AT CELLS AND TISSUES

With very few exceptions, all living things are either a single cell or an assembly of cells. This chapter will begin to describe what a cell is, and further chapters will say much more. However, to begin with, we can briefly describe a cell as an **aqueous** (watery) droplet enclosed by a lipid (fatty) membrane. Cells are, with a few notable exceptions, small (Figure 1.1), with dimensions measured in micrometers (μm, 1 μm = 1/1000 mm). They are more or less self-sufficient: a single cell taken from a human being can survive for many days in a dish of nutrient broth, and many human cells can grow and divide in such an environment. In 1838 the botanist Matthias Schleiden and the zoologist Theodor Schwann formally proposed that all living organisms are composed of cells. Their "cell theory," which nowadays seems so obvious, was a milestone in the development of modern biology. Nevertheless, general acceptance took many years, in large part because the **plasma membrane** (Figure 1.2), the membrane surrounding the cell that divides the living inside from the nonliving **extracellular medium,** is too thin to be seen using a light microscope. **Microorganisms** such as bacteria, yeast, and protozoa exist as single cells. In contrast, the adult human is made up of about 30 trillion cells (1 trillion = 10^{12}), which are mostly organized into collectives called **tissues.**

 ONLY TWO TYPES OF CELL

Superficially at least, cells exhibit a staggering diversity. Some have defined, geometric shapes; others have flexible boundaries; some lead a solitary existence; others live in communities; some swim, some crawl, and some are sedentary. Given these differences, it is perhaps surprising that there are only two types of cell (Figure 1.2). **Prokaryotic** (Greek for "before nucleus") cells have very little visible internal organization so that, for instance, the genetic material, stored in the molecule **deoxyribonucleic acid** (DNA), is free within the cell. These cells are especially small, the vast majority being 1–2 μm in length. The prokaryotes are made up of two broad groups of organisms, the bacteria and the archaea (Figure 1.3). The archaea were originally thought to be an unusual group of bacteria but we now know that they are a distinct group of prokaryotes with an independent evolutionary history. The cells of all other organisms, from yeasts to plants to worms to humans, are **eukaryotic** (Greek for "with a nucleus"). These are generally larger (5–100 μm, although some eukaryotic cells are large enough to be seen with the naked eye; Figure 1.1) and structurally more complex. Eukaryotic cells contain a variety of specialized

Cell Biology: A Short Course, Fourth Edition. Stephen Bolsover, Andrea Townsend-Nicholson, Greg FitzHarris, Elizabeth Shephard, Jeremy Hyams and Sandip Patel.
© 2022 John Wiley & Sons Ltd. Published 2022 by John Wiley & Sons Ltd.
Companion website: www.wiley.com/go/bolsover/cellbiology4

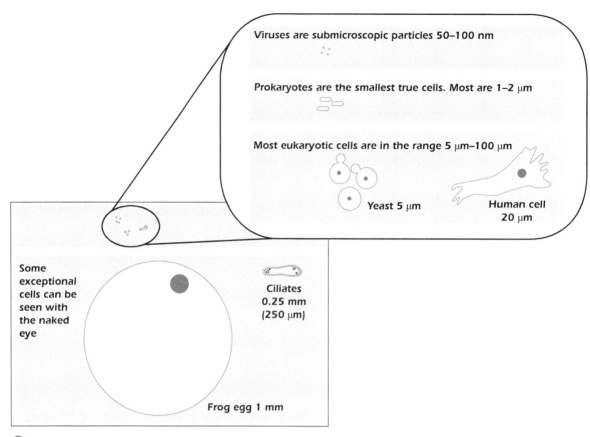

 Figure 1.1. Dimensions of some example cells. 1 mm = 10^{-3} m; 1 μm = 10^{-6} m; 1 nm = 10^{-9} m.

structures known collectively as **organelles**, embedded within a viscous substance called **cytosol.** Their DNA is held within the largest organelle, the **nucleus.** The structure and function of organelles will be described in detail in subsequent chapters. Table 1.1 summarizes the differences between prokaryotic and eukaryotic cells.

Cell Division

One of the major distinctions between prokaryotic and eukaryotic cells is their mode of division. In prokaryotes the circular chromosome is duplicated from a single replication origin by a group of proteins that reside on the inside of the plasma membrane. At the completion of replication the old and new copies of the chromosome lie side by side on the plasma membrane which then pinches inwards between them. This process, which generates two equal, or roughly equal, daughter cells is described as **binary fission.** In eukaryotes the large, linear chromosomes, housed in the nucleus, are duplicated from multiple origins of replication by enzymes located in the nucleus. Sometime later the **nuclear envelope** breaks down and the replicated chromosomes are compacted so that they can be segregated without damage during **mitosis.** We will deal with mitosis in detail in Chapter 14. For the moment we should be aware that

although it is primarily about changes to the nucleus, mitosis is accompanied by dramatic changes in the organization of the rest of the cell. A new structure, the **mitotic spindle,** is assembled specifically to move the chromosomes apart whilst other structures are dismantled so that their components can be divided among the two daughter cells following cell division.

VIRUSES

Viruses occupy a unique position between the living and nonliving worlds. On the one hand they are made of the same molecules as living cells. On the other they are incapable of independent existence, being completely dependent on a host cell for reproduction. Almost all living organisms have viruses that infect them. Human viruses include polio, influenza, herpes, rabies, smallpox, chickenpox, HIV, and SARS-CoV-2, the causative agent of COVID-19. Viruses are submicroscopic particles consisting of a core of genetic material enclosed within a protein coat called the capsid. Some have an extra membrane layer called the envelope. Viruses are inert until they enter a host cell, whereupon their genetic material directs the host cell machinery to produce viral protein and viral genetic

(a) Bacterium, prokaryotic

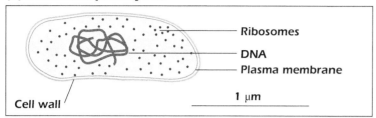

(b) Animal cell, eukaryotic

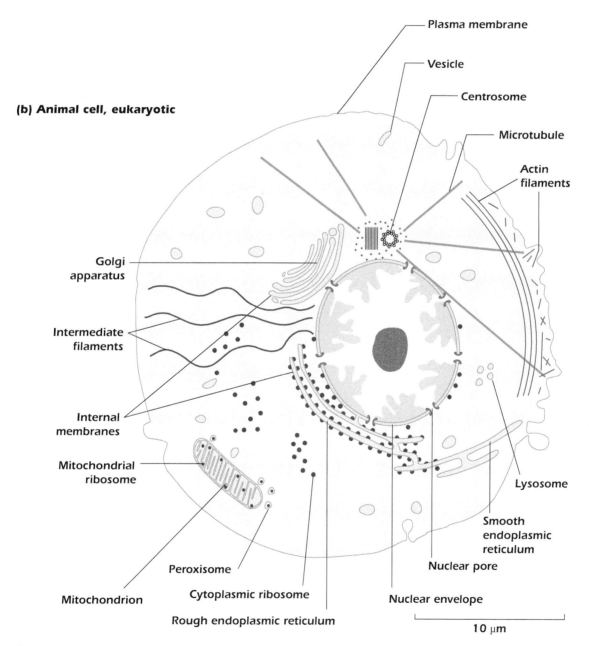

Figure 1.2. Organization of prokaryotic and eukaryotic cells.

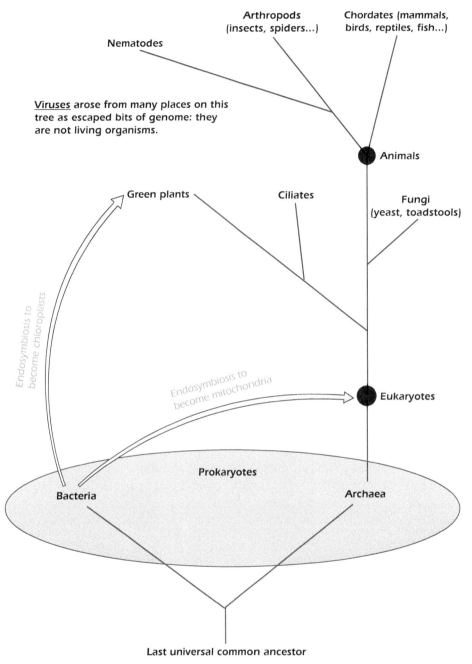

 Figure 1.3. The tree of life. The diagram shows the currently accepted view of how the different types of organism arose from a common ancestor. Many minor groups have been omitted. Distance up the page should not be taken as indicating complexity or how "advanced" the organisms are. All organisms living today represent lineages that have had the same amount of time to evolve and change from the last universal common ancestor.

material. Viruses often insert their genome into that of the host, an ability that is widely made use of in molecular biology research (Chapter 8). Bacterial viruses, **bacterio-phages,** are used by scientists to transfer genes between bacterial strains. As we will see, human viruses are used as vehicles for gene therapy.

ORIGIN OF EUKARYOTIC CELLS

Prokaryotic cells are simpler in their organization than eukaryotic cells and are assumed to be more primitive. According to the fossil record, prokaryotic organisms

◉ TABLE 1.1. Differences between prokaryotic and eukaryotic cells.

	Prokaryotes	Eukaryotes
Size	Usually 1–2 μm	Usually 5–100 μm
Nucleus	Absent	Present
DNA	Usually a single circular molecule (= chromosome)	Multiple linear molecules (chromosomes)[a]
Cell division	Simple fission	Mitosis or meiosis
Internal membranes	Rare	Complex
Ribosomes	70S[b]	80S (70S in mitochondria and chloroplasts)
Cytoskeleton	Rudimentary	Microtubules, microfilaments, intermediate filaments
Motility	Rotary motor (drives bacterial flagellum)	Dynein (drives cilia and flagella); kinesin, myosin
First appeared	3.5×10^9 years ago	2×10^9 years ago

[a] The tiny chromosomes of mitochondria and chloroplasts are exceptions; like prokaryotic chromosomes they are often circular.
[b] The S value, or Svedberg unit, is a sedimentation rate. It is a measure of how fast a molecule moves in a gravitational field, and therefore in an ultracentrifuge.

Example 1.1   Sterilization by Filtration

Because even the smallest cells are larger than 1 μm, harmful bacteria and other organisms can be removed from drinking water by passing it through a filter with holes 200 nm in diameter. These filters can vary in size from huge, such as those used in various commercial processes, to small enough to be easily transportable by backpackers. Filtering drinking water greatly reduces the chances of bringing back an unwanted souvenir from your camping trip!

---

precede, by at least 1.5 billion years, the first eukaryotes that appeared some 2 billion years ago. It seems highly likely that eukaryotes evolved from prokaryotes, and the most likely explanation of this process is the **endosymbiotic theory.** The basis of this theory is that some eukaryotic organelles originated as free-living bacteria that were engulfed by larger cells in which they established a mutually beneficial relationship. For example, **mitochondria** would have originated as free-living aerobic bacteria and **chloroplasts** as photosynthetic cyanobacteria. The endosymbiotic theory provides an attractive explanation for the fact that mitochondria and chloroplasts contain their own DNA and ribosomes both of which are more closely related to those of bacteria than to all the other DNA and ribosomes in the same cell. The case for the origin of other eukaryotic organelles is less persuasive. Nevertheless, while it is clearly not perfect, most biologists are now prepared to accept that the endosymbiotic theory provides at least a partial explanation for the evolution of the eukaryotic cell from prokaryotic ancestors.

IN DEPTH 1.1   OUR ANCESTOR, THE ARCHAEON

When we say prokaryote we usually mean bacterium. The prokaryotes we use to make yogurt and kimchi, those which give us diseases, and those that we use in genetic engineering (Chapter 8) are all bacteria. However, from the very origin of life on earth a second group of prokaryotes called the archaea lived alongside bacteria. Archaea are found throughout nature, for example each of our guts contains at least a trillion cells of the archaeon *Methanobrevibacter smithii* that help to break down complex sugars. It is now thought that the cell that incorporated bacteria to become the ancestor of the eukaryotes was an archaeon related to a present-day group called the Asgard archaea. Asgard archaea contain several genes and proteins that are otherwise found only in eukaryotes, including the building blocks of microtubules and microfilaments (Table 1.1). When a cell similar to a present-day Asgard archaeon engulfed an oxygen-using bacterium it could then begin the slow process of evolving into the multitude of eukaryotes that exist today.

CELL SPECIALIZATION IN ANIMALS

Animals are multicellular communities of individual cells. Lying between and supporting the cells is the **extracellular matrix** (Figure 1.4) of different types of fiber around which the fluids and solute of the **interstitial fluid** can easily pass. All the body cells that comprise a single organism share the same set of genetic instructions in their nuclei. Nevertheless, the cells are not all identical. Rather, they form a variety of **tissues,** groups of cells that are specialized to carry out a common function. This specialization occurs because different cell types read out different parts of the DNA blueprint and therefore make different proteins. In animals there are four major tissue types: epithelium, connective tissue, nervous tissue, and muscle. Some examples of the cells that make up these tissues are shown in Figure 1.5.

Epithelia are sheets of cells that cover the surface of the body and line its internal cavities, such as the lungs and intestine. The cells may be **columnar,** meaning taller than they are broad (Figure 1.5b), or **squamous,** meaning flat

(e.g. the capillary epithelial cell in Figure 1.4). They are often **polarized,** meaning that one surface of the cell is distinct in its organization, composition, and appearance from the other. In the intestine, the single layer of columnar cells lining the inside, or **lumen,** has an absorptive function that is increased by the folding of the surface into villi (Figure 1.6). The luminal surfaces of these polarized cells have **microvilli** that increase the surface area even further. The basal (bottom) surface sits on a thin planar sheet of specialized extracellular matrix called the **basement membrane** or **basal lamina.** Many of the epithelial cells of the airways, for instance those lining the trachea and bronchioles, have **cilia** on their surfaces (Figure 1.7). These are hairlike appendages that actively beat back and forth, moving a layer of mucus away from the lungs (Chapter 13). Particles and bacteria are trapped in the mucus layer, preventing them from reaching the delicate membranes in the lung that carry out air exchange. In the case of the skin, the epithelium is said to be **stratified** because it is composed of several layers.

Connective tissues provide essential support for the other tissues of the body. They include bone, cartilage, and

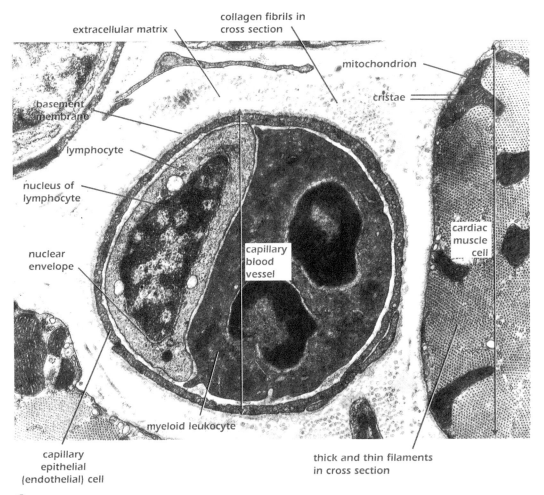

Figure 1.4. Transmission electron micrograph of a capillary blood vessel running between cardiac muscle cells. Source: Image by Giorgio Gabella, Department of Cell and Developmental Biology, University College London. Reproduced by permission.

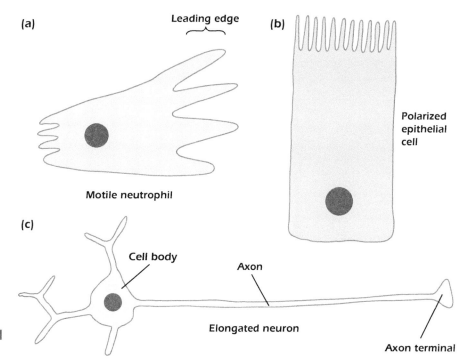

(a) Leading edge

Motile neutrophil

(b) Polarized epithelial cell

(c) Cell body

Axon

Elongated neuron

Axon terminal

Figure 1.5. Different types of animal cells.

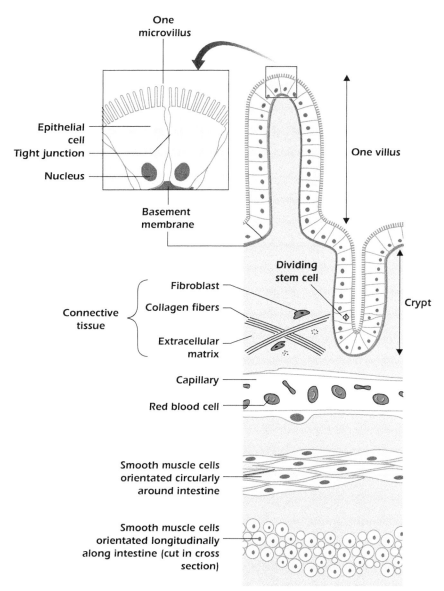

One microvillus

Epithelial cell

Tight junction

Nucleus

Basement membrane

One villus

Dividing stem cell

Crypt

Connective tissue

Fibroblast

Collagen fibers

Extracellular matrix

Capillary

Red blood cell

Smooth muscle cells orientated circularly around intestine

Smooth muscle cells orientated longitudinally along intestine (cut in cross section)

Figure 1.6. Tissues and structures of the intestine wall.

Figure 1.7. Scanning electron micrograph of airway epithelium. Source: Image by Giorgio Gabella, Department of Cell and Developmental Biology, University College London. Reproduced by permission.

adipose (fat) tissue. Unlike other tissues, connective tissue contains relatively few cells within a large volume of extracellular matrix that consists of different types of fiber embedded in **amorphous** ground substance (Figure 1.6). The most abundant of the fibers is **collagen,** a protein with the tensile properties of steel that accounts for about a third of the protein of the human body. Other fibers have elastic properties that permit the supported tissues to be displaced and then to return to their original position. The amorphous ground substance contains large quantities of water, facilitating the diffusion of metabolites, oxygen, and carbon dioxide to and from the cells in other tissues and organs. Of the many cell types found in connective tissue, two of the most important are **fibroblasts,** which make and secrete the ground substance and fibers, and **macrophages,** which remove foreign, dead, and defective material. A number of inherited diseases are associated with defects in connective tissue. Osteogenesis imperfecta or Brittle Bone Disease, for example, is characterized by short height, loose joints, and bones that break easily. These characteristics result from a defect in the organization of the collagen fibers (see Medical Relevance 3.2 on page 47).

Nervous tissue is a highly modified epithelium that is composed of several cell types. Principal among these are the **nerve cells,** also called **neurons** (Figure 1.5c), along with a variety of supporting cells that help maintain them. Neurons extend processes called **axons,** which can be over a meter in length. Neurons constantly monitor what is occurring inside and outside the body. They integrate and summarize this information and mount appropriate responses to it (Chapters 9–11). Another type of cell, **glia,** has other roles

in nervous tissue, including forming the electrical insulation around axons.

Muscle tissue can be of two types, **smooth** or **striated.** Smooth muscle cells are long and slender and are usually found in the walls of tubular organs such as the intestine and many blood vessels. In general, smooth muscle cells contract slowly and can maintain the contracted state for a long time. There are two classes of striated muscle: **cardiac** and **skeletal.** Cardiac muscle cells (Figure 1.4) make up the walls of the heart chambers. These are branched cells that are connected electrically by **gap junctions** (page 26), and their automatic rhythmical contraction powers the beating of the heart. Each skeletal muscle is a bundle of hundreds to thousands of fibers, each fiber being a giant single cell with many nuclei. This rather unusual situation is the result of an event that occurs in the embryo when the cells that give rise to the fibers fuse together, pooling their nuclei in a common **cytoplasm** (the term cytoplasm is historically a crude term meaning the semi-viscous ground substance of cells; we use the term to mean everything inside the plasma membrane except the nucleus). The mechanism of muscle contraction will be described in Chapter 13.

STEM CELLS AND TISSUE REPLACEMENT

Cells multiply by division. In the human body an estimated 25 million cell divisions occur every second! These provide new cells for the blood and immune systems, for the repair of wounds and the replacement of dead cells. In complex tissues such as those described above, division is restricted to a small number of undifferentiated **stem cells** that are capable of dividing many times; some of their daughter cells then differentiate to become all of the other cells of the tissue. In the case of the intestine, folds in the surface epithelium form crypts, each of which contains ~250 cells (Figure 1.6). Mature cells at the top die and must be replaced by the division of between four and six stem cells near the base of the crypt. Each stem cell divides roughly twice a day, the resulting cells moving up the crypt to replace those lost at the surface. Benign (non-cancerous) polyps can be formed in the intestine if this normal balance between birth and death is disturbed.

As in the intestine, stem cells in other tissues exist in specific locales, called **niches,** with environments that support their special and vital functions. In many tissues the requirement to replace dead cells is much less than it is in the intestine and in such cases the stem cell niche must maintain its occupants in a quiescent (nondividing) state until needed (for more on stem cells see In Depth 14.1 on page 234).

● THE CELL WALL

Many types of cell, particularly bacteria and plant cells, create a rigid case around themselves called a **cell wall.** For cells that live in an extracellular medium more dilute than their own cytosol, the cell wall is critical in preventing the cell bursting. For example, penicillin and many other **antibiotics** block the synthesis of bacterial cell walls with the result that the bacteria burst. Within trees, plant cells modify the cell wall to generate the woody trunk. Animal cells do not have cell walls.

● MICROSCOPES REVEAL CELL STRUCTURE

Many different techniques have contributed to our understanding of the structure of cells but nothing can compare to actually seeing what is there. Microscopy, the visualization of small objects, began with Robert Hooke (1635–1703) who described the *cella* (open spaces) of plant tissues. But the colossus of this era of discovery was Anton van Leeuwenhoek (1632–1723), a Dutchman with no scientific training but with unrivaled talents as both a microscope maker and as an observer and recorder of the microscopic living world. Van Leeuwenhoek's microscope was a single glass lens that bent light rays to form a magnified image so it, and all the later instruments that use visible light to image small structures, are called light microscopes.

The Modern Light Microscope

Today's light microscopes, such as those one would find in a school laboratory, consist of a light source, which may be the sun or an artificial light, plus three glass lenses: a **condenser lens** to focus light on the specimen, an **objective lens** to form the magnified image, and a **projector lens,** usually called the eyepiece, to convey the magnified image to the eye (Figures 1.8 and 1.9). Since the image is formed by light passing through the specimen, this is a **transmission light microscope.** Depending on the focal length of the various lenses and their arrangement, a given magnification is achieved. In **bright-field microscopy,** the image that reaches the eye consists of the colors of white light minus those absorbed by the cell. Most living cells have little color and are therefore largely transparent to light.

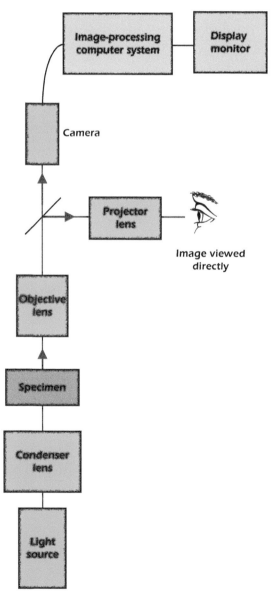

● Figure 1.8. Basic design of a light microscope.

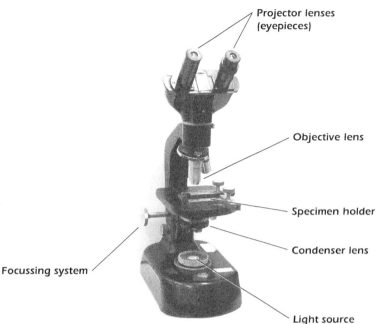

Projector lenses
(eyepieces)

Objective lens

Specimen holder

Condenser lens

Focussing system

Light source

Figure 1.9. A simple upright light microscope.

This problem can be overcome by **cytochemistry,** the use of colored stains to selectively highlight particular structures and organelles. However, many of these compounds are highly toxic and to be effective they often require that the cell or tissue is subjected to a series of harsh chemical treatments. A different approach, and one that can be applied to living cells, is the use of **phase contrast microscopy.** This relies on the fact that light travels at different speeds through regions of the cell that differ in composition. The phase contrast microscope converts these differences in refractive index into differences in contrast, and considerably more detail is revealed (Figure 1.10). Transmitted light microscopes can distinguish objects as small as about half the wavelength of the light used, so

about 250 nm (nm, 1 nm = 1/1000 μm). They can therefore be used to visualize the smallest cells and the major intracellular structures and organelles (Figure 1.11a).

The Transmission Electron Microscope

In principle the smaller the wavelength of the radiation used to image a structure, the better the resolution. This fact led to the invention in 1931 of the transmission electron microscope by Max Knoll and Ernst Ruska. An electron gun generates a beam of electrons by heating a thin, V-shaped piece of tungsten wire to 3000 °C. A large voltage accelerates the beam down the microscope column, which is under vacuum because the electrons are slowed and scattered if they collide

Bright field   **Phase contrast**

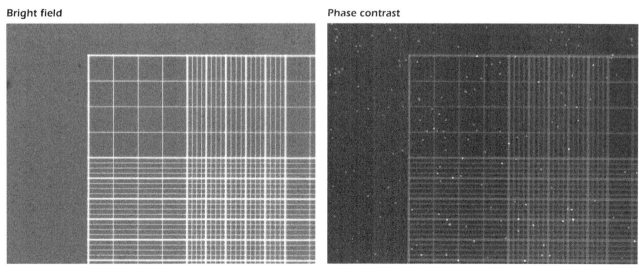

Figure 1.10. Cultured human cells on a hemocytometer grid under bright-field and phase contrast.

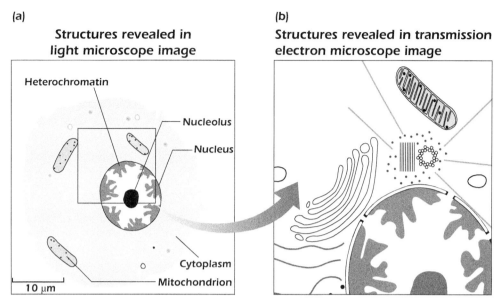

(a)
Structures revealed in light microscope image

Heterochromatin

Nucleolus

Nucleus

Cytoplasm

Mitochondrion

10 μm

(b)
Structures revealed in transmission electron microscope image

Figure 1.11. Cell structure as seen through light and transmission electron microscopes. For the identity of the structures revealed by the electron microscope, see Figure 1.2.

with air molecules. The beam passes through the specimen and is bent by powerful magnets to form a highly magnified image (Figure 1.11b). This image can be viewed on a fluorescent screen that emits light when struck by electrons. While the electron microscope offers enormous resolution, electron beams are potentially highly destructive, and biological material must be subjected to a complex processing schedule before it can be examined. The preparation of cells for electron microscopy is summarized in Figure 1.12.

The transmission electron microscope produces a detailed image but one that is static, two-dimensional, and highly processed (Figure 1.4). Often, only a small region of what was once a dynamic, living, three-dimensional cell is revealed. Moreover, the picture revealed is essentially a snapshot taken at the particular instant that the cell was killed ("fixed"). Clearly, such images must be interpreted with great care. Also, electron microscopes are large, expensive, and require a skilled operator. Nevertheless, since they

A small piece of tissue (~1 mm³) is immersed in glutaraldehyde and osmium tetroxide. These chemicals bind all the component parts of the cells together; the tissue is said to be **fixed**. It is then washed thoroughly.

The tissue is **dehydrated** by soaking in acetone or ethanol.

The tissue is **embedded** in resin which is then baked hard.

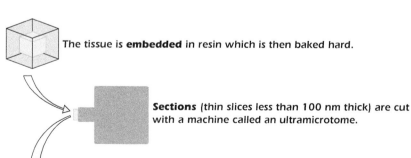

Sections (thin slices less than 100 nm thick) are cut with a machine called an ultramicrotome.

The sections are placed on a small copper grid and **stained** with uranyl acetate and lead citrate. When viewed in the electron microscope, regions that have bound lots of uranium and lead will appear dark because they are a barrier to the electron beam.

Figure 1.12. Preparation of tissue for electron microscopy.

can resolve objects down to about 0.2 nm in size, they are the main source of information on the organization of the cell at the nanometer scale, the **ultrastructure** of the cell.

The Scanning Electron Microscope

Whereas the image in a transmission electron microscope is formed by electrons transmitted through the specimen, in the scanning electron microscope it is formed from electrons that are reflected back from the surface of a specimen as the electron beam scans rapidly back and forth over it. These reflected electrons are detected and used to generate a picture on a display monitor. The scanning electron microscope operates over a wide magnification range, from 10 times to 100 000 times, and has a wide **depth of focus.** The images created give an excellent impression of the three-dimensional shape of objects (Figure 1.7). The scanning electron microscope is therefore particularly useful for providing topographical information on the surfaces of cells or tissues. Modern instruments have a resolution of about 1 nm.

 ## FLUORESCENCE MICROSCOPY

A major advance in light microscopy was the development of the **fluorescence microscope. Fluorescent** molecules emit light when they are illuminated with light of a shorter wavelength. Familiar examples are the hidden signature in bank passbooks, which is written in fluorescent ink that glows blue (wavelength about 450 nm) when illuminated with ultraviolet light (UV) (wavelength about 360 nm), and the whitener in fabric detergents that causes your white shirt to glow blue when illuminated by the ultraviolet light in a club. The fluorescent dye Hoechst 33342 has a similar wavelength dependence: it is excited by UV light and emits blue light. However, it differs from the dyes used in ink or detergent in that it binds tightly to the DNA in the nucleus and only fluoresces when so bound. Figure 1.13a shows the optical path through a microscope set up to look at a preparation stained with Hoechst. White light from an arc lamp passes through an excitation filter that allows only UV light to pass. This light then strikes the heart of the fluorescence

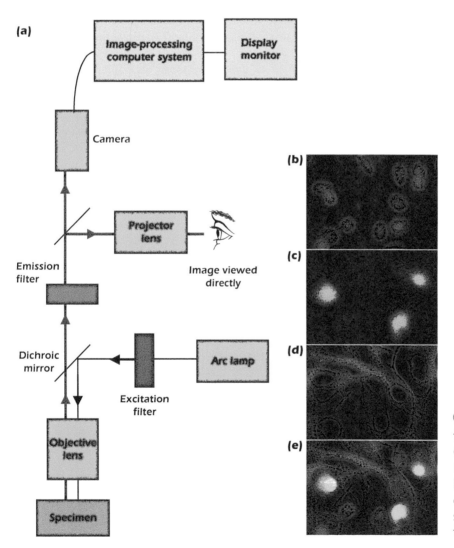

Figure 1.13. (a) Basic design of a fluorescence light microscope. (b–d) Cultured brain cells labeled with (b) Hoechst to label DNA, (c) fluorescently labeled antibody against ELAV, (d) fluorescently labeled antibody against glial-specific intermediate filament. (e) Three fluorescence images merged.

microscope: a special mirror called a dichroic mirror that reflects light of wavelengths shorter than a designed cutoff but transmits light of longer wavelength. To view Hoechst, we use a dichroic mirror of cutoff wavelength 400 nm, which therefore reflects the UV excitation light down through the objective lens and onto the specimen. Any Hoechst bound to DNA in the preparation will emit blue light. Some of this will be captured by the objective lens and, because its wavelength is greater than 400 nm, will not be reflected by the dichroic mirror but will instead pass through. An emission filter, set to pass only blue light, cuts out any scattered UV light. The blue light now passes to the eye or camera in the usual way. Image (b) shows a field of cells cultured from rat brain after staining with Hoechst. Only the nuclei are seen, as bright ovals.

Although some of the structures and chemicals found in cells can be selectively stained by specific fluorescent dyes such as Hoechst, others are most conveniently revealed by using antibodies. In this technique an animal (usually a mouse, rabbit, or goat) is injected with a protein or other molecule of interest. The animal's immune system recognizes the chemical as foreign and generates antibodies that bind to (and therefore help neutralize) the chemical. Some blood is then taken from the animal and the antibodies purified. The antibodies can then be labeled by attaching a fluorescent dye. Images (c) and (d) show the same field of brain cells but with the excitation filter, dichroic mirror, and emission filter changed so as to reveal in (c) a protein called ELAV that is found only in neurons; then in (d) an intermediate filament protein (page 228) found only in glial cells. The antibody that binds to ELAV is labeled with a fluorescent dye that is excited by blue light and fluoresces green. The antibody that binds to the glial filaments is labeled with a dye that is excited by green light and fluoresces red. Because these wavelength characteristics are different, the location of the three chemicals – DNA, ELAV, and intermediate filament – can be revealed independently in the same specimen. Panel (e) shows the three images superimposed.

The technique just described is **primary immunofluorescence** and requires that the antibody to the chemical of interest be labeled with a dye. Only antibodies to chemicals that many laboratories study are so labeled. In order to reveal other chemicals, scientists use **secondary immunofluorescence.** In this approach, a commercial company injects an animal (e.g. a goat) with an antibody from another animal (e.g. a rabbit). The goat then makes "goat anti-rabbit" antibody. This, the **secondary antibody,** is purified and labeled with a dye. All the scientist has to do is make or buy a rabbit antibody that binds to the protein of interest. No further modification of this specialized **primary antibody** is necessary. Once the primary antibody has bound to the specimen and excess antibody rinsed off, the specimen is then exposed to the fluorescent secondary antibody that binds selectively to the primary antibody. Viewing the stained preparation in a fluorescence microscope then reveals the location of the

chemical of interest. The same dye-labeled secondary antibody can be used in other laboratories or at other times to reveal the location of many different proteins because the specificity is determined by the unlabeled primary antibody.

Increasing the Resolution of Fluorescence Microscopes

The resolution and precision of fluorescence microscopes have steadily improved with time. In 1979 a team in Amsterdam invented the **confocal light microscope** that scanned a point of excitation light across the specimen to markedly reduce the contribution of out-of-focus light. In 1987 a team in Cambridge developed a prototype of a commercial system and within a couple of years all major microscope manufacturers began offering confocal light microscopes.

A series of advanced microscopy techniques start with confocal microscopy and then use clever optical approaches to dramatically improve the resolution so that objects considerably smaller than the wavelength of light are revealed. Collectively these techniques are known as **super-resolution microscopy.** One such technique is Stimulated Emission Depletion Microscopy (STED), in which the excitation spot is surrounded by a doughnut of light of a different wavelength that actually de-excites the dye. Figure 1.14 shows one use of STED. Over 100 proteins are associated with the nuclear pores that perforate the nuclear envelope (Figure 1.2). Göttfert and coworkers used an antibody that recognizes one of these proteins, called gp210, labeled with a dye that emits red light, and a second antibody, labeled with a green emitter, against the transport machinery inside the pore. Figure 1.14 shows how an uninterpretable fuzz of red and green fluorescence in the standard confocal image is resolved into beautiful images by STED, revealing how eight gp210 molecules surround the pore. Data like these contribute to our present understanding of the eightfold symmetry of the structure that braces the pore to keep it open (Figure 12.10 on page 211). Super-resolution microscopes can resolve objects down to tens of nanometers in size and can visualize biological structures previously thought to be unresolvable using light.

Fluorescent Proteins

Instead of introducing fluorescent dyes into fixed or living cells, we can cause cells to make them. This technology began with the description in 1962 of **Green Fluorescent Protein** (GFP), a protein from the jellyfish *Aequorea victoria* that glows green when excited with blue light. Since the original description of GFP, a family of fluorescent proteins of different colors has become available, some through artificially mutating the original GFP, some found in other organisms. Using recombinant DNA technology (Chapter 8),

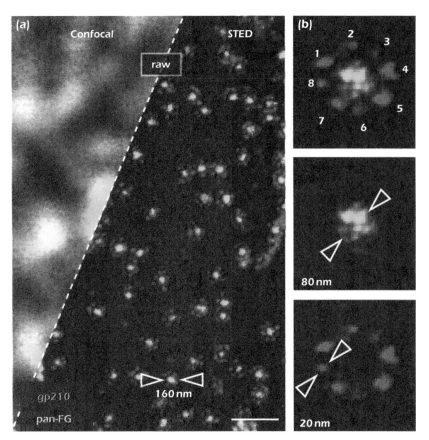

Figure 1.14. Super-resolution microscopy. Fluorescence image of the surface of a eukaryotic nucleus by standard confocal microscopy and by STED. Source: Göttfert et al. (2013). Coaligned Dual-Channel STED Nanoscopy and Molecular Diffusion Analysis at 20 nm Resolution. Biophysical Journal, 105(1), L01 -L03. doi:10.1016/j.bpj.2013.05.029

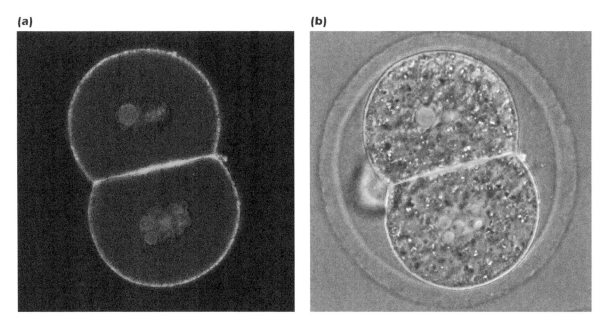

Figure 1.15. (a) Fluorescence microscope image of two-cell mouse embryo expressing GFP and RFP chimeras. Two images were acquired in succession using optical configurations suitable for GFP and RFP respectively. The two images were then combined in a computer. (b) Image (a) combined with a phase contrast transmitted light image; the clear envelope around the embryo is the zona pellucida. Source: Images by Lia Paim and Adelaide Allais, University of Montreal.

the gene for a fluorescent protein can be introduced into a living cell, which then makes the protein.

This basic technique is useful in itself for labeling a population of cells. However a battery of more and more sophisticated techniques has been developed from this starting point. Most use the approach of fusing the gene for a fluorescent protein with the gene for another protein of interest so that the cell makes a chimera – a single protein comprising the protein of interest plus the fluorescent protein. For example, Figure 1.15 shows a fluorescence image of a mouse embryo at the two-cell stage engineered to express a GFP chimera that concentrates at the plasma membrane together with a chimera of red fluorescent protein (RFP, from coral) and histone H2B (page 39) that concentrates in the nuclei.

In this example the two parts of the chimeras worked independently, so that the GFP and RFP simply showed where their respective partners were located. However, clever protein design has created more complex chimeras of the calcium-binding protein calmodulin (page 115) with GFP mutants so that the fluorescence changes according to the concentration of calcium. Calcium concentrations change dramatically as cells respond to stimuli (Chapter 10) and these fluorescent calmodulin chimeras can be used to report these changes. Even more clever, if the calcium-measuring chimera is fused to a third protein with a known specific location in the cell, then the protein can be used to report the calcium concentration in that specific location.

BrainBox 1.1   Osamu Shimomura, Martin Chalfie, and Roger Tsien

Osamu Shimomura, Martin Chalfie, and Roger Tsien. Source: The Nobel Foundation. Photo: U. Montan.

Many proteins are colored, but in most cases the color is generated by a prosthetic group (for example the heme group in hemoglobin (page 118) and in chlorophyll). However, in 1979 Osamu Shimomura, working at Princeton University in the USA, showed that the colored moiety in a GFP made by the jellyfish *Aequorea aequorea* was a reaction product of the amino acids themselves. This opened up the possibility of using the cell's own machinery to make genetically encoded labeling proteins that could be targeted to precise tissues and even specific sites within the cell. However, the suspicion was that one or more specialized enzymes in the jellyfish cells would be needed to carry out the conversion of the amino acids to the fluorophore, so that simply introducing the *gfp*

gene would do nothing. In 1994 Martin Chalfie, working at Columbia University in New York, showed that this was not the case: the *gfp* gene product, on its own, converted itself into fluorescent GFP. The next leap in technology was to engineer GFP and GFP chimeras to be more than markers, and instead to be reporters of cell behavior. From 1992 onward, working at the University of California at San Diego, Roger Tsien and his lab engineered an ever-increasing family of fluorescent proteins that are now used universally by cell biologists and drug companies to study almost all aspects of cell behavior, creating both beautiful science and beautiful images, such as Figure 1.15. Shimomura, Chalfie, and Tsien were awarded the Nobel Prize in Chemistry in 2008.

SUMMARY

1. All living organisms are made of cells.

2. There are only two types of cells, prokaryotic and eukaryotic.

3. Prokaryotic cells have little visible internal organization. They are usually 1–2 μm in size.

4. Eukaryotic cells usually measure from 5–100 μm. They contain a variety of specialized internal organelles, the largest of which, the nucleus, contains the genetic material.

5. The endosymbiotic theory proposes that some eukaryotic organelles, such as mitochondria and chloroplasts, originated as free-living prokaryotes.

6. In multicellular organisms, cells are organized into tissues. In animals there are four tissue types: epithelium, connective tissue, nervous tissue, and muscle.

7. The extracellular matrix is found on the outside of animal cells.

8. In tissues, specialized cells arise from unspecialized stem cells.

9. Light microscopy revealed the diversity of cell types and the existence of the major organelles, such as the nucleus and mitochondria.

10. The electron microscope revealed the detailed structure of the larger organelles and resolved the cell ultrastructure, the fine detail, at the nanometer scale.

11. Fluorescence microscopy can reveal the location of specific molecules within the cell by using antibodies or genetically encoded fluorescent chimeras.

12. Super-resolution microscopy is an advanced form of fluorescence microscopy that can reveal objects tens of nanometers in size.

13. Cells can be engineered to express fluorescent proteins that reveal many aspects of cell organization and behavior.

FURTHER READING

Gest, H. (2004). The discovery of microorganisms by Robert Hooke and Antoni van Leeuwenhoek, fellows of the royal society. *Notes and Records of the Royal Society of London* 58: 187–201.

Kubitscheck, U. (2017). *Fluorescence Microscopy: From Principles to Biological Applications*, 2e. Weinheim: Wiley-VCH.

Lane, N. (2009). *Life Ascending: The Ten Great Inventions of Evolution*. London: Profile Books.

⬤ REVIEW QUESTIONS

We use the same format of review questions throughout the book. For each of the numbered questions choose the best response from the lettered list. The same response may apply to more than one numbered question. Unless specifically told to do so, you should not refer back to the chapter text or figures in answering the questions. Answers are at the back of the book, starting on page 265.

1.1 Theme: Dimensions in cell biology

A 0.025 nm
B 0.2 nm
C 20 nm
D 250 nm
E 2000 nm
F 20 000 nm

Answer to thought question: Only transmission electron microscopy reveals all the structures present in a particular volume of the cell at sufficient resolution to determine whether it is malformed. Super-resolution microscopy has the resolution to reveal individual molecules on or within the Golgi, but only those individual molecules that the scientist chose to study are revealed, not the overall structure. Malformation of the endoplasmic reticulum and Golgi apparatus is thought to underlie one type of inherited spastic paraplegia.

G   200 000 nm
H   5 000 000 nm
I   1 000 000 000 nm
J   20 000 000 000 nm

From the above list, select the dimension most appropriate for each of the descriptions below.

1. a typical bacterium
2. a typical eukaryotic cell
3. the longest cell in the human body
4. resolution of a transmission light microscope
5. resolution of a transmission electron microscope

1.2   Theme: Types of cell

A   bacterium
B   epithelial cell
C   fibroblast
D   macrophage
E   glial cell
F   skeletal muscle cell
G   stem cell

From the above list of cell types, select the cell corresponding to each of the descriptions below.

1. a cell that synthesizes collagen
2. a cell type found in nervous tissue
3. a cell type that forms sheets, e.g. to separate different spaces in the body
4. a cell whose role is to remove dead and foreign material

5. a cell with no nuclear envelope
6. large cells with multiple nuclei
7. an undifferentiated cell capable of multiple rounds of cell division

1.3   Theme: Revealing cell organization

A   bright-field light microscopy
B   fluorescence microscopy using a DNA stain such as Hoechst
C   fluorescence microscopy using a GFP chimera
D   fluorescence microscopy using a specific antibody
E   phase contrast light microscopy
F   scanning electron microscopy
G   transmission electron microscopy

From the above list of microscopical techniques, select one appropriate for each of the studies or questions below.

1. to count the number of nuclei in a single skeletal muscle cell
2. to use in an assay for steroid hormones, which cause a particular protein (the hormone receptor) to move from the cytosol to the nucleus
3. to reveal the mitotic spindles of cells in a biopsy of a tumor from a patient
4. to look for cultured animal cells in a culture flask
5. to study the density of cilia on the surface of lung epithelial cells
6. to demonstrate that ribosomes can be found on the surface of the nucleus as well as on the endoplasmic reticulum

---

⬤ THOUGHT QUESTION

Each chapter also has a thought question. For these you are encouraged to refer back to the text and diagrams within the chapter to formulate your response. Answers appear earlier in the relevant chapter, printed upside down.

You wish to test the hypothesis that a particular inherited human disease is characterized by malformation of the Golgi apparatus (see Figure 1.2). What type of microscopy would you use to examine a tissue biopsy from a patient?

2

MEMBRANES AND ORGANELLES

In much the same way that our homes are divided into rooms that are adapted for particular activities, so eukaryotic cells contain distinct compartments or **organelles** to house specific functions. Organelles, like the cell itself, are delimited by **membranes.** It is therefore first necessary to consider some of the fundamental properties of cell membranes before discussing different organelles.

 BASIC PROPERTIES OF CELL MEMBRANES

It is difficult to overstate the importance of membranes to living cells; without them life as we know it could not exist. The **plasma membrane,** also known as the **cell membrane** or **plasmalemma,** defines the boundary of the cell. It regulates the movement of materials into and out of the cell and facilitates electrical and chemical signaling between cells. Other membranes define the boundaries of organelles and provide a matrix upon which complex chemical reactions can occur. In the following section, the basic structure of the cell membrane will be outlined.

The basic structure of a biological membrane is shown in Figure 2.1. Approximately half the mass is **phospholipid,** molecules comprising an electrically charged **head group** and long **hydrocarbon** tails. The head groups readily associate with water, a property called being **hydrophilic.** In contrast the tails are **hydrophobic** and avoid contact with water. Phospholipid molecules spontaneously organize to form a bilayer about 4 nm thick with hydrophilic surfaces formed of their heads and a hydrophobic interior formed of their tails. All the membranes of the cell, including the plasma membrane, also contain proteins. These may be embedded within the membrane and extracted from it only with great difficulty, in which case they are **integral proteins** (e.g. connexin, Figure 2.6); or they may be associated with the inner or outer surface and thus separated with relative ease, in which case they are **peripheral proteins** (e.g. Ras, Medical Relevance 7.1 on page 108). Membrane proteins are free to move laterally, within the plane of the membrane. Integral plasma membrane proteins are often **glycosylated:** they have sugar residues attached on the side facing the extracellular medium. In addition to phospholipids and proteins, eukaryotic cell membranes also contain **cholesterol.** Cholesterol makes the membrane more fluid. Cholesterol is essential for life but excess cholesterol in the bloodstream is strongly implicated in the development of atherosclerosis and heart disease.

Molecules of oxygen are uncharged. Although they dissolve readily in water, they are also able to dissolve in the hydrophobic interior of lipid bilayers. Oxygen molecules can therefore pass from the extracellular medium into the interior of the plasma membrane, and from there on into the cytoplasm, in a simple diffusion process (Figure 2.2). Water itself also passes across the plasma membrane in a process

Cell Biology: A Short Course, Fourth Edition. Stephen Bolsover, Andrea Townsend-Nicholson, Greg FitzHarris, Elizabeth Shephard, Jeremy Hyams and Sandip Patel.
© 2022 John Wiley & Sons Ltd. Published 2022 by John Wiley & Sons Ltd.
Companion website: www.wiley.com/go/bolsover/cellbiology4

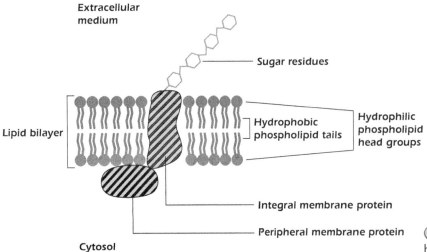

Figure 2.1. Membranes comprise a lipid bilayer plus integral and peripheral proteins.

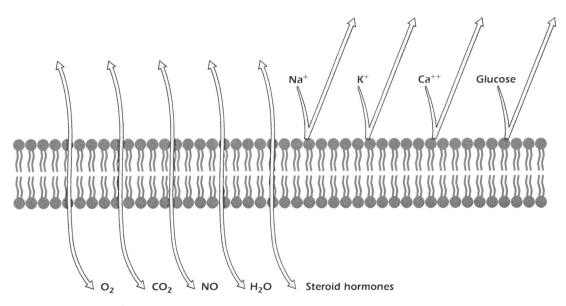

Figure 2.2. Small uncharged molecules can pass through membranes by simple diffusion, but hydrophilic solutes cannot.

known as **osmosis** (In Depth 2.1), as do the uncharged hormones of the steroid family. Indeed, any foreign chemical that is not strongly hydrophilic will readily pass into and out of the cell. In contrast, simple ions and charged molecules are strongly hydrophilic. They cannot dissolve in the hydrophobic interior of the membrane and therefore cannot cross membranes by simple diffusion (Figure 2.2). In this case, specialized proteins are present in our cells to facilitate their entry and exit. We will cover these in detail in Chapter 9.

ORGANELLES BOUNDED BY DOUBLE-MEMBRANE ENVELOPES

The nucleus, mitochondrion, and chloroplast (in plants) are enclosed within an envelope consisting of two parallel

membranes. These major cell organelles all contain the genetic material, DNA.

The Nucleus

The nucleus is often the most prominent cell organelle. It contains the **genome,** the cell's database, which is encoded in molecules of the **nucleic acid,** DNA. The nucleus is bounded by a **nuclear envelope** composed of two membranes separated by an intermembrane space (Figure 2.3). The inner membrane of the nuclear envelope is lined by the **nuclear lamina,** a meshwork of **lamin** proteins that provide rigidity to the nucleus and anchorage for the DNA. A two-way traffic of proteins and nucleic acids between the nucleus and the cytoplasm passes through holes in the nuclear envelope called **nuclear pores.** The nucleus of a

IN DEPTH 2.1 WATER, WATER (AND AQUAPORINS) EVERYWHERE

Our bodies are ~70% water. Water can readily diffuse into and out of cells through the process of osmosis. It is thus essential that we maintain correct water balance otherwise our cells would distort or even lyse. Osmosis can be defined as the movement of water across a membrane down its concentration gradient from a solution of low **osmolarity** to a solution of high osmolarity. Osmolarity is calculated by summing the molar concentration of all the solutes in a particular solution. The more concentrated a solution is, the lower its water concentration and the higher its osmolarity.

For most mammalian fluids, the osmolarity is approximately 300 mOsm/l. Hypertonic solutions have an osmolarity that is higher, whereas hypotonic solutions have an osmolarity that is lower. If cells were placed in a hypertonic

solution, water would move out of the cell from the cytosol where the osmolarity is lower than the bathing medium. Cells shrink under these conditions. Conversely, water would move into the cell if placed in hypotonic solutions because the osmolarity of the cytosol is now higher relative to the bathing medium. This can cause lysis. Changes in osmolarity can occur in pathological situations.

Water is a hydrophilic molecule, obviously! Yet it can cross the hydrophobic membrane relatively quickly. Why? The answer is that membranes have water channels known as aquaporins. They facilitate the diffusion of water across the membrane. Aquaporins are expressed in all cells but are particularly abundant in red blood cells and kidney tubules. Consequently, the plasma membranes of these cells are highly permeable to water.

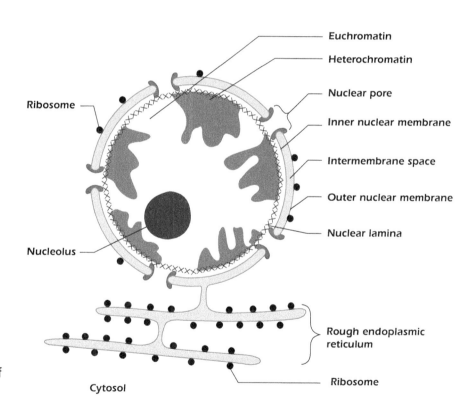

Figure 2.3. The nucleus and the relationship of its membranes to those of the endoplasmic reticulum.

cell that is synthesizing proteins at a low level will have few nuclear pores. In cells that are undergoing active protein synthesis, however, virtually the whole nuclear surface is perforated.

Within the nucleus, it is usually possible to recognize discrete areas. Much of it is occupied by chromatin, a complex of DNA and certain DNA-binding proteins such as **histones** (page 39). In most cells, it is possible to discern two

types of chromatin. A central region of lightly staining **euchromatin** is that portion of the cell's DNA database that is being actively read out by being **transcribed** into **RNA,** another nucleic acid (Chapter 5). In contrast the peripheral, darkly staining **heterochromatin** is the inactive portion of the genome where no RNA synthesis is occurring. The DNA in heterochromatin is densely packed, leading to its dark appearance.

Example 2.1   DNA Destruction in the Cytosol

An animal cell's own DNA should remain in the nucleus, except for the tiny amount that is within mitochondria. DNA in the cytosol will likely belong to a pathogen such as an invading virus. Cells therefore contain active

DNAses in the cytosol that rapidly destroy DNA, while leaving RNA intact. It is to evade this defense mechanism that many viruses use RNA as their genetic material, even though RNA is a much less stable molecule than is DNA.

Unlike DNA, RNA is also found in the cytoplasm associated with particles called ribosomes whose function is to make proteins. Ribosomes are made in the nucleus, in specialized regions called **nucleoli** that form at specific **nucleolar organizer region** sites on the DNA. These contain blocks of genes that code for the ribosomal RNA. Nuclear pores allow ribosomal subunits to exit the nucleus.

It should be stressed that the appearance of the nucleus we have described thus far relates to the cell in **interphase,** the period between successive rounds of cell division. As the cell enters mitosis (Chapter 14) the organization of the nucleus changes dramatically. The DNA becomes more and more tightly packed and is revealed as a number of separate rods called **chromosomes,** of which there are usually 46 in human cells. The nucleolus disperses, and the nuclear envelope fragments. Upon completion of mitosis, these structural rearrangements are reversed and the nucleus resumes its typical interphase organization.

Mitochondria

Like nuclei, mitochondria are encapsulated by an outer and inner membrane (Figure 2.4). Perhaps the most distinctive feature of mitochondria is that the inner membrane is markedly elaborated and folded to increase its surface area. These shelf-like projections, named **cristae,** make mitochondria

among the most easily recognizable organelles (e.g. Figure 1.4 on page 8). The number of cristae, like the number of mitochondria themselves, depends upon the energy budget of the cell in which they are found. In muscle cells, which must contract and relax repeatedly over long periods of time, there are many mitochondria that contain numerous cristae; in fat cells, which generate little energy, there are few mitochondria and their cristae are less well developed. This gives a clue as to the function of mitochondria: they are the cell's power stations. Mitochondria produce the molecule **adenosine triphosphate (ATP)** (page 35), the cell's main energy currency that provides the energy to drive a host of cellular reactions and mechanisms. Mitochondria make ATP through the process of oxidative phosphorylation whereby oxygen is used to pass electrons from energy intermediates to a series of protein complexes on the inner mitochondrial membrane known as the electron transport chain. This results in the transfer of H^+ out of the mitochondria and the generation of a concentration and voltage gradient. This gradient is subsequently tapped into by a protein known as ATP synthase which, as its name suggests, produces ATP. This process is essential for aerobic life and is the reason we breathe. We will return to ion gradients and the uses the cell puts them to in Chapter 9.

The great majority of proteins of the mitochondrion are encoded by nuclear genes and synthesized in the cytoplasm. But some of the information necessary for the function of mitochondria is stored within the organelle itself. Mitochondria contain many small circular DNA molecules (Table 1.1 on page 7) that are very different from the long, linear DNA molecules in the nucleus. This is strong evidence for the endosymbiotic theory of the origin of mitochondria (page 7), which proposes that the small circular DNA molecules found in mitochondria are all that is left of the chromosomes of the original symbiotic bacteria. Mitochondria also contain ribosomes (again, more like those of bacteria than the ribosomes in the cytoplasm of their own cell) which allows synthesis of a small subset of mitochondrial proteins.

⬤ ORGANELLES BOUNDED BY SINGLE MEMBRANES

Eukaryotic cells contain many sacs and tubes bounded by a single membrane. These are often rather similar in appearance and indeed if roughly spherical are lumped together as

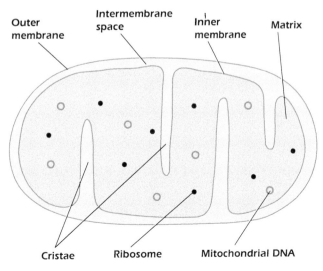

Figure 2.4. The mitochondrion.

vesicles. Nevertheless, there are in fact distinct types specialized to carry out distinct functions.

Peroxisomes

Mitochondria are frequently found close to another membrane-bound organelle, the peroxisome (Figure 1.2 on page 5). In human cells peroxisomes have a diameter of about 500 nm and their dense matrix contains a heterogeneous collection of proteins concerned with a variety of metabolic functions, some of which are only now beginning to be understood. Peroxisomes are so named because they are frequently responsible for the conversion of the highly reactive molecule hydrogen peroxide (H_2O_2), which is formed as a by-product of the reactions in the mitochondrion, into water. This reaction is carried out by a protein called catalase, which sometimes forms an obvious crystal within the peroxisome. Catalase is an **enzyme** – a protein catalyst that increases the rate of a chemical reaction. In fact, it was one of the first enzymes to be discovered. In humans, peroxisomes are primarily associated with lipid metabolism. Understanding peroxisome function is important for a number of inherited human diseases such as X-linked adrenoleukodystrophy where peroxisome malfunction and the consequent inability to metabolize lipid properly typically leads to death in childhood or early adulthood unless dietary lipid is extremely restricted.

Endoplasmic Reticulum

The endoplasmic reticulum (**ER**) is a network of membrane-enclosed tubes that run throughout the cell, forming a continuous mesh whose lumen (interior) is at all points separated from the cytosol by a single membrane. The membrane of the ER is continuous with the outer nuclear membrane (Figure 2.3). Two regions can be recognized in most cells, known as smooth ER and rough ER (Figure 1.2 on page 5). The basic difference is that the rough ER is covered in ribosomes, which gives it its rough appearance in the electron microscope.

The function of the smooth ER varies from tissue to tissue. In the ovaries, testes, and the adrenal gland it is where steroid hormones are made; in the liver it is the site of detoxication of foreign chemicals including drugs. Probably the most universal role of the smooth ER is the storage and sudden release of calcium ions. Calcium ions are pumped from the cytosol into the lumen of the smooth ER to more than 1000 times the concentration found in the cytosol. Many stimuli can cause this calcium to be released back into the cytosol, where it activates myriad cell processes (Chapter 10).

The rough ER is where cells make the proteins that will end up as integral membrane proteins in the plasma membrane, and the proteins that the cells will export (secrete) to the extracellular medium (such as the proteins of the extracellular matrix, page 8).

Golgi Apparatus

The Golgi apparatus, named after its discoverer, 1906 Nobel prize winner Camillo Golgi, is a distinctive stack of flattened sacks called **cisternae** (Figure 1.2 on page 5). The Golgi apparatus is the distribution point of the cell where proteins made within the rough ER are further processed and then directed to their final destination, the interior of the cell or the cell surface (see Chapter 12). Appropriately, given this central role, the Golgi apparatus is situated at the so-called cell center, a point immediately adjacent to the nucleus that is also occupied by a structure called the **centrosome.** The centrosome helps to organize the **cytoskeleton,** the supporting framework of the cell (Chapter 13).

Lysosomes

Lysosomes vary in shape and size but are often spherical, with a diameter of 250–500 nm. They are particularly plentiful in cells that digest and destroy other cells, such as the

---

Medical Relevance 2.1   Lysosomal Storage Disorders

A number of inherited diseases are characterized by cells filled with very large lysosomes. Many of these diseases involve abnormalities of development of the skeleton and connective tissues, as well as of the nervous system. The severity varies with the particular disease, but they often lead to death in infancy. In the majority of these diseases one lysosomal enzyme is missing or defective. Lysosomes work to degrade cellular components that are damaged or no longer needed and their enzymes function under the acidic conditions in the lysosome. If one of these enzymes is defective then the substrate will accumulate, filling the lysosome. The distended lysosomes eventually fill and damage the cell. Some of the best understood lysosomal storage diseases involve deficiencies in one or other of the enzymes required to degrade the complex glycosylated proteins and lipids found on cell surfaces.

Tay-Sachs disease involves severe mental retardation and blindness, with death by the age of three. In this case an enzyme required to break down a particular complex membrane lipid called a ganglioside is missing and undegraded ganglioside accumulates, swelling the lysosomes. Gangliosides are especially important in neuronal membranes so neurons are particularly damaged.

white blood cells called macrophages. This relates to their primary function in degrading material. Indeed, lysosomes are sometimes called "cell stomachs" because they contain a battery of enzymes that digest cellular components. Over 50 such enzymes exist, responsible for the breakdown of proteins, lipids, and even damaged organelles. Many of the enzymes show an acidic pH optimum, meaning that they function most efficiently at low pH. For this reason, lysosomes maintain a pH of 4–5. This aids breakdown of the material. We shall return to lysosomes in Chapter 12.

THE CONNECTED CELL

Although organelles compartmentalize cell activity they do not work in isolation. Similarly, cells often function as a collective. Organelles and cells form both intracellular and intercellular junctions to facilitate this. We will consider both these in turn.

Organelle Junctions

Organelles form junctions with other organelles and the plasma membrane via specialized structures known as **membrane contact sites.** Membrane contact sites are regions of close apposition whereby the membranes on either side of the junction are separated by <30 nm. They are stabilized by tethering proteins that span the junction.

The ER is a particularly well-connected organelle. It forms membrane contact sites with most organelles, including lysosomes (Figure 2.5), as well as with the plasma membrane. ER membrane contact sites are often tethered by a protein called VAP (VAP is short for the cumbersome name "vesicle-associated membrane protein associated protein"!). VAPs are integral proteins of the ER that recognize and bind to specific target proteins on the apposing membrane (page 120). This is an example of a protein–protein interaction and we will meet many other examples throughout this book.

Membrane contact sites serve a variety of functions. They are important for the trafficking of lipids around the cell and in the transfer of calcium between compartments, for example between the ER and mitochondria.

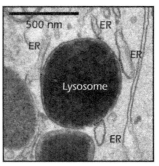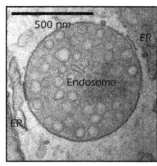

Figure 2.5. Electron micrographs showing contact sites between the endoplasmic reticulum (ER) and lysosomes (left) and endosomes (right). Endosomes will be described in Chapter 12. Source: Images by Bethan S. Kilpatrick, Clare E. Futter, and Sandip Patel, University College London. Reproduced by permission.

Cell Junctions

In multicellular organisms, and particularly in their epithelia, it is often necessary for neighboring cells within a tissue to be connected. This function is provided by cell junctions. In animal cells there are three types of junction. Those that form a tight seal between adjacent cells are known as **tight junctions;** those that allow communication between cells are known as **gap junctions.** A third class of cell junction anchors cells together, allowing the tissue to be stretched without tearing. These are called **anchoring junctions.**

Tight junctions are found wherever the flow of extracellular medium is to be restricted and are particularly common in epithelial cells, such as those lining the small intestine. The plasma membranes of adjacent cells are pressed together so tightly that no intercellular space exists between them (Figure 1.6 on page 9). Tight junctions between the epithelial cells of the intestine ensure that the only way that molecules can get from the lumen of the intestine to the blood supply that lies beneath is by passing through the cells, a route that can be selective.

Gap junctions are specialized structures that allow cell-to-cell communication in animals (Figure 2.6). When two cells form a gap junction, ions and small molecules can pass directly from the cytosol of one cell to the cytosol of the other

---

IN DEPTH 2.2   MY OLD MAM

The existence of membrane contact sites between the ER and mitochondria has been known for decades from electron microscope studies. Relatively recently, these contacts have been isolated biochemically from cell extracts for study. The resulting fractions are referred to as mitochondria-associated membranes (**MAMs**) because

they contain proteins not only typically found in mitochondrial membranes but also in the ER. In live cells, ER-mitochondria contact sites provide a restricted space such that when calcium ions are released from the ER (Chapter 10), they achieve a very high local concentration that facilitates their uptake by adjacent mitochondria.

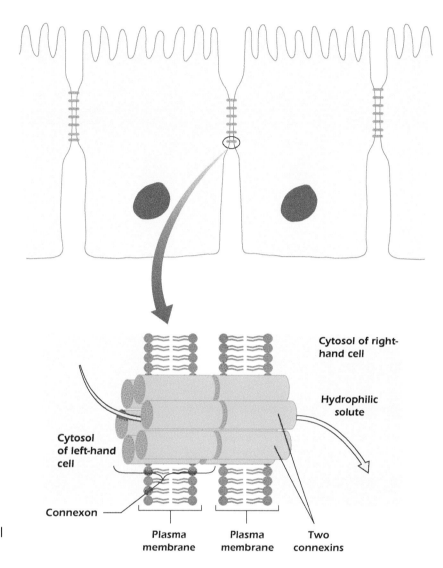

Figure 2.6. Gap junctions allow solute and electrical current to pass from the cytosol of one cell to the cytosol of its neighbor.

Cytosol of right-hand cell

Hydrophilic solute

Cytosol of left-hand cell

Connexon

Plasma membrane

Plasma membrane

Two connexins

Example 2.2 Gap Junctions Keep Eggs Healthy

In the days leading up to ovulation, oocytes develop within structures called follicles, in which they are connected to surrounding granulosa cells by gap junctions. During their development oocytes are not yet themselves capable of performing several fundamental homeostatic processes, such as regulating intracellular pH. However, the surrounding granulosa cells have ample ability to regulate pH, and H⁺ ions can pass through the gap junctions, such that the granulosa cells effectively regulate the pH of the oocyte on its behalf. By the time the oocyte is fully grown and ready to be ovulated it can finally regulate its own pH, at which time it jettisons the granulosa cells and becomes ready to be fertilized by a spermatozoon.

cell without going into the extracellular medium. Since ions can move through the junction, changes in electrical voltage are also rapidly transmitted from cell to cell by this route. In vertebrates, the structure that makes this possible is the **connexon.** When two compatible connexons meet, they form a tube, about 1.5 nm in diameter, that runs through the plasma membrane of the first cell, across the small gap between the cells, and through the plasma membrane of the second cell. This hole is large enough to allow through small organic molecules as well as ions, but it is too small for proteins or nucleic

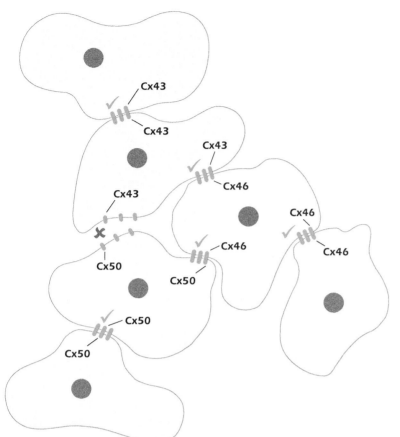

Figure 2.7. Not all connexins are compatible. A
√ indicates a working gap junction, x indicates that
gap junction channels cannot form.

acids. The limit is a relative molecular mass (Mr) of about
1000. Gap junctions are especially important in the heart,
where they allow an electrical signal to pass rapidly between
all the cardiac muscle cells, ensuring that they all contract at
the appropriate time. Each connexon is composed of six protein subunits called connexins that can twist against each
other to open and close the central channel in a process called
gating (page 147). This allows the cell to control the degree
to which it shares material with its neighbor.

There are over 20 different members of the connexin
gene family in humans. Each can, if made in a population of

cells, generate complete, working gap junction channels of
12 identical connexins. However, not all connexin **isoforms**
are compatible. A cell with connexin 43 cannot form gap
junctions with another cell that makes connexin 50
(Figure 2.7). However, a connexon made from connexin 43
is able to form a good working channel if it meets a connexon made from connexin 46, and a connexon made from
connexin 46 is in turn compatible with connexin 50.

Anchoring junctions bind cells tightly together and are
found in tissues such as the skin and heart that are subjected
to mechanical stress. They are described later (page 229).

Answer to thought question: This can only be speculation, but one obvious effect of cells having incompatible connexin isoforms is that the intracellular route for cell–cell communication is lost. Consider a tissue whose cells make connexin 43 that lies
adjacent to another tissue whose cells use connexin 50. The cells of each tissue can coordinate their activity by passing signals
via gap junctions, for example by intracellular messengers (page 166) or as voltage changes. However, the signals will remain
private to each tissue and will not be shared with the neighboring tissue. The two tissues can still communicate when necessary,
for instance by transmitter chemicals that the cells release into the extracellular medium (see Chapters 10 and 11).

SUMMARY

1. Membranes are made of phospholipids, protein, and cholesterol.

2. Cells are bounded by membranes, while cell functions are compartmentalized into membrane-bound organelles.

3. Solutes with a significant solubility in hydrophobic solvents can pass across biological membranes by simple diffusion. Charged molecules cannot.

4. The nucleus and mitochondrion are bounded by double-membrane envelopes. In the case of the nucleus, this is perforated by nuclear pores. Both organelles contain DNA.

5. Mitochondria produce the energy currency ATP.

6. Peroxisomes carry out a number of reactions, including the destruction of hydrogen peroxide.

7. The ER is a major site of protein synthesis. Cell stimulation can often cause calcium ions stored in the ER to be released into the cytosol.

8. The Golgi apparatus is concerned with the modification of proteins after they have been synthesized.

9. Lysosomes contain powerful degradative enzymes that degrade material.

10. Organelles form membrane contact sites with other organelles and the plasma membrane to allow information flow.

11. Tight junctions prevent the passage of extracellular water or solute between the cells of an epithelium.

12. Gap junctions allow solute and electrical current to pass from the cytosol of one cell to the cytosol of its neighbor.

13. Anchoring junctions form a strong physical link between cells.

FURTHER READING

Balda, M. S. & Matter, K. (2008). Tight junctions at a glance. Journal of Cell Science 121: 3677–3682.

Duchen, M. R. (2004). Mitochondria in health and disease: perspectives on a new mitochondrial biology. Molecular Aspects of Medicine 25: 365–451.

Dundr, M. & Misteli, T. (2001). Functional architecture in the cell nucleus. Biochemical Journal 356: 297–310.

Gall, J. G. & McIntosh, J. R. (2001). Landmark Papers in Cell Biology. New York: Cold Spring Harbor Laboratory Press. 544 pp.

Lamond, A. I. & Earnshaw, W. C. (1998). Structure and function in the nucleus. Science 280: 547–553.

Platt, F. M. et al. (2018). Lysosomal storage diseases. *Nature Reviews Disease Primers* 4 (27).

Short, B. & Barr, F. A. (2000). The Golgi apparatus. Current Biology 10: R583–585.

van der Klei, I. & Veenhuis, M. (2002). Peroxisomes: flexible and dynamic organelles. Current Opinion in Cell Biology 14: 500–505.

Wu, H, Carvalho, P, & Voeltz, G. K. 2018). Here, there, and everywhere: the importance of ER membrane contact sites. Science 361 (6401): eaan5835.

REVIEW QUESTIONS

2.1 Theme: Membranes

A the membrane or set of membranes surrounding the ER

B the membrane or set of membranes surrounding the Golgi apparatus

C the membrane or set of membranes surrounding the lysosome

D the membrane or set of membranes surrounding the mitochondrion

E the membrane or set of membranes surrounding the peroxisome

F the plasma membrane

From the above list, select the membrane that best fits the descriptions below.

1. a double layer comprising an inner and an outer membrane
2. is continuous with a membrane surrounding the nucleus
3. forms a set of flattened sacks called cisternae
4. often contains connexons as integral membrane proteins
5. surrounds a space containing calcium ions at a much higher concentration than in the cytosol. This store of calcium ions can be released into the cytosol upon cell stimulation. (Other such releasable calcium stores may exist in cells, but this is the largest.)

2.2 Theme: Organelles in eukaryotic cells

A   endoplasmic reticulum
B   Golgi apparatus
C   lysosome
D   mitochondrion
E   nucleus
F   peroxisome

From the above list of organelles, select the organelle described by each of the descriptions below.

1. a major site of protein synthesis
2. contains many powerful digestive enzymes
3. contains small circular chromosomes
4. contains the enzyme catalase
5. filled with chromatin
6. made up of flattened sacks called cisternae
7. most of the cell's ATP is made here
8. usually found at the cell center

2.3 Theme: Transport across membranes

A   can move from the cytosol of one cell to the cytosol of a neighboring cell, crossing the lipid bilayer component of each cell's plasma membrane as it does so
B   cannot cross lipid bilayers, but can move from the cytosol of one cell to the cytosol of a neighboring cell via gap junctions
C   cannot move to a neighboring cell either by crossing the lipid bilayer component of the plasma membrane or via gap junctions

Above we list three different possible constraints on the movement of a cytosolic solute. For each of the molecules below, state which of the three conditions apply.

1. an RNA molecule of $Mr = 10\,000$. As we will describe later, in Chapter 5, RNA molecules bear many negative charges
2. inositol trisphosphate, a small charged molecule of $Mr = 649$
3. K^+ (atomic weight = 39)
4. nitric oxide (NO) ($Mr = 30$)

Review question covering chapters 1 and 2: Some basic components of the eukaryotic cell

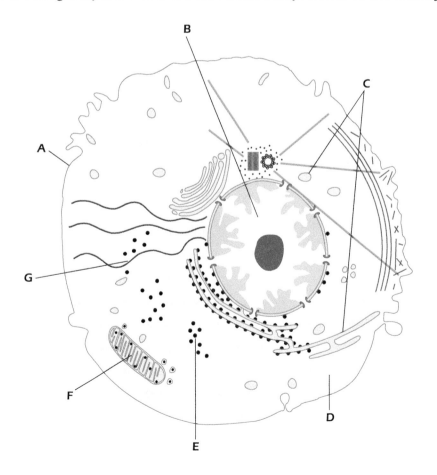

Identify each of the cellular components below from the figure above.

1. cytosol
2. internal membranes
3. mitochondrion
4. nucleus
5. plasma membrane

 THOUGHT QUESTION

Why might it be useful for the genome to code for many connexin isoforms, some of which are incompatible with each other?

SECTION 2

THE MOLECULAR BIOLOGY OF THE CELL

The central dogma of molecular biology is "DNA makes RNA makes protein." That central concept defines the structure of this section of the book, which moves from DNA through RNA to the synthesis of proteins. Single-celled organisms change their behavior by altering the spectrum of RNAs and proteins that they make, while the cells of an animal or plant differentiate into different cell types by selecting different RNAs and proteins to synthesize. We will therefore describe the control mechanisms that operate to allow selective synthesis and readout of RNA. We will describe the ribosome, the machine for making protein using the instructions on DNA, and then describe the many and varied structures and behaviors of proteins. Lastly, in Chapter 8, we will describe some of the techniques that have made molecular biology such a powerful technology for both manipulating and investigating cells and organisms.

3

DNA STRUCTURE AND THE GENETIC CODE

Our genes are made of deoxyribonucleic acid (**DNA**). This remarkable molecule contains all of the information needed to make a cell and to pass on this information when a cell divides. This chapter describes the structure and properties of DNA molecules, the way in which our DNA is packaged into chromosomes, and how the information stored within DNA is retrieved via the genetic code.

THE STRUCTURE OF DNA

Deoxyribonucleic acid is an extremely long polymer made from monomeric units called **deoxyribonucleotides (dNTPs),** which are often simply called **nucleotides.** Nucleotides are made up of three components: a base, a sugar, and a phosphate group. Figure 3.1 shows a deoxyribonucleotide, deoxyadenosine triphosphate, on the left. On the right is the corresponding ribonucleotide, adenosine triphosphate or ATP. As mentioned in Chapter 2, ATP is the cell's primary energy currency. As we will see in Chapter 5, ATP also plays a critical role as one of the four nucleotides in RNA, taking the place of the deoxyadenosine triphosphate in DNA. Note that deoxyribose, unlike ribose, has no **hydroxyl** (OH) group on its $2'$ carbon.

Four bases are found in DNA; they are the two purines **guanine** (G) and **adenine** (A) and the two pyrimidines **thymine** (T) and **cytosine** (C) (Figure 3.2). The lines represent covalent bonds formed when atoms share electrons, each seeking the most stable structure.

The combined base and sugar is known as a **nucleoside** to distinguish it from the phosphorylated form, which is called a nucleotide. Four different nucleotides are used to make DNA. They are $2'$-deoxyguanosine-$5'$-triphosphate (dGTP), $2'$-deoxyadenosine-$5'$-triphosphate (dATP), $2'$-deoxythymidine-$5'$-triphosphate (dTTP), and $2'$-deoxycytidine-$5'$-triphosphate (dCTP).

DNA molecules are very large. The single chromosome of the bacterium *Escherichia coli* is made up of two strands of DNA that are hydrogen-bonded together to form a single circular molecule comprising 9 million nucleotides. DNA molecules in eukaryotes are even larger: the DNA molecules in humans comprise on average 260 million nucleotides, and a cell has 46 of these massive molecules, each forming one chromosome. We inherit 23 chromosomes from each parent. Each set of 23 chromosomes encodes a complete copy of our **genome** and is made up of 6×10^9 nucleotides (or 3×10^9 **base pairs** – see below).

Figure 3.3 illustrates the structure of the DNA chain. As nucleotides are added to the chain by the enzyme **DNA polymerase** (Chapter 4), they lose two phosphate groups. The last (the α phosphate) remains and forms a **phosphodiester bond** between successive

Cell Biology: A Short Course, Fourth Edition. Stephen Bolsover, Andrea Townsend-Nicholson, Greg FitzHarris, Elizabeth Shephard, Jeremy Hyams and Sandip Patel.
© 2022 John Wiley & Sons Ltd. Published 2022 by John Wiley & Sons Ltd.
Companion website: www.wiley.com/go/bolsover/cellbiology4

IN DEPTH 3.1   WE HAVE A SECOND GENOME IN OUR CELLS

The set of 23 chromosomes that we inherit from each parent encodes a complete copy of our nuclear genome that resides in the nucleus of our cells. We have a second genome that resides in our mitochondria – the energy-producing organelles inside our cells. Unlike our large nuclear genome, which is organized in linear chromosomes, our mitochondrial genome is circular and only 16 569 base pairs in length. The mitochondrial genome contains 37 genes but only 13 of these encode proteins.

These proteins are all involved in mitochondrial energy production. While nuclear genomes are inherited from both parents, our mitochondrial genome is always inherited maternally. This has allowed a prediction of "mitochondrial Eve," the most recent female ancestor from whom all living humans descend in the matrilineal line, estimated to have lived between 165 000 and 190 000 years ago.

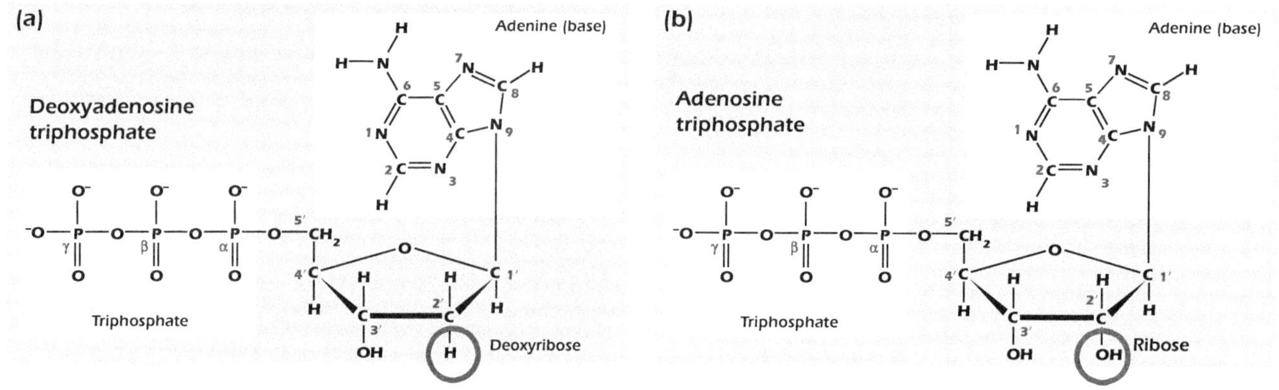

Figure 3.1. Adenine nucleotides. (a) Deoxyadenosine triphosphate. The H on the 2′ carbon of the ribose ring is circled. (b) Adenosine triphosphate. The OH group on the 2′ carbon of the ribose ring is circled.

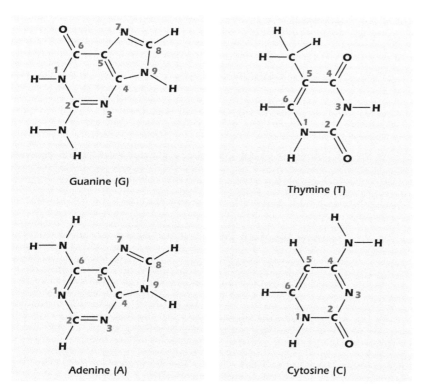

Figure 3.2. The four bases found in DNA.

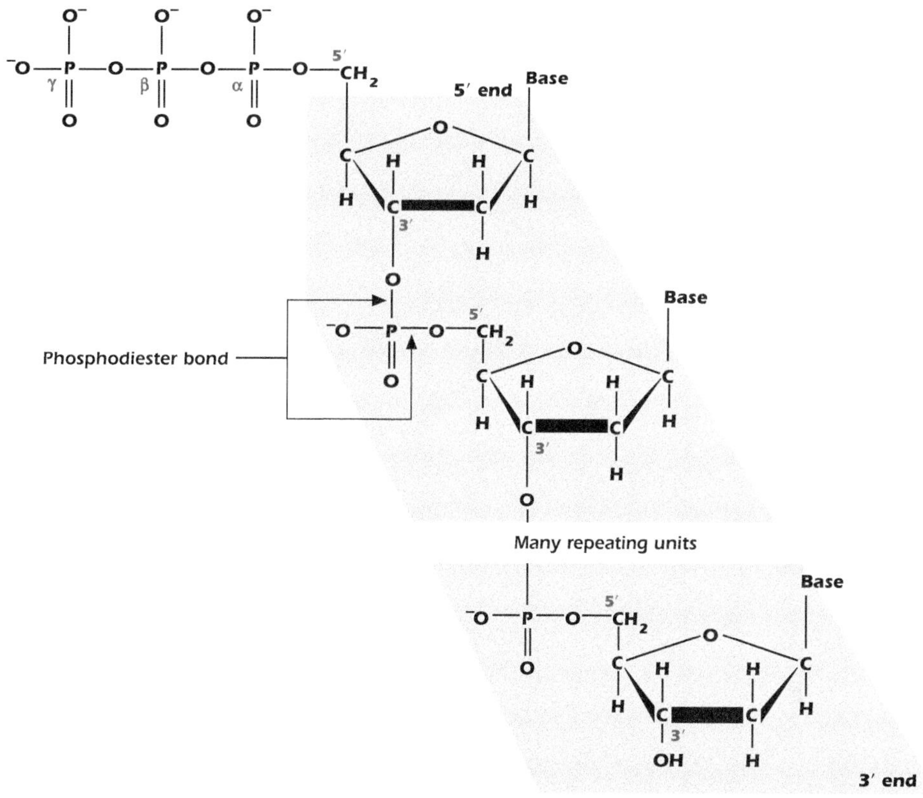

Figure 3.3. The phosphodiester bond and the sugar-phosphate backbone of DNA.

Sugar-phosphate backbone of DNA

deoxyribose residues. This bond is formed between the hydroxyl group on the 3′ carbon of the deoxyribose of the last nucleotide in the DNA chain and the α-phosphate group attached to the 5′ carbon of the nucleotide that the polymerase will add to the chain. The linkage gives rise to the sugar-phosphate backbone of a DNA molecule. A DNA chain has polarity because its two ends are different. In the first nucleotide in the chain, the 5′ carbon of the deoxyribose is phosphorylated but otherwise free. This is called the 5′ end of the DNA chain. At the other end is a deoxyribose with a free hydroxyl group on its 3′ carbon. This is the 3′ end.

The DNA Molecule Is a Double Helix

In 1953 Rosalind Franklin used X-ray diffraction to show that DNA was a helical (i.e. twisted) polymer. James Watson and Francis Crick demonstrated, by building three-dimensional models, that the molecule is a double helix (Figure 3.4). Two hydrophilic sugar-phosphate backbones lie on the outside of the molecule and the purine and pyrimidine bases lie on the inside of the molecule. There is just enough space for one purine and one pyrimidine in the center of the double helix. The Watson–Crick model showed that the purine guanine (G) would fit nicely with the pyrimidine cytosine (C). The purine adenine (A) would fit nicely with the pyrimidine thymine (T). Thus A always base pairs with T and G always base pairs with C.

Hydrogen Bonds Form Between Base Pairs

A **hydrogen bond** forms when a hydrogen atom is shared. The hydrogen bonds in an A–T and a G–C base pair (Figure 3.4) form when the hydrogen attached to a nitrogen in one base gets close to an electron-grabbing oxygen or nitrogen in the other base of the pair. The hydrogen bonds formed between the base pairs hold the DNA helix together. The three hydrogen bonds formed between G and C produce a relatively strong base pair. Because only two hydrogen bonds are formed between A and T, this weaker base pair is more easily broken. The difference in strengths between a G–C and an A–T base pair is important in the initiation of DNA replication (page 51) and in the initiation and termination of RNA synthesis (page 69).

DNA Strands Are Antiparallel

The two strands of DNA are said to be antiparallel because they lie in the opposite orientation with respect to one another, with the 3′-hydroxyl terminus of one strand opposite the 5′-phosphate terminus of the second strand. The sugar-phosphate backbones do not completely conceal the bases inside. There are two grooves along the surface of the DNA molecule. One is wide and deep – the major groove – and the other is narrow and shallow – the minor groove (Figure 3.4). DNA-binding proteins can use the grooves to gain access to the bases and bind to specific sequences. This is important in initiating replication (page 51) and transcription (page 69) and is also used when manipulating DNA in the laboratory.

Schematic representation of the double-helical structure of DNA

Cytosine-guanine base pair (C≡G)

Thymine-adenine base pair (T=A)

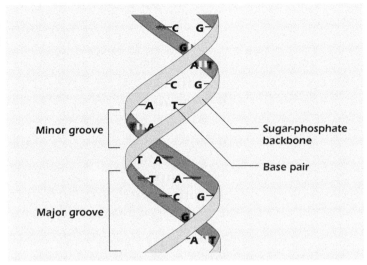

Figure 3.4. The DNA double helix is held together by hydrogen bonds.

Example 3.1   Erwin Chargaff's Puzzling Data

In a key discovery of the 1950s, Erwin Chargaff analyzed the purine and pyrimidine content of DNA isolated from many different organisms and found that the amounts of A and T were always the same, as were the amounts of G and C. Such an identity was inexplicable at the time but helped James Watson and Francis Crick build their double-helix model in which every A on one strand of the DNA helix has a matching T on the other strand and every G on one strand has a matching C on the other.

The Two DNA Strands Are Complementary

A consequence of the base pairing that joins the two strands of DNA is that if the base sequence of one strand is known, then that of its partner can be inferred. A **G** in one strand will always be paired with a **C** in the other. Similarly an **A** will always pair with a **T.** The two strands are therefore said to be **complementary.**

 ## DNA AS THE GENETIC MATERIAL

Deoxyribonucleic acid carries the genetic information encoded in the sequence of the four bases – guanine, adenine, thymine, and cytosine. The information in DNA is transferred to its daughter molecules through **replication** (the duplication of DNA molecules) and subsequent cell division. DNA directs the synthesis of proteins through the intermediary molecule **messenger RNA (mRNA).** The DNA code is transferred to mRNA by a process known as **transcription** (Chapter 5). The mRNA code is then **translated** into a sequence of amino acids during protein synthesis (Chapter 6). This is the central dogma of molecular biology: DNA makes RNA makes protein.

Retroviruses such as the human immunodeficiency virus, the cause of AIDS, are an exception to this rule. As their name suggests, they reverse the normal order of data transfer. Inside the virus coat is a molecule of RNA plus an enzyme that can make DNA from an RNA template by the process known as **reverse transcription.**

We do not yet know the exact number of genes that encode messenger RNA in the human genome. The current estimate is 19 116. Table 3.1 compares the number of predicted messenger RNA genes in the genomes of different organisms. In each organism, there are also a small number of genes (about 100 in humans) that code for ribosomal RNAs and transfer RNAs. The roles these three types of RNA play in protein synthesis is described in Chapter 6.

 ## PACKAGING OF DNA MOLECULES INTO CHROMOSOMES

Eukaryotic Chromosomes and Chromatin Structure

A human cell contains 46 chromosomes (23 pairs), each of which is a single DNA molecule bundled up with various proteins. On average, each human chromosome contains about 1.3×10^8 base pairs (bp) of DNA. If the DNA in a human chromosome were stretched as far as it would go without breaking it would be about 5 cm long, so the 46 chromosomes in all represent about 2 m of DNA. The nucleus in which this DNA must be contained has a diameter of only about 10 μm, so large amounts of DNA must be packaged into a small space. This represents a formidable problem that is dealt with by binding the DNA to proteins to form chromatin. As shown in Figure 3.5, the DNA double helix is packaged at both small and larger scales. In the first stage, shown on the right of the figure, the DNA double helix with a diameter of 2 nm is bound to proteins known as **histones.** Histones are positively charged because they contain large amounts of the amino acids arginine and lysine (page 104) and bind tightly to the negatively charged phosphates on DNA. A 146 bp length of DNA is wound around a protein complex composed of two molecules each of four different histones – H2A, H2B, H3, and H4 – to form a **nucleosome.** Because each nucleosome is separated from its neighbor by about

TABLE 3.1. Numbers of predicted protein-coding genes in various organisms.

Organism	Number of predicted genes
Bacterium – *Escherichia coli*	4288
Yeast – *Saccharomyces cerevisiae*	6091
Fruit fly – *Drosophila melanogaster*	14 133
Worm – *Caenorhabdites elegans*	19 735
Plant – *Arabidopsis thaliana*	27 029
Human – *Homo sapiens*	19 116

Medical Relevance 3.1   Anti-Viral Drugs for HIV

It is hard to find drugs that will inhibit replication of the HIV virus, which causes the disease AIDS, without damaging the host cell because the HIV virus uses the host cell's synthetic machinery. However, the first act of the virus is to make DNA from its RNA genome using an enzyme contained within its viral envelope, called reverse transcriptase. Azidothymidine (**AZT**), the most widely used anti-AIDS drug, inhibits this enzyme.

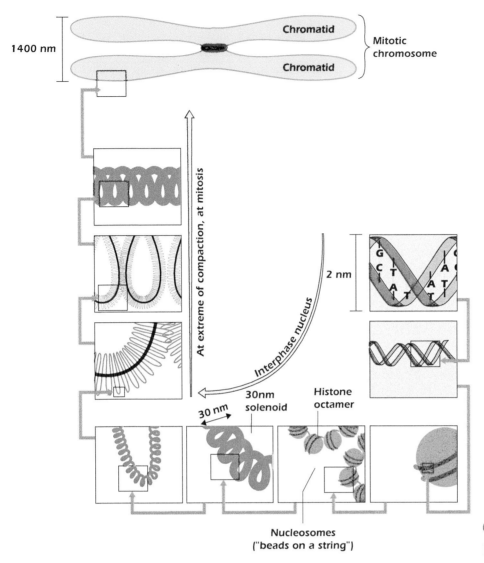

Figure 3.5. How DNA is packaged into chromosomes.

50 bp of linker DNA, this unfolded chromatin state looks like beads on a string when viewed in an electron microscope. Nucleosomes undergo further packaging. A fifth type of histone, H1, binds to the linker DNA and pulls the nucleosomes together, helping to further coil the DNA into chromatin fibers 30 nm in diameter, which are referred to as 30-nm solenoids. The fibers then form loops with the help of a class of proteins known as nonhistones and this further condenses the DNA (panels on left-hand side of Figure 3.5) into a higher order set of coils in a process of **supercoiling.**

In a normal interphase cell about 10% of the chromatin is in the highly compacted form and is visible in the light microscope as darkly staining heterochromatin (Figure 2.3 on page 23). Heterochromatin is the portion of the genome where there is no RNA synthesis taking place. Euchromatin, which is chromatin that is being transcribed into RNA, is wholly or partly unpacked from the histones to allow it to be read and has a less dense appearance in the microscope. Chromatin is in its most compacted form when the cell is preparing for mitosis, as shown at the top left of Figure 3.5. The chromatin

folds and condenses further to form the 1400 nm-wide chromosomes we see under the light microscope. Because the cell is to divide, the DNA has been replicated, so that each chromosome is now formed by two chromatids, each one a DNA double helix. This means each daughter cell will receive a full set of 46 chromosomes. Figure 3.6 is a photograph of human chromosomes as they appear at cell division.

Prokaryotic Chromosomes

The chromosome of the bacterium *E. coli* is a single circular DNA molecule of about 4.6×10^6 base pairs. It has a circumference of 1 mm, yet must fit into the 1 μm cell so, like eukaryotic chromosomes, it is coiled, supercoiled, and packaged with basic proteins that are similar to eukaryotic histones. However, an ordered nucleosome structure similar to the "beads on a string" seen in eukaryotic cells is not observed in prokaryotes. Prokaryotes do not have nuclear envelopes so the condensed chromosome, together with its associated proteins, lies free in the cytoplasm, forming a mass that is

BrainBox 3.1 Marie Maynard Daly

Marie Maynard Daly working in her lab, c. 1960.
Source: Archives of the Albert Einstein College of
Medicine. Photographer Ted Burrows.

Nucleosomes are the first level of DNA packaging, helping to compact the nuclear genome so that it can fit inside the cell. Nucleosomes form when negatively charged DNA binds to positively charged histone proteins. Histones are rich in the positively charged amino acids arginine and lysine. Marie Maynard Daly was a biochemist who studied histones and was able to demonstrate that there were lysine-rich histones, in addition to the classical arginine-rich histones that had previously been described in the literature. In 1947, Marie Maynard Daly became the first African American woman to receive a PhD in chemistry in the United States.

called the **nucleoid** to emphasize its functional equivalence to the eukaryotic nucleus.

Plasmids

Plasmids are small circular minichromosomes found in bacteria and some eukaryotes. They are several thousand base pairs long and are probably tightly coiled and supercoiled inside the cell. Plasmids often code for proteins that confer a selective advantage to a bacterium, such as resistance to a particular antibiotic. In Chapter 8 we describe how plasmids are used by scientists to artificially introduce foreign DNA molecules into bacterial cells.

IN DEPTH 3.2   DNA – A GORDIAN KNOT

At the start of his career Alexander the Great was shown the Gordian Knot, a tangled ball of knotted rope, and told that whoever untied the knot would conquer Asia. Alexander cut through the knot with his sword. A similar problem occurs in the nucleus, where the 46 chromosomes form 2 m of tangled, knotted DNA. How does the DNA ever untangle at mitosis? The cell adopts Alexander's solution – it cuts the rope. At any place where the DNA helix is under strain, for instance, where two chromosomes press against each other, an enzyme called **topoisomerase** II cuts one chromosome double helix so that the other can pass through the gap. Then, surpassing Alexander, the enzyme rejoins the cut ends. Topoisomerases are active all the time in the nucleus, relieving any strain that develops in the tangled mass of DNA.

Concerns that a terrorist organization might release large amounts of anthrax spores have caused several governments to stockpile large amounts of the antibiotic Cipro. This works by inhibiting the prokaryotic form of topoisomerase II (sometimes called gyrase) hence preventing cell replication.

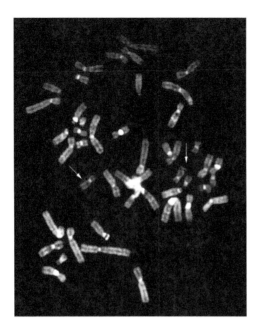

 Figure 3.6. A spread of human chromosomes (at metaphase – see page 236). The red dots (generated using a technique called fluorescence in-situ hybridization or FISH) reveals the COL1A1 gene which codes for collagen 1A1 (see Medical Relevance 3.2 on page 47). Source: Image by Mariano Rocchi, Resources for Molecular Cytogenetics, Department of Biology, University of Bari. Reproduced by permission.

Viruses

Viruses (page 4) rely on the host cell to make more virus. Once viruses have entered cells, the cells' machinery is used to copy the viral genome. Depending on the virus type, the genome may be single- or double-stranded DNA, or even RNA. A viral genome is packaged within a protective protein coat. Viruses that infect bacteria are called bacteriophages. One of these, lambda, has a fixed-size DNA molecule of 4.8×10^4 base pairs. In contrast, the bacteriophage M13 can change its chromosome size, its protein coat expanding in parallel to accommodate the chromosome.

THE GENETIC CODE

Amino Acids and Proteins

Proteins are made up of individual amino acid building blocks. Amino acids contain both a **carboxyl group,** which readily gives an H^+ to water and is therefore acidic, and a basic **amino group** ($-NH_2$), which readily accepts H^+ to become $-NH_3^+$. Figure 3.7a shows two amino acids, leucine and γ-**aminobutyric acid (GABA),** in the form in which they are found at normal pH: the carboxyl groups have each lost an H^+ and the amino groups have each gained one, so that the molecules bear both a negative and a positive charge.

We name organic acids by labeling the carbon adjacent to the carboxyl group α, the next one β, and so on. When we add an amino group, making an amino acid, we state the letter of the carbon to which the amino group is attached. Hence leucine is an α-**amino acid** while GABA stands for *gamma*-aminobutyric acid. α-Amino acids are the building blocks of proteins. They have the general structure shown in Figure 3.7b where R is the **side chain.** Leucine has a simple side chain of carbon and hydrogen. Other amino acids have different side chains and so have different properties. It is the diversity of amino side chains that give proteins their characteristic properties (page 104).

α-Amino acids can link together to form long chains through the formation of a **peptide bond** between the carboxyl group of one amino acid and the amino group of the next. Figure 3.7c shows the generalized structure of such a chain of α-amino acids. If there are fewer than about 50 amino acids in a polymer we tend to call it a **peptide.** More and it is a **polypeptide.** Polypeptides that fold into a specific shape are **proteins.**

Reading the Genetic Code

It is the sequence of bases along the DNA strand that determines the sequence of the amino acids in proteins. There are four different bases in DNA (G, A, T, and C). Each amino acid is specified by a **codon,** a group of three bases. Because there are four bases in DNA, a three-letter code gives 64 ($4 \times 4 \times 4$) possible codons. These 64 codons form the **genetic code** – the set of instructions that tells a cell the order in which amino acids are to be joined together to form a protein (Figure 3.8). Despite the fact that the linear sequence of codons in DNA determines the linear sequence of amino acids in proteins, the DNA helix does not itself play a role in protein synthesis. The translation of the sequence from codons into amino acids occurs through the intervention of members of a third class of molecule – mRNA. Messenger RNA acts as a template, guiding the assembly of amino acids into a polypeptide chain. Messenger RNA uses the same code as the one used in DNA with one difference: in mRNA the base uracil (U) is used in place of thymine (T). When we write the genetic code we usually use the RNA format, that is, we use U instead of T.

The code is read in sequential groups of three, codon by codon. Adjacent codons do not overlap and each triplet of bases specifies one particular amino acid. This discovery was made by Sydney Brenner, Francis Crick, and their colleagues by studying the effect of various mutations (changes in the DNA sequence) on the bacteriophage T4, which infects the common bacterium *E. coli*. If a mutation caused either one or two nucleotides to be added or deleted from one end of the T4 DNA, then a defective polypeptide was produced, with a completely different sequence of amino acids. However, if three bases were added or deleted, then the protein made often retained its normal function. These

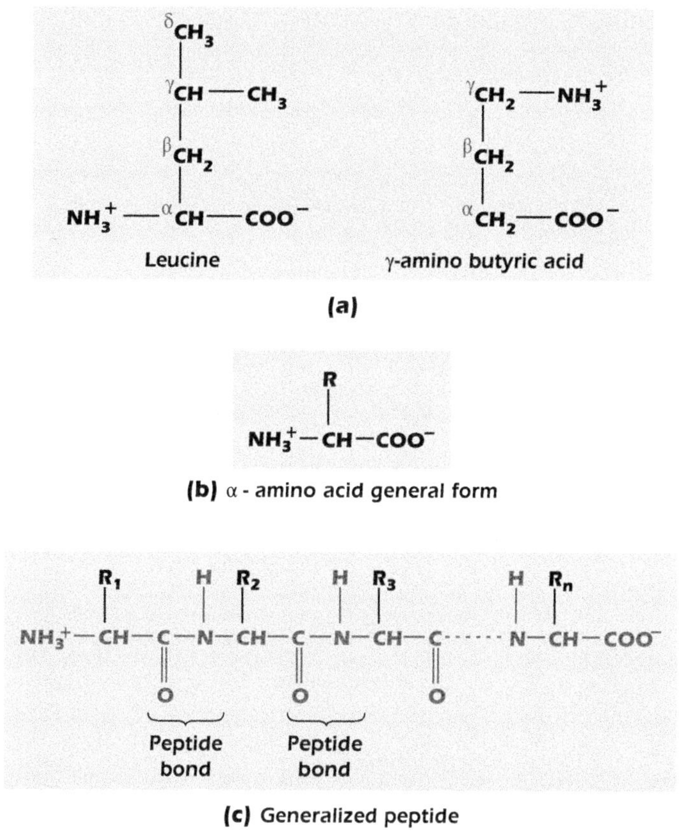

Figure 3.7. Amino acids and the peptide bond.

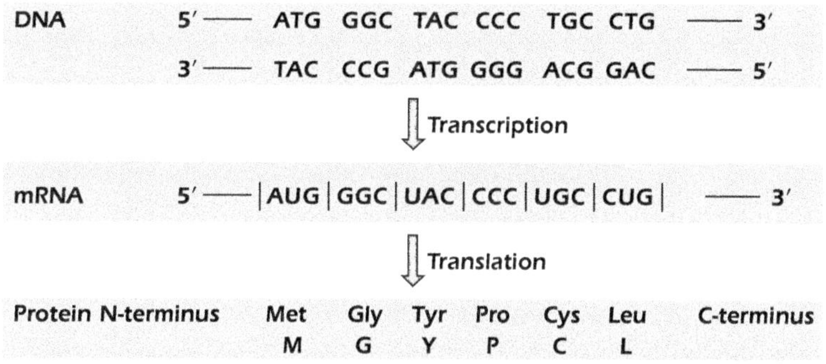

(a)

(b) α - amino acid general form

(c) Generalized peptide

Figure 3.8. DNA makes RNA makes protein: the central dogma of molecular biology.

proteins were found to be identical to the original protein, except for the addition or loss of one amino acid.

The identification of the triplets encoding each amino acid began in 1961. This was made possible by using a cell-free protein synthesis system prepared by breaking open *E. coli* cells. Synthetic RNA polymers, of known sequence, were added to the cell-free system together with the 20 amino acids. When the RNA template contained only uridine residues (poly-U) the polypeptide produced contained only phenylalanine – therefore codon UUU must specify phenylalanine. A poly(A) template produced a polypeptide of lysine and poly-C one of proline: AAA and CCC must therefore specify lysine and proline, respectively. Synthetic RNA polymers containing all possible combinations of the bases G, A, U, and C, were added to the cell-free system to determine the codons for the other amino acids. A template made of the repeating unit CU gave a polypeptide with the alternating sequence leucine–serine. Because the first amino acid in the chain was found to be leucine, CUC must code for leucine and UCU must code for serine. Although much of the genetic code was read in this way, the amino acids defined by some codons were particularly hard to determine. Only when specific **transfer RNA** molecules (page 85) were used was it possible to demonstrate that GUU codes for valine. The genetic code was finally solved by the combined efforts of several research teams. The leaders of two of these, Marshall Nirenberg and Har Gobind Khorana, received the Nobel prize in 1968 for their part in cracking the code.

Amino Acid Names Are Abbreviated

To save time we usually write an amino acid as either a three-letter abbreviation, for example, glycine is written as Gly and leucine as Leu, or as a one-letter code, for example, glycine is G and leucine is L. Figure 3.9 shows the full name and the three- and one-letter abbreviations used for each of the 20 amino acids found in proteins.

The Code Is Degenerate but Unambiguous

To introduce the terms *degenerate* and *ambiguous*, consider the English language. English shows considerable degeneracy, meaning that the same concept can be indicated using a number of different words – think, for example, of *lockup, cell, pen, pound, brig,* and *dungeon.* English also shows ambiguity, so that it is only by context that one can tell

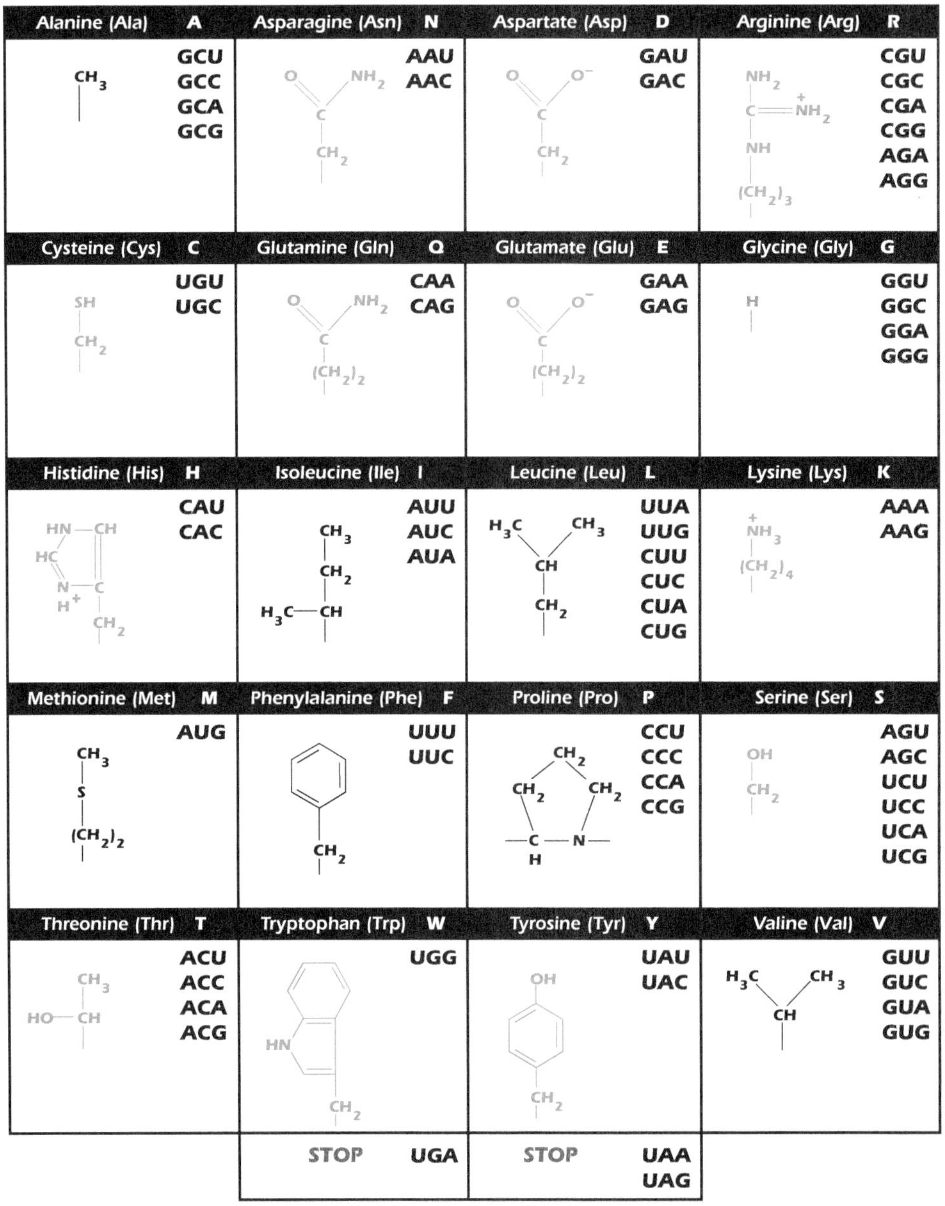

Figure 3.9. The genetic code. Amino acid side chains are shown in alphabetical order together with the three- and one-letter amino acid abbreviations. Hydrophilic side chains are shown in green, hydrophobic side chains are black, and the minimal side chain of glycine is shown in gray. The significance of this distinction is discussed in Chapter 7. To the right of each amino acid we show the corresponding mRNA codons.

whether *cell* means *a lockup* or *a living aqueous droplet enclosed by a membrane*. Like the English language the genetic code shows degeneracy but, unlike language, the genetic code is unambiguous.

The 64 codons of the genetic code are shown in Figure 3.9 together with the side chains of the amino acids for which each codes. Amino acids with hydrophilic side chains are shown in green while those with hydrophobic side chains are in black. Glycine, which has a hydrogen for a side chain, is shown in gray. The importance of these distinctions will be discussed in Chapter 7. Methionine is encoded by a single codon: AUG. Tryptophan is also encoded by a single codon, but the other 18 amino acids are encoded by more than one codon and so the code is **degenerate.** Although there are 64 possible codons, there are only 20 amino acids. Sixty-one codons specify an amino acid and the remaining three act as **stop signals** for protein synthesis (Figure 3.9). No triplet codes for more than one amino acid and so the code is **unambiguous.** Notice that when two or more codons specify the same amino acid, they usually only differ in the third base of the triplet. Thus **single base substitutions** in the third base can often leave the amino acid sequence unaltered. Perhaps degeneracy evolved in the triplet system to avoid a situation in which 20 codons each meant one amino acid and 44 specified none. If this were the case, then most mutations would stop protein synthesis dead.

Start and Stop Codons and the Reading Frame

The order of the codons in DNA and the amino acid sequence of a protein are colinear. The **start signal** for protein synthesis is the codon AUG, specifying the incorporation of methionine. Because the genetic code is read in blocks of three, there are three potential **reading frames** in any

mRNA. Figure 3.10 shows that only one of these results in the synthesis of the correct protein. When we look at a sequence of bases, it is not obvious which of the reading frames should be used to code for the protein. As we shall see later (page 89), the ribosome scans along the mRNA until it encounters an AUG. This both defines the first amino acid of the protein and the reading frame used from that point on. A mutation that inserts or deletes a nucleotide will change the normal reading frame and is called a **frameshift mutation** (Figure 3.11).

The codons UAA, UAG, and UGA are stop signals for protein synthesis. A base change that causes an amino acid codon to become a **stop codon** is known as a **nonsense mutation** (Figure 3.11). If, for example, the codon for tryptophan UGG changes to UGA, then a premature stop signal will have been introduced into the messenger RNA template. A shortened protein, usually without function, is produced.

The Code Is Nearly Universal

The code shown in Figure 3.9 is the one used by organisms as diverse as *E. coli* and humans for their nuclear-encoded proteins. It was originally thought that the code would be

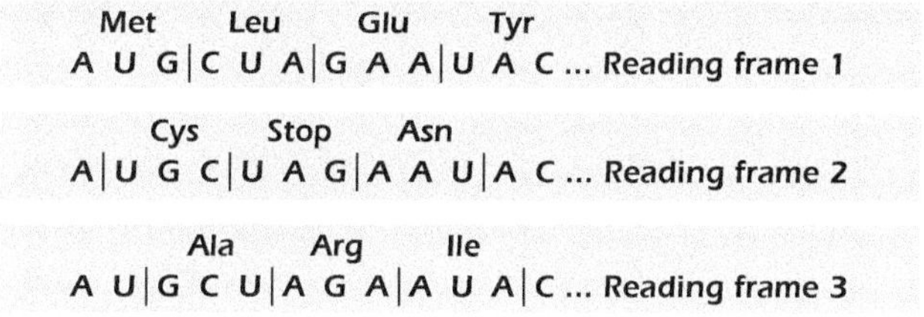

Figure 3.10. Reading frames. The genetic code is read in blocks of three.

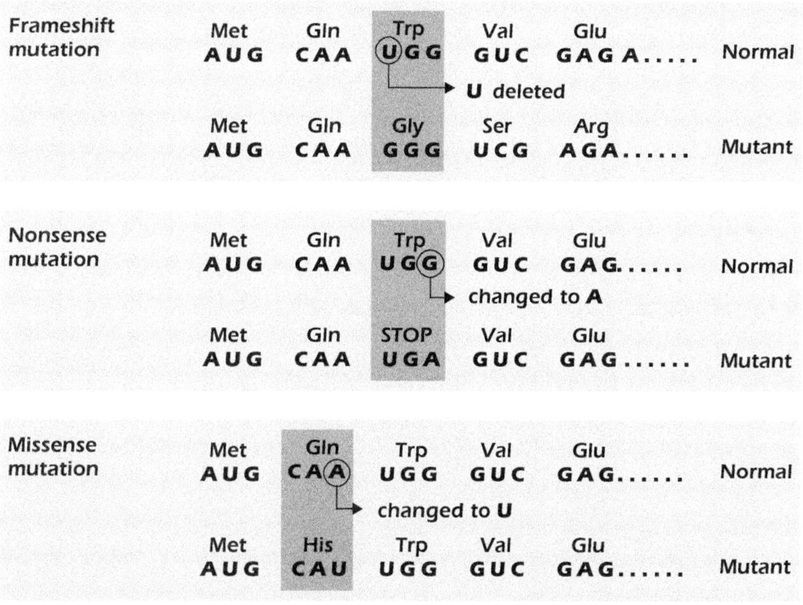

Figure 3.11. Mutations that alter the sequence of bases.

universal. However, several mitochondrial genes use UGA to mean tryptophan rather than *stop*. The nuclear code for some unicellular eukaryotes uses UAA and UAG to code for glutamine rather than *stop*.

Missense Mutations

A mutation that changes the codon from one amino acid to that for another by substitution of one base for another is a **missense mutation** (Figure 3.11). As shown in Figure 3.9, the second base of each codon shows the most consistency with the chemical nature of the amino acid it encodes. Amino acids with hydrophobic side chains, shown in black in Figure 3.9, have a U or a C – a pyrimidine – in the second position. With two exceptions, serine and threonine, amino acids with hydrophilic side chains, shown in green in Figure 3.9, have a G or an A – a purine – in the second

position. This has implications for mutations of the second base. Substitution of a purine for a pyrimidine is very likely to change the chemical nature of the amino acid side chain significantly and can therefore seriously affect the protein. Sickle cell anemia is an example of such a mutation. At position 6 in the β-globin chain of hemoglobin, the mutation in DNA changes a glutamate residue encoded by GAG to a valine residue encoded by GTG (GUG in RNA). The shorthand notation for this mutation is E6V, meaning that the glutamate (E) at position 6 of the protein becomes a valine (V). This change in amino acid alters the overall charge of the chain and the hemoglobin tends to precipitate in the red blood cells of those affected. The cells adopt a sickle shape and therefore tend to block blood vessels, causing sickle cell anemia with painful cramp-like symptoms and progressive damage to vital organs.

 BrainBox 3.2   William Warrick Cardozo

William Warrick Cardozo. Source: AAREG. Image from https://aaregistry.org/story/sickle-cell-pioneer-willliam-w-cardozo/.

The peculiar shape of red blood cells in patients with sickle cell anemia was first described in 1910 but little experimental investigation had been conducted until William Warrick Cardozo published a paper in 1937 reporting a comprehensive study of the largest number of patients ever tested for the disease. Cardozo was a pediatrician whose research on sickle cell anemia was conducted during a two-year fellowship in pediatrics at the Children's Memorial Hospital and Provident Hospital in Chicago. Cardozo's findings confirmed the heritability of the disorder and revealed that "the sickling factor remains within the cell, no matter how long preserved, as long as the cell itself remains intact." He concluded that future therapeutic interventions would need to be interventions on the cell itself. Today, the only cure for sickle cell anemia is a stem cell or bone marrow transplant that replaces the damaged red blood cells with healthy ones.

Medical Relevance 3.2   Osteogenesis Imperfecta

Collagen is the most abundant protein in the body and a major component of the extracellular matrix. The triple helix of Type I collagen is made up of two alpha1 chains and an alpha2 chain encoded by the COL1A1 and COL1A2 genes, respectively. The repetition of glycine, the smallest amino acid, at every third position in the triple helical domain of collagen allows the formation of the triple helix. A change of one base in one of these codons in COL1A1 could cause a missense mutation and the introduction of any one of eight different amino acids at this position. These altered amino acid residues affect the structural properties of collagen, disrupting the extracellular matrix and causing the disease Osteogenesis Imperfecta. The clinical severity of the disease ranges from mild to fatal and correlates with the identity of the amino acid that replaces glycine.

Answer to thought question: Glycine is coded for by four codons: GGU, GGC, GGA, and GGG. A single base substitution in the third base will have no effect on the protein since the codon will still encode glycine. For each of the four codons GGU, GGC, GGA, and GGG we can tabulate the effect of a single base substitution in the first or second base; the unmutated codon is shown in green and the mutated base in red:

Base 1	Base 2	Base 3	
G	G	U	coding for glycine
A	G	U	coding for serine
U	G	U	coding for cysteine
C	G	U	coding for arginine
G	A	U	coding for aspartate
G	U	U	coding for valine
G	C	U	coding for alanine

Base 1	Base 2	Base 3	
G	G	C	coding for glycine
A	G	C	coding for serine
U	G	C	coding for cysteine
C	G	C	coding for arginine
G	A	C	coding for aspartate
G	U	C	coding for valine
G	C	C	coding for alanine

Base 1	Base 2	Base 3	
G	G	A	coding for glycine
A	G	A	coding for arginine
U	G	A	coding for STOP
C	G	A	coding for arginine
G	A	A	coding for glutamate
G	U	A	coding for valine
G	C	A	coding for alanine

Base 1	Base 2	Base 3	
G	G	G	coding for glycine
A	G	G	coding for arginine
U	G	G	coding for tryptophan
C	G	G	coding for arginine
G	A	G	coding for glutamate
G	U	G	coding for valine
G	C	G	coding for alanine

so the eight possible substituted amino acids are alanine, arginine, aspartate, cysteine, glutamate, serine, tryptophan, and valine. What other kind of mutation could also arise from a single base substitution of a codon for glycine? The answer is a nonsense mutation since a single base substitution that converts GGA to UGA creates a STOP codon.

SUMMARY

1. DNA, the cell's database, contains the genetic information necessary to encode RNA and protein.

2. The information is stored in the sequence of four bases. These are the purines – guanine and adenine – and the pyrimidines – thymine and cytosine. Each base is attached to the 1′-carbon atom of the sugar deoxyribose. A phosphate group is attached to the 5′-carbon atom of the sugar. The base + sugar + phosphate is called a nucleotide.

3. The enzyme DNA polymerase joins nucleotides together by forming a phosphodiester bond between the 5′-phosphate group of one nucleotide and the hydroxyl group on the 3′ carbon of deoxyribose of another. This gives rise to the sugar-phosphate backbone structure of DNA.

4. The two strands of DNA are held together in an antiparallel double-helical structure because guanine hydrogen bonds with cytosine and adenine hydrogen bonds with thymine. This means that if the sequence of one strand is known, that of the other can be inferred. The two strands are complementary in sequence.

5. DNA binds to histone and nonhistone proteins to form chromatin. DNA is wrapped around histones to form a nucleosome structure. This is then folded again and again. This packaging compresses the DNA molecule to a size that fits into the cell.

6. The genetic code specifies the sequence of amino acids in a polypeptide. The code is transferred from DNA to mRNA and is read in groups of three bases (a codon) during protein synthesis. There are 64 codons; 61 specify an amino acid and 3 are the stop signals for protein synthesis.

FURTHER READING

Annunziato, A.T. (2008). DNA packaging: nucleosomes and chromatin. *Nature Education* 1 (1): 2008. http://www.nature.com/scitable/topicpage/DNA-Packaging-Nucleosomes-and-Chromatin-310.

DiGuilo, M. (1997). The origin of the genetic code. *Trends in Biochemical Sciences* 22: 49–50.

Franklin, R.E. and Gosling, R.G. (1953)). Molecular configuration in sodium thymonucleate. *Nature* 171: 740.

Lappalainen, T., Scott, A.J., Brandt, M., and Hall, I.M. (2019). Genomic analysis in the age of human genome sequencing. *Cell* 177 (1): 70–84. https://doi.org/10.1016/j.cell.2019.02.032.

Maddox, B. (2002). *Rosalind Franklin: The Dark Lady of DNA*. New York: Harper Collins.

Ravichandran, S., Subramani, V.K., and Kim, K.K. (2019 Jun). Z-DNA in the genome: from structure to disease. *Biophysical Reviews* 11 (3): 383–387. https://doi.org/10.1007/s12551-019-00534-1. Epub 2019 May 22. PMID: 31119604; PMCID: PMC6557933.

Sivakumar, A., de Las Heras, J.I., and Schirmer, E.C. (2019). Spatial genome organization: from development to disease. *Frontiers in Cell and Development Biology* 7: 18. https://doi.org/10.3389/fcell.2019.00018.

Travers, A. & Muskhelishvili, G. (2015 Jun). DNA structure and function. The FEBS Journal 282 (12):2279–2295. doi: https://doi.org/10.1111/febs.13307. Epub 2015 Jun 2. PMID: 25903461.

Wang, J.C. (2002 Jun). Cellular roles of DNA topoisomerases: a molecular perspective. *Nature Reviews. Molecular Cell Biology* 3 (6): 430–440. https://doi.org/10.1038/nrm831. PMID: 12042765.

Watson, J.D. and Crick, F.H.C. (1953). A structure for deoxyribose nucleic acid. *Nature* 171: 737.

● REVIEW QUESTIONS

3.1 Theme: Mutations

A frameshift
B missense
C nonsense
D none of the above

Consider the mRNA strand 5′ ACU AUC UGU AUU AUG UUA CAC CCA 3′ coding for the amino acid sequence TICIMLHP. For each of the errors described below, choose the appropriate description from the list above. Refer to Figure 3.9 on page 44 while answering this question.

1. a change of a U to a A in the 6th codon of the sequence, generating the sequence
   5′ ACUAUCUGUAUUAUGUAACACCCA 3′

2. a change of a U to a C in the 6th codon of the sequence, generating the sequence
   5′ ACUAUCUGUAUUAUGCUACACCCA 3′

3. a change of a U to a G in the 2nd codon of the sequence, generating the sequence
   5′ ACUAGCUGUAUUAUGUUACACCCA 3′
4. deletion of a U in the 3rd codon, generating the sequence
   5′ ACUAUCUGAUUAUGUUACACCCA 3′
5. deletion of an A in the 4th codon, generating the sequence
   5′ ACUAUCUGUUUAUGUUACACCCA 3′

3.2 Theme: Bases and amino acids

A   adenine
B   alanine
C   arginine
D   aspartate
E   cytosine
F   glutamate
G   glycine
H   guanine
I   thymine
J   uracil
K   aline

From the above list of compounds, select the one described by each of the descriptions or questions below.

1. the base that is found in RNA but not in DNA
2. a positively charged amino acid that is found in large amounts in chromatin, where it neutralizes the negative charge on the phosphodiester bonds of DNA

3. a protein is described as having the mutation G5E. Which amino acid is present in this protein in place of the amino acid present in the normal protein?
4. the base that pairs with guanine in double-stranded DNA
5. the base that pairs with thymine in double-stranded DNA

3.3 Theme: Structures associated with DNA

A   30 nm solenoid
B   codon
C   euchromatin
D   gene
E   heterochromatin
F   nucleoid
G   nucleosome

From the above list of structures, select the one described by each of the descriptions below.

1. a highly compacted, darkly staining substance comprising DNA and protein found at the nuclear periphery
2. a mass of DNA and associated proteins lying free in the cytoplasm
3. a structure formed when a 146 base-pair length of DNA winds around a complex of histone proteins
4. the form adopted by those parts of chromosomes that are being transcribed into RNA

⬤ THOUGHT QUESTION

In Medical Relevance 3.2 on page 47 we state that a single base substitution in a codon for glycine could result in the substitution of any one of eight different amino acids at this position. Explain this statement and list the eight amino acids. What other kind of mutation could also arise from a single base substitution in a codon for glycine?

4

DNA AS A DATA STORAGE MEDIUM

The genetic material DNA must be faithfully replicated every time a cell divides to ensure that the information encoded in it is passed unaltered to the daughter cells. DNA molecules have to last a long time compared to RNA and protein. The sugar-phosphate backbone of DNA is a very stable structure because there are no free hydroxyl groups on the sugar – they are all used up in bonds, either to the base or to phosphate. The bases themselves are protected from chemical attack because they are hidden within the DNA double helix. Nevertheless, chemical changes – **mutations** – do occur in the DNA molecule and cells have had to evolve mechanisms to ensure that mutation is kept to a minimum. Repair systems are essential for both cell survival and to ensure that the correct DNA sequence is passed on to daughter cells. This chapter describes how new DNA molecules are made during chromosome duplication and how the cell acts to correct base changes in DNA.

 ## DNA REPLICATION

During replication the two strands of the double helix unwind. Each then acts as a template for the synthesis of a new strand. This process generates two double-stranded daughter DNA molecules, each of which is identical to the parent molecule. The base sequences of the new strands are complementary in sequence to the template strands upon which they were built. This means that G, A, T, and C in the old strand cause C, T, A, and G, respectively, to be placed in the new strand.

The DNA Replication Fork

Replication of a new DNA strand starts at specific sequences known as **origins of replication.** The small circular chromosome of *Escherichia coli* has only one of these, whereas eukaryotic chromosomes, which are usually linear and much larger, have many. At each origin of replication, the parental strands of DNA untwist to give rise to a structure known as the **replication fork** (Figure 4.1). This unwinding permits each parental strand to act as a template for the synthesis of a new strand. The structure of the double helix and the nature of DNA replication pose a mechanical problem. How do the two strands unwind and how do they stay unwound so that each can act as a template for a new strand?

 ## PROTEINS OPEN UP THE DNA DOUBLE HELIX DURING REPLICATION

The DNA molecule must be opened up before replication can proceed. The helix is a very stable structure, and in a test tube the two strands separate only when the temperature reaches about 90 °C. In the cell, the combined actions of

Cell Biology: A Short Course, Fourth Edition. Stephen Bolsover, Andrea Townsend-Nicholson,
Greg FitzHarris, Elizabeth Shephard, Jeremy Hyams and Sandip Patel.

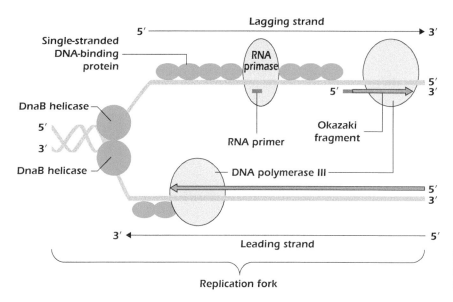

Figure 4.1. DNA replication. The helicases, and the replication fork, are moving to the left.

several proteins help to separate the two strands. Much of our knowledge of replication comes from studying *E. coli*, but similar systems operate in all organisms, bacteria, archaea, and eukaryotes. The proteins *E. coli* uses to open up the double helix during replication include DnaA, DnaB, DnaC, and single-stranded DNA-binding proteins.

DnaA Protein

Several copies of the protein DnaA, which is activated by a molecule of ATP, bind to four sequences of nine base pairs within the *E. coli* origin of replication (*ori C*). This causes the two strands to begin to separate (or "melt") because the hydrogen bonds in DNA are broken near to where the DnaA protein binds. The DNA is now in the **open complex** formation and has been prepared for the next stage in replication, which is to open up the helix even further.

DnaB and DnaC Proteins

DnaB is a **helicase.** It moves along a DNA strand, breaking hydrogen bonds, and in the process unwinds the helix (Figure 4.1). Two molecules of DnaB are needed, one for each strand of DNA. One DnaB attaches to one of the template strands and moves in the 5′ to 3′ direction; the second DnaB attaches to the other strand and moves in the 3′ to 5′ direction. The unwinding of the DNA double helix by DnaB is an ATP-dependent process. DnaB is escorted to the DNA strands by another protein, DnaC. However, having delivered DnaB to its destination, DnaC plays no further role in replication.

Single-Stranded DNA-Binding Proteins

As soon as DnaB unwinds the two parental strands, they are engulfed by single-stranded DNA-binding proteins. These proteins bind to adjacent groups of 32 nucleotides. DNA

covered by single-stranded DNA-binding proteins is rigid, without bends or kinks. It is, therefore, an excellent template for DNA synthesis. Single-stranded binding proteins are sometimes called helix-destabilizing proteins.

BIOCHEMISTRY OF DNA REPLICATION

In prokaryotes the synthesis of a new DNA molecule is catalyzed by the enzyme **DNA polymerase III.** Its substrates are the four deoxyribonucleoside triphosphates, dGTP, dATP, dTTP, and dCTP. DNA polymerase III catalyzes the formation of a phosphodiester bond (Figure 3.3 on page 37) between the 3′ hydroxyl of the sugar residue on the most recently added nucleotide and the 5′ phosphate of the incoming nucleotide. The elongation of the new DNA molecule takes place in the 5′ to 3′ direction (Figure 4.2a). The base sequence of a newly synthesized DNA strand is dictated by the base sequence of its parental strand. If the sequence of the template strand is 3′ CATCGA 5′, then that of the daughter strand is 5′ GTAGCT 3′. In eukaryotes, DNA replication is performed by three isoforms, DNA polymerases α, δ, and ε, but the mechanism is much the same.

DNA polymerase III can only add a nucleotide to a free 3′-hydroxyl group and therefore synthesizes DNA in the 5′ to 3′ direction. The template strand is read in the 3′ to 5′ direction. However, the two strands of the double helix are antiparallel. They cannot be synthesized in the same direction because only one has a free 3′-hydroxyl group, the other has a free 5′-phosphate group. No DNA polymerase has been found that can synthesize DNA in the 3′ to 5′ direction, that is, by attaching a nucleotide to a 5′ phosphate, so the synthesis of the two daughter strands must differ. One strand, the **leading strand,** is synthesized continuously while the other, the **lagging strand,** is synthesized discontinuously.

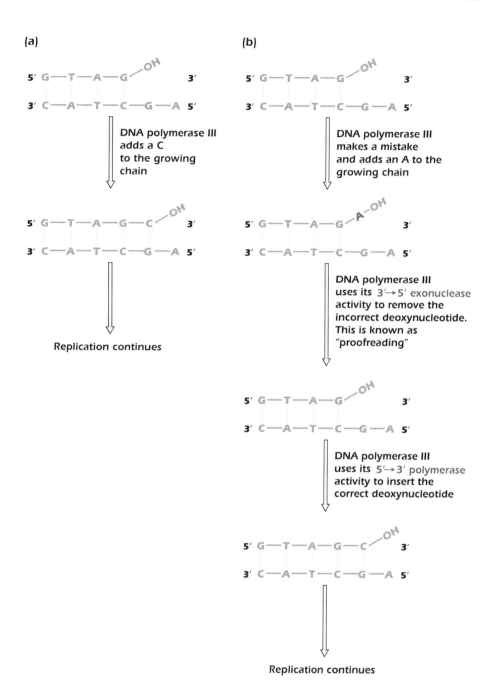

Figure 4.2. DNA polymerase III can correct its own mistakes.

DNA polymerase III can synthesize both daughter strands, but must make the lagging strand as a series of short 5′ to 3′ sections (Figure 4.1). These fragments of DNA, called **Okazaki fragments** after Reiji Okazaki who discovered them in 1968, are then joined together by **DNA ligase.**

Medical Relevance 4.1   Inhibiting DNA Polymerase Fights Cancer

Drugs that inhibit DNA replication prevent cells dividing. Cytarabine (also known as cytosine arabinoside and arabinofuranosyl cytidine) is sold as Cytosar-U, Tarabine PFS, Depocyt, and AraC. It inhibits eukaryote DNA polymerases α, δ, and ε, and is therefore widely used to treat cancers. It is on the World Health Organization's model list of essential medicines.

Example 4.1   The Meselson and Stahl Experiment

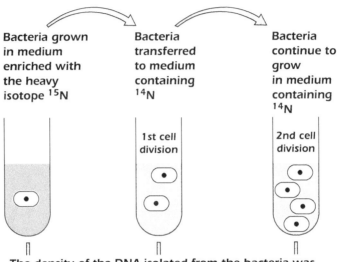

Bacteria grown in medium enriched with the heavy isotope ^{15}N

Bacteria transferred to medium containing ^{14}N

Bacteria continue to grow in medium containing ^{14}N

1st cell division

2nd cell division

The density of the DNA isolated from the bacteria was analyzed by high speed centrifugation

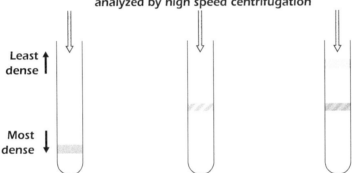

Least dense

Most dense

This experiment demonstrated that the two strands of the parental DNA unwind and each acts as template for a new daughter strand

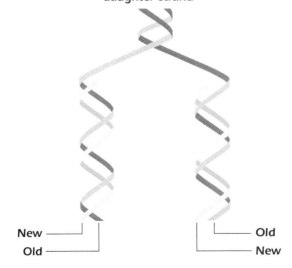

New

Old

Old

New

In 1958 Matthew Meselson and Franklin Stahl designed an ingenious experiment to test whether each strand of the double helix does indeed act as a template for the synthesis of a new strand. They grew the bacterium *E. coli* in a medium containing the heavy isotope ^{15}N that could be incorporated into new DNA molecules. After several cell divisions they transferred the bacteria, now containing "heavy" DNA, to a medium containing only the lighter, normal, isotope ^{14}N. Any newly synthesized DNA molecules would therefore be lighter than the original parent DNA molecules containing ^{15}N. The difference in density between the heavy and light DNAs allows their separation using very high-speed centrifugation. The results of this experiment are illustrated in the figure. DNA isolated from cells grown in the ^{15}N medium had the highest density and migrated the furthest during centrifugation.

The lightest DNA was found in cells grown in the ^{14}N medium for two generations, whereas DNA from bacteria grown for only one generation in the lighter ^{14}N medium had a density halfway between these two. This is exactly the pattern expected if each strand of the double helix acts as a template for the synthesis of a new strand. The two heavy parental strands separated during replication, with each acting as a template for a newly synthesized light strand, which remained bound to the heavy strand in a double helix. The resulting DNA was therefore of intermediate density. Only in the second round of DNA replication, when the light strands created during the first round of replication were allowed to act as templates for the construction of complementary light strands, did DNA double helices composed entirely of ^{14}N-containing building blocks appear.

DNA Synthesis Requires an RNA Primer

DNA polymerase III cannot itself initiate the synthesis of DNA. The enzyme **primase** is needed to catalyze the formation of a short stretch of RNA complementary in sequence to the DNA template strand (Figure 4.1). This RNA chain, the primer, is needed to prime (or start) the synthesis of the new DNA strand. DNA polymerase III catalyzes the formation of a phosphodiester bond between the 3′-hydroxyl group of the RNA primer and the 5′-phosphate group of the appropriate deoxyribonucleotide. Several RNA primers are made along the length of the lagging strand template. Each is extended in the 5′ to 3′ direction by DNA polymerase III until it reaches the 5′ end of the next RNA primer. In prokaryotes the lagging strand is primed about every 1000 nucleotides whereas in eukaryotes this takes place every 200 nucleotides.

RNA Primers Are Removed

Once the synthesis of the DNA fragment is complete, the RNA primers must be replaced by deoxyribonucleotides. In prokaryotes the enzyme DNA polymerase I removes ribonucleotides using its 5′ to 3′ **exonuclease** activity and then uses its 5′ to 3′ polymerizing activity to incorporate deoxyribonucleotides. In this way, the entire RNA primer is replaced by DNA. Synthesis of the lagging strand is completed by the enzyme DNA ligase, which joins the DNA fragments together by catalyzing the formation of phosphodiester bonds between adjacent fragments.

Eukaryotic organisms use an enzyme called **ribonuclease H** to remove their RNA primers. These enzymes break phosphodiester bonds in an RNA strand that is hydrogen-bonded to a DNA strand. Within each cell, ribonuclease H2 is involved in removing RNA primers during replication of nuclear genomic DNA and ribonuclease H1 is involved in removing RNA primers during replication of mitochondrial genomic DNA.

The Self-Correcting DNA Polymerase

The genome of *E. coli* consists of about 4.6×10^6 base pairs of DNA. DNA polymerase III makes a mistake about every 1 in 10^5 bases and joins an incorrect deoxyribonucleotide to the growing chain. If unchecked, these mistakes would lead to a catastrophic mutation rate. Fortunately, DNA polymerase III has a built-in proofreading mechanism that corrects its own errors. If an incorrect base is inserted into the newly synthesized daughter strand, the enzyme recognizes the change in shape of the double-stranded molecule, which arises through incorrect base pairing, and DNA synthesis stops (Figure 4.2b). DNA polymerase III then uses its 3′ to 5′ exonuclease activity to remove the incorrect deoxyribonucleotide and replace it with the correct one. DNA synthesis then continues. DNA polymerase III hence functions as a self-correcting enzyme.

Mismatch Repair Backs Up the Proofreading Mechanism

The proofreading function of DNA polymerase III improves the accuracy of DNA replication about 100-fold. However, sometimes the enzyme does miss a nucleotide that has been incorrectly inserted into the newly synthesized DNA strand. Cells have evolved a backup mechanism, **mismatch repair,** that detects when an incorrect nucleotide has been inserted into the daughter strand (Figure 4.3). The repair mechanism relies on the cell being able to distinguish, within the double helix, between the template strand (the parental strand) and the newly synthesized strand (the daughter strand).

(a)

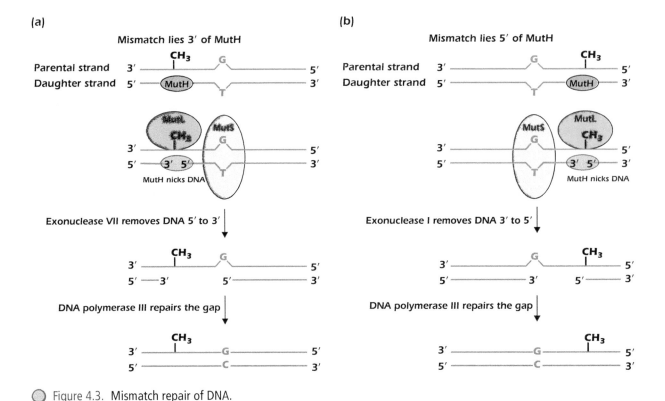

(b)

 Figure 4.3. Mismatch repair of DNA.

We best understand this repair process in *E. coli*. The bacterium has an enzyme called Dam methylase that adds a -CH₃ group, called a methyl group, onto the **A** of the sequence 5′ GATC3′. This sequence occurs very frequently in DNA, about once every 256 bp. The methylation of DNA happens very soon after a DNA strand has been replicated. However, for a short time during replication the double-stranded DNA molecule will have one strand methylated (the parental strand) and one strand not methylated (the daughter strand). The DNA molecule is said to be hemi-methylated (half methylated). Because the newly synthesized strand has not yet been methylated the cell knows that if a mismatch in base pairing has occurred between the two strands it is the nonmethylated, newly synthesized, strand that must carry the mistake.

A protein called MutH binds on the newly synthesized strand at a site opposite a methylated **A** in the template strand. If there is no mismatched base pair nearby then MutH does nothing. However, if two other proteins called MutL and MutS have detected a mismatched base pair then MutH, which is an **endonuclease,** is activated and nicks (cleaves a phosphodiester bond between two nucleotides in) the unmethylated newly synthesized strand. This allows a stretch of DNA containing the mismatched base pair to be removed. Two different proteins are involved in removing the stretch of DNA. If MutH nicks the DNA 5′ to the mismatch (Figure 4.3a), then exonuclease VII degrades the DNA strand in the 5′ to 3′ direction. However, if MutH nicks the DNA 3′ to the mismatch (Figure 4.3b), then the DNA strand

is removed by exonuclease I in the 3′ to 5′ direction. In either case, the gap in the daughter strand is then replaced by DNA polymerase III.

DNA REPAIR AFTER REPLICATION

Deoxyribonucleic acid can be damaged by a number of agents, which include oxygen, water, naturally occurring chemicals in our diet, and radiation. Because damage to DNA can change the sequence of bases, a cell must be able to repair alterations in the DNA code if it is to survive and pass on the DNA database unaltered to its daughter cells.

Spontaneous and Chemically Induced Base Changes

The most common damage suffered by a DNA molecule is **depurination** – the loss of an adenine or guanine because the bond between the purine base and the deoxyribose sugar to which it is attached spontaneously hydrolyzes (Figure 4.4). Within each human cell about 5000–10 000 depurinations occur every day.

Deamination is a less frequent event; it happens about 100 times a day in every human cell. Collision of H_3O^+ ions with the bond linking the amino group to carbon number 4 in cytosine sets off a spontaneous deamination that produces uracil (Figure 4.4). Cytosine base pairs with guanine, whereas uracil pairs with adenine. If this change were not

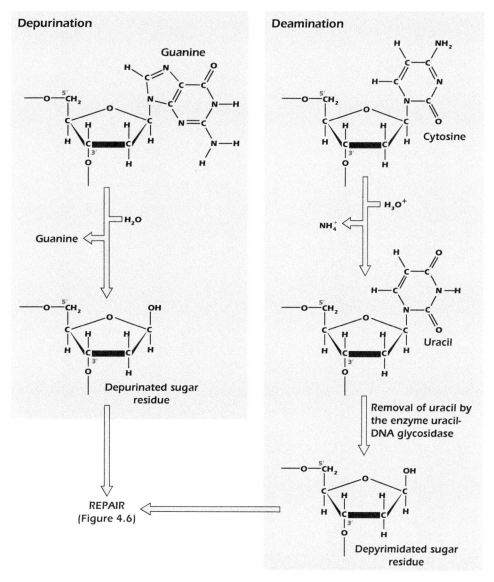

Figure 4.4. Spontaneous reactions corrupt the DNA database.

corrected, then a CG base pair would mutate to a UA base pair the next time the DNA strand was replicated, introducing a mutation at this position in one of the two copies of the DNA double helix that are obtained post-replication.

Ultraviolet light or chemical carcinogens such as benzopyrene, present in cigarette smoke, can also disrupt the structure of DNA. The absorption of ultraviolet light can cause two adjacent thymine residues to link and form a thymine dimer (Figure 4.5). If uncorrected, thymine dimers create a distortion in the DNA helix known as a **bulky lesion.** This inhibits normal base pairing between the two strands of the double helix and blocks the replication process. Ultraviolet light has a powerful germicidal action and is widely used to sterilize equipment. One of the reasons why bacteria are killed by this treatment is because the formation of large numbers of thymine dimers prevents replication.

Repair Processes

If there were no way to correct altered DNA, the rate of mutation would be intolerable. **DNA excision** and **DNA repair enzymes** have evolved to detect and to repair altered DNA. The role of the repair enzymes is to cut out (excise) the damaged portion of DNA and then to repair the base sequence. Much of our knowledge of DNA repair has been derived from studies on *E. coli,* but the general principles apply to other organisms such as ourselves. Repair is possible because DNA comprises two complementary strands. If the repair mechanisms can identify which of the two strands is the damaged one, it can be repaired to be as good as new by rebuilding it so that is again complementary to the undamaged strand.

Two types of excision repair are described in this section: base excision repair and nucleotide excision repair.

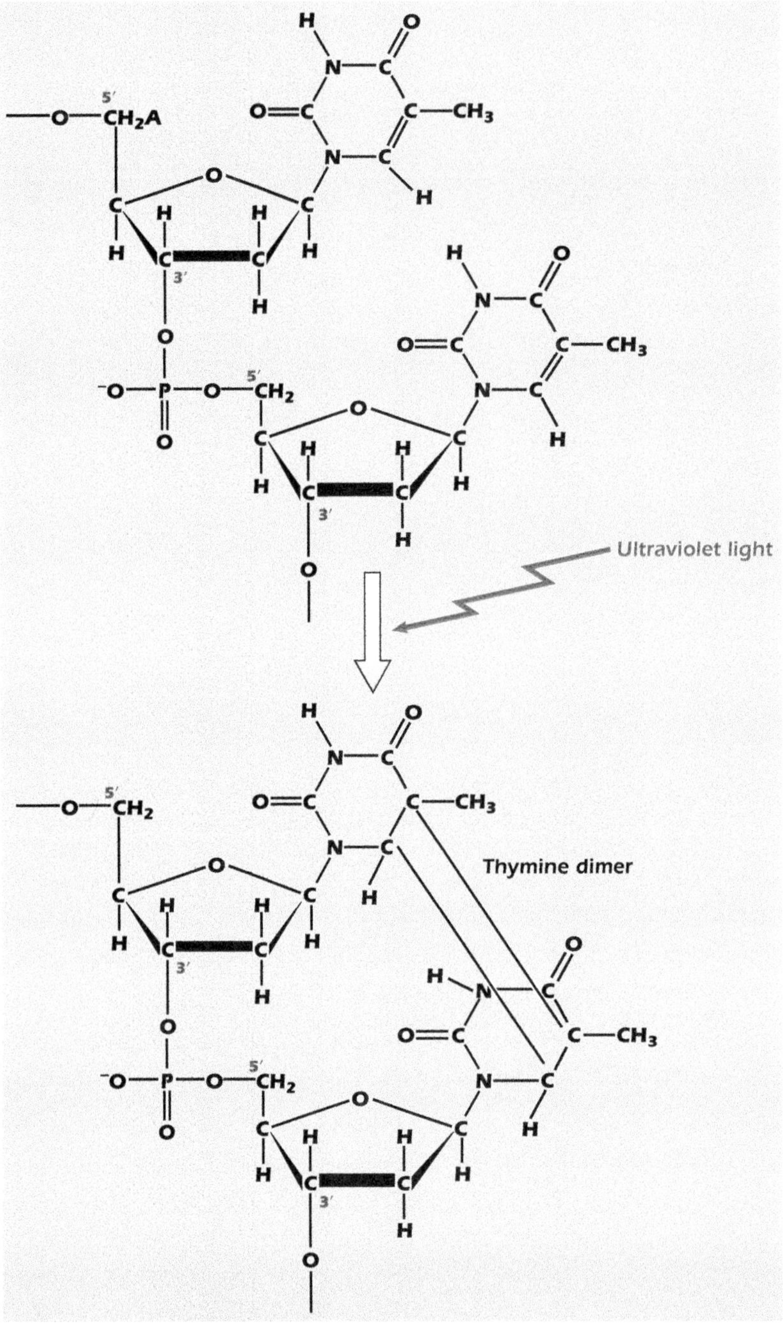

Figure 4.5. Formation of a thymine dimer in DNA.

The common themes for each of these repair mechanisms are: (i) an enzyme recognizes the damaged DNA, (ii) the damaged portion is removed, (iii) DNA polymerase inserts the correct nucleotide(s) into position (according to the base sequence of the second DNA strand), and (iv) DNA ligase joins the newly repaired section to the remainder of the DNA strand.

Base excision repair is needed to repair DNA that has lost a purine (depurination), or where a cytosine has been deaminated to uracil (U). Although uracil is a normal constituent of RNA, it does not form part of undamaged DNA and is recognized and removed by the repair enzyme **uracil – DNA glycosidase** (Figure 4.4). This leaves a gap in the DNA where the base had been attached to deoxyribose. There is no enzyme that can simply reattach a C into the vacant space on the sugar. Instead, an enzyme called **AP endonuclease** recognizes the gap and removes the sugar by breaking the phosphodiester bonds on either side (Figure 4.6). When DNA has been damaged by the loss of a purine (Figure 4.4), AP endonuclease also removes the sugar that has lost its base. The AP in the enzyme's name

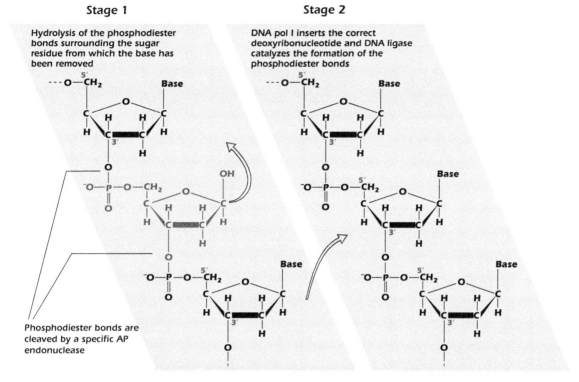

Stage 1

Hydrolysis of the phosphodiester bonds surrounding the sugar residue from which the base has been removed

Phosphodiester bonds are cleaved by a specific AP endonuclease

Stage 2

DNA pol I inserts the correct deoxyribonucleotide and DNA ligase catalyzes the formation of the phosphodiester bonds

Figure 4.6. Base excision repair.

means apyrimidinic (without a pyrimidine) or apurinic (without a purine).

The repair process for reinserting a purine or a pyrimidine into DNA is now the same (Figure 4.6). DNA polymerase I replaces the appropriate deoxyribonucleotide into position. DNA ligase then seals the strand by catalyzing the reformation of a phosphodiester bond.

Nucleotide excision repair is required to correct a thymine dimer. The thymine dimer, together with some 30 surrounding nucleotides, is excised from the DNA. Repairing damage of this bulky type requires several proteins because the exposed, undamaged, DNA strand must be protected from nuclease attack while the damaged strand is repaired by the actions of DNA polymerase I and DNA ligase.

Even with all these protection systems in place, the cell divisions that create and repair our bodies generate errors, so that the adult human contains many cells with **somatic mutations.** Most are irrelevant to the specialized operation of that cell, or merely reduce its ability to function. Some, however, can cause cancer (page 241).

GENE STRUCTURE AND ORGANIZATION IN EUKARYOTES

Introns and Exons – Additional Complexity in Eukaryotic Genes

Genes that code for proteins should be simple things: DNA makes RNA makes protein, and a gene codes for the amino acids of a protein by the three-base genetic code. In prokaryotes, indeed, a gene is a continuous series of bases that, read

---

Medical Relevance 4.2   Bloom's Syndrome and Xeroderma Pigmentosum

DNA helicases are essential proteins required to open up the DNA helix during replication. In Bloom's syndrome, mutations give rise to a defective helicase. The result is excessive chromosome breakage, and affected people are predisposed to many different types of cancers when they are young.

People who suffer from the genetic disorder xeroderma pigmentosum are deficient in one of the enzymes

for excision repair. As a result, they are very sensitive to ultraviolet light. They contract skin cancer even when they have been exposed to sunlight for very short periods because thymine dimers produced by ultraviolet light are not excised from their genomes.

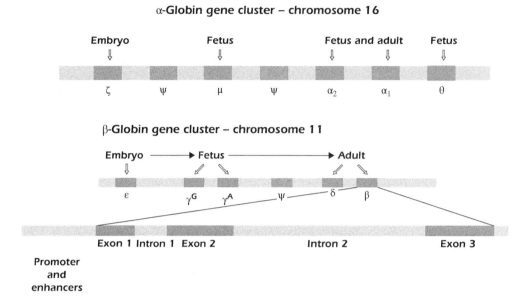

Figure 4.7. The human α-and β-globin gene family clusters. Ψ indicates a pseudogene. Adults only express α, β, and δ, and of these the expression of δ is very low. The exon/intron boundaries of the β-globin gene are indicated at the bottom.

in threes, code for the protein. This simple and apparently sensible system does not apply in eukaryotes. Instead, the protein-coding regions of almost all eukaryotic genes are organized as a series of separate bits interspersed with non-coding regions. The protein-coding regions of the split genes are **exons.** The regions between are called **introns,** short for intervening sequences. At the bottom of Figure 4.7 we show the structure of the β-globin gene, which contains three exons and two introns. Introns are often very long compared to exons. As happens in prokaryotes, messenger RNA complementary to the DNA is synthesized, but then the introns are spliced out before the mRNA leaves the nucleus (page 22). This means that a gene is much longer than the mRNA that ultimately codes for the protein. The name exon derives from the fact that these are the regions of the gene that, when transcribed into mRNA, exit from the nucleus.

In fact, there is an evolutionary rationale to this apparently perverse arrangement. As we will see (page 114), a single protein is often composed of a series of **domains,** with each domain performing a different role. The breaks between exons often correspond to domain boundaries. During evolution, reordering of exons has created new genes that have some of the exons of one gene, and some of the exons of another, and hence generates novel proteins composed of new arrangements of domains, each of which still does its job.

The Major Classes of Eukaryotic DNA

We do not yet fully understand the construction of our nuclear genome. Only about 1.1% of the human genome codes for exonic sequences (i.e. makes protein) with about 24% coding for introns. Most protein-coding genes occur only once in the genome and are called single-copy genes.

Many genes have been duplicated at some time during their evolution. Mutation over the succeeding generations causes the initially identical copies to diverge in sequence and produce what is known as a gene family. Members of a gene family usually have a related function, for example the products of the globin family transport oxygen from our lungs to our tissues. These genes generate related proteins or **isoforms,** which are often distinguished by placing a Greek letter after the protein name, for example, hemoglobin α and hemoglobin β. Different members of a family sometimes encode proteins that carry out the same specialized function but at different times during development. The α- and β-globin gene families, illustrated in Figure 4.7, are an example. The α-globin gene cluster is on human chromosome 16 while the β-globin gene cluster is on human chromosome 11. Hemoglobin is composed of two α globins and two β globins. The different globin proteins are produced at different stages of gestation, embryo to fetus to adult, to cope with the different oxygen transport requirements at each step. The duplication of genes and their subsequent divergence allows the expansion of the gene repertoire, the production of new protein molecules and the elaboration of ever more specialized gene functions during evolution.

Some sections of DNA are very similar in sequence to other members of their gene family but do not produce mRNA. These are known as **pseudogenes.** There are two in the α-globin gene cluster and one in the β-globin gene cluster (Ψ in Figure 4.7). Pseudogenes may be former genes that

IN DEPTH 4.1 THERE ARE MORE PROTEINS THAN GENES IN MULTI-CELLULAR ORGANISMS

As genomes of more and more organisms were sequenced, the most surprising feature to emerge was just how few genes supposedly complex organisms possess (Table 3.1 on page 39). The first eukaryotic genome to be sequenced was that of the budding yeast, *Saccharomyces cerevisiae,* the simple unicellular fungus that we use to make bread and beer. *S. cerevisiae* has 6091 genes. The fruit fly, *Drosophila melanogaster,* a much more complex organism with a brain, nervous and digestive systems, and the ability to fly and navigate, on the other hand, has 14 133 genes, or roughly twice the number in a yeast. Even more surprising was the finding that humans have only about 19 116 protein-coding genes. However, humans make many more than 19 116 proteins and it is these that contribute to the complexity of an organism such as ourselves.

How is it possible to have so few genes and yet make 100 000s of different proteins? It is the arrangement of human genes into exons and introns (page 59) that provides the solution. **Alternative splicing** (page 76) allows the cell to "cut and paste" exons in different ways to produce many different mRNAs from the same gene. The most extreme case known is the human gene called *SLO,* which encodes a protein found in some potassium channels (page 152). This gene has 35 exons, which can produce 40 320 different combinations of exons from a single gene. Estimates are that something like 50% of human genes show alternative splicing with the pattern of splicing (the range of proteins produced) varying from tissue to tissue. *Drosophila* genes also show alternative splicing but those of yeast, which contain few introns, do not.

---

have mutated to such an extent that they can no longer be transcribed into RNA. Some pseudogenes have arisen because an mRNA molecule has been copied back into DNA by an enzyme called reverse transcriptase found in some viruses (Medical Relevance 3.1 on page 39). Such pseudogenes are immediately recognizable because some or all of their introns were spliced out before the integration occurred. Some have the poly(A) tail characteristic of intact mRNA present in the gene (page 76). These are called processed pseudogenes.

Sometimes DNA that encodes RNA is repeated as a series of copies that follow one after the other along the chromosome. Such genes are said to be **tandemly repeated.** The genes that code for ribosomal RNAs (about 250 copies/cell), transfer RNAs (50 copies/cell), and histone proteins (20–50 copies/cell) are tandemly repeated. The products of these genes are required in large amounts.

This still leaves about 75% of our nuclear genome that lacks a very clearly understood function. A large proportion of this **extragenic** DNA is made up of **repetitive DNA** sequences that are repeated many times in the genome. Some sequences are repeated more than a million times and are called **satellite DNA.** The repeating unit is usually several hundred base pairs long, and many copies are often lined up next to each other in tandem repeats. Most of the satellite DNA is found in a region called the **centromere,** which plays a role in the physical movement of the chromosomes that occurs at cell division (page 235), and one theory is that it has a structural function.

Our genome also contains **minisatellite DNA** where the tandem repeat is about 25 bp long. Minisatellite DNA stretches can be up to 20 000 bp in length and are often found near the ends of chromosomes, a region called the **telomere.** **Microsatellite DNA** has an even smaller repeat unit of about

4 bp or less. Again, the function of these repeated sequences is unknown but microsatellites, because their number varies between different individuals, have proved very useful in DNA testing (page 130). Other extragenic sequences, known as LINEs (long interspersed nuclear elements) and SINEs (short interspersed nuclear elements) occur in our genome. There are about 50 000 copies of LINEs in a mammalian genome and they make up about 17% of the human genome.

GENE NOMENCLATURE

One of the great difficulties that has arisen out of genome-sequencing projects is how to name the genes and the proteins they encode. This has not been easy and a number of committees have been set up to deal with this problem. In general, each gene is designated by an abbreviation, written in capitalized italics. For example, type 1 collagen (the commonest form in the human body) is a trimer formed of two molecules of collagen 1 α1 and one molecule of collagen 1 α2. The abbreviated names of these proteins are COL1A1 and COL1A2 respectively, using normal capitals, while the names of the genes coding for these proteins use capitalized italics: *COL1A1* and *COL1A2.* It is mutations in *COL1A1* that give rise to osteogenesis imperfecta (Medical Relevance 3.2 on page 47).

There are many instances where for historical reasons the correlation between the protein and gene names are not so simple. For example the proteins connexin 43, 46, and 50 (page 28) are named for their relative molecular masses (43 kDa, etc.) and have the abbreviated names Cx43, Cx46, and Cx50. However, the genes that encode these proteins are called GJA1, GJA3, and GJA8 respectively, where GJ stands for gap junction.

IN DEPTH 4.2   GENOME PROJECTS

The publication in 1996 of the sequence of the genome of the single-celled yeast *S. cerevisiae* was a milestone in biology. Not only did scientists have before them the complete genetic blueprint of a eukaryotic organism, but the technology for obtaining and curating huge amounts of genetic data was established. The genomes of other simple organisms such as the tiny nematode worm *Caenorhabditis elegans,* with just 959 body cells, and the fruit fly *D. melanogaster*, were published soon after, followed by more complex organisms such as the mouse and, of course, humans. Today, the sequence of the genomes of nearly 60 000 organisms, including 15 000 eukaryotic species, has been determined. Genomes from every branch of the tree of life are now available for study, including the platypus, our most distant mammalian relative, and both the nuclear and mitochondrial genomes of the Neanderthal, the hominid most closely related to present-day humans.

Sophisticated databases have been created to store and analyze base sequence information from the various genome projects. Computer programs analyze the data for exon sequences and compare the sequence of one genome to that of another. In this way sequences encoding related proteins (proteins that share stretches of similar amino acids) can be identified. The genome data from patients can be used to identify mutations and inform clinical decision-making. Some important programs that can be easily accessed through the internet are BLASTN for the comparison of a nucleotide sequence to other sequences stored in a nucleotide database and BLASTP, which compares an amino acid sequence to protein sequence databases. Programs such as Clustal, MAFFT, MUSCLE, and T-Coffee can be used to compare multiple DNA or multiple protein sequences simultaneously. 3D-Coffee is a version of T-Coffee that can combine data from sequence and protein structure databases in the analysis.

The Human Genome Project, completed in 2003, was a 13-year international effort that was described at the time as the biological equivalent of putting a man on the moon. As more and more genomes were sequenced, the technology became quicker and, more importantly, cheaper. Using Next Generation Sequencing (NGS) technologies, it is currently possible to have our genomes sequenced at a cost that ranges between a few hundred and a thousand dollars per person. As an increasing number of us have our genomes sequenced, this inexpensive but informative resource is bringing personalized medicine closer to our everyday lives. Soon, clinicians will routinely tailor treatment for a wide range of diseases to our own unique genetic makeup.

In 2012 the 100 000 genomes project was set in motion through a collaboration between scientists and the government of the United Kingdom. Under the direction of Genomics England, the remit was to sequence 100 000 complete genomes from NHS (National Health Service) patients. The aim of this large-scale project was to analyze DNA from patients with cancer or who had a rare disorder to try to provide an understanding of the causes of a condition and inform best treatments. In a cancer patient the genome sequence from both tumor and normal tissue was compared. For patients with a rare disease, the genomes of two relatives were also sequenced. In December 2018 the project met its target of sequencing 100 000 genomes, a remarkable achievement of progress in DNA sequence technology and analysis. To date, the project has generated over 21 petabytes of genome data and is already delivering valuable insights into how DNA sequences inform an individual's medical condition. In response to the COVID-19 pandemic, genome sequencing of both patients and SARS-CoV-2 samples has provided us with information on both the way in which an individual's genome influences their susceptibility to COVID-19 infection and on the spread of new variants through the population.

daughters.
This answer neglects the fact that a bacterium infected by a bacteriophage will likely die without generating any
as arising from a deamination.
and the daughter cell that inherits this, and all its daughters in turn, will have a mutation that can no longer be easily identified

3′ATTT 5′

5′TAAA 3′

3′ will generate a chromosome with the structure
The lower strand, 3′ AUTT 5′, will as before have the matching strand 5′ TAAA 3′ synthesized on it. The upper strand 5′ TAAA
When this cell replicates its DNA the strands will separate and each will act as a template for the synthesis of a new strand.
fact that a deamination event has occurred.
with each base pair now nicely hydrogen bonded to its partner. Only the presence of uracil in the DNA molecule betrays the

BrainBox 4.1   Elizabeth Blackburn, Carol Greider, and Jack Szostak

Elizabeth Blackburn, Carol Greider, and Jack Szostak. Source: Nobel organization website.

Maintaining the information encoded within DNA is essential for the health of cells and organisms, and since the 1930s it had been speculated that the ends of chromosomes, called the telomeres, must be highly specialized to protect DNA strands from being degraded. In the late 1970s Elizabeth Blackburn, studying the model organism tetrahymena, discovered that the ends of chromosomes contain repeats of the sequence CCCCAA. Together with colleague Jack Szostak she also found that a minichromosome injected into cells would be rapidly degraded, unless this CCCCAA sequence was located at its ends – in which case the minichromosome was protected. Later,

a PhD student named Carol Greider working in Blackburn's lab discovered the enzyme telomerase, which is responsible for depositing the CCCCAA repeats at chromosome ends. It is now understood that loss of telomeres contributes to the loss of cells' ability to divide and, conversely, that the unwanted ability of many cancer cells to over-proliferate is linked to the overactivity of telomerase. Blackburn, Szostak, and Greider together won the Nobel Prize in Physiology and Medicine in 2009 for explaining "how chromosomes are protected by telomeres and the enzyme telomerase."

Answer to thought question: Guanine cannot base pair with uracil, so there is certainly a mismatch in the DNA. However, mismatch repair cannot correct the error, because the mutation has occurred in a mature chromosome in which both DNA strands are methylated.

When the bacterium replicates its DNA, the strands will separate and each will act as a template for the synthesis of a new strand. The unmodified strand, 5′ TGAA 3′ will have the matching strand 3′ ACTT 5′ synthesized on it, and the resulting newly synthesized strand will be unmutated, as will its own daughter strands.

However, the modified strand 3′ AUTT 5′ will have the matching strand 5′ TAAA 3′ synthesized on it, since adenine base pairs with uracil. The daughter cell that inherits this chromosome, assuming that it is still infected with PBS2 and therefore allows the uracil to remain, will now have a chromosome with the structure

5′ TAAA 3′

3′ AUTT 5′

SUMMARY

1. During replication each parent DNA strand acts as the template for the synthesis of a new daughter strand. The base sequence of the newly synthesized strand is complementary to that of the template strand.

2. Replication starts at specific sequences called origins of replication. The two strands untwist and form the replication fork. Helicase enzymes unwind the double helix, and single-stranded DNA-binding proteins keep it unwound during replication. In prokaryotes, DNA polymerase III synthesizes the leading strand continuously in the 5′ to 3′ direction. The lagging strand is made discontinuously in short pieces in the 5′ to 3′ direction. These are joined together by DNA ligase. DNA polymerase is a self-correcting enzyme. It can remove an incorrect base using its 3′- to 5′-exonuclease activity and then replace it with the correct base.

3. DNA repair enzymes can correct mutations. Uracil in DNA, resulting from the spontaneous deamination of cytosine, is removed by uracil – DNA glycosidase. The depyrimidinated sugar is cleaved from the sugar-phosphate backbone by AP endonuclease, and DNA polymerase then inserts the correct nucleotide. The phosphodiester bond is reformed by DNA ligase.

4. In eukaryotes protein-coding genes are split into exons and introns. Only exons code for protein. The human genome has a large amount of DNA whose function is not obvious. This includes much repetitious DNA, whose sequence is multiplied many times.

5. Protein-coding genes may be found in repeated groups of slightly diverging structure called gene families, either close together or scattered over the genome. Some of the family members have lost the ability to operate – they are pseudogenes.

FURTHER READING

Brenner, S., Elgar, G., Sandford, R. et al. (1993). Characterization of the pufferfish (Fugu) genome as a compact model vertebrate genome. *Nature* 366: 265–268.

Chatterjee, N. and Walker, G.C. (2017). Mechanisms of DNA damage, repair, and mutagenesis. *Environmental and Molecular Mutagenesis* 58: 235–263.

O'Donnell, M., Langston, L., and Stillman, B. (2013). Principles and concepts of DNA replication in bacteria, archaea, and eukarya. *Cold Spring Harbor Perspectives in Biology* 5 (7): a010108.

Radman, M. and Wagner, R. (1988). The high fidelity of DNA duplication. *Scientific American* 259 (2): 40–46.

Venter, J.C. et al. (2001). The sequence of the human genome. *Science* 291: 1304–1351.

● REVIEW QUESTIONS

4.1 Theme: Normal or damaging chemical change in DNA

A depurination
B deamination
C excision
D methylation
E priming
F thymine dimer

From the list of processes on the left, select the one described in each of the statements below.

1. a hydrolysis reaction creating a uracil base in a DNA strand
2. addition of a -CH$_3$ group to adenine bases on a DNA strand
3. hydrolysis of the bond between the deoxyribose moiety and a guanine base
4. hydrolysis of the bond between the deoxyribose moiety and an adenine base
5. loss of the NH$_2$ group on the 4th carbon of a cytosine base
6. the formation of a covalent bond between adjacent bases on a DNA strand

4.2   Theme: DNA replication

A   breaking of base–base hydrogen bonds
B   connection of adjacent Okazaki fragments
C   formation of an RNA sequence complimentary to the DNA chain
D   hydrolysis and removal of a DNA strand
E   incorporation of deoxyribonucleotide monomers into both the leading and lagging strand
F   incorporation of deoxyribonucleotide monomers into the lagging strand but not into the leading strand
G   in eukaryotes, removal of an RNA sequence complimentary to the DNA chain

From the above list of processes, select the one performed by each of the enzymes below.

1. DNA ligase
2. DNA polymerase I
3. DNA polymerase III
4. exonuclease I
5. exonuclease VII
6. helicase
7. primase
8. ribonuclease H

4.3   Theme: Regions within eukaryotic chromosomes

A   exon
B   gene family
C   intron
D   long interspersed nuclear element
E   pseudogene
F   satellite DNA
G   tandemly repeated DNA

From the above list of DNA regions within eukaryotic chromosomes, select the region described by each of the descriptions below.

1. a section of DNA that, read in triplets of bases, encodes successive amino acids in a polypeptide chain
2. a section of DNA with a sequence similar to a working gene but which no longer encodes a functional protein
3. a section of DNA within a gene that is transcribed into RNA but which does not encode amino acids and which must be removed from the RNA before it leaves the nucleus
4. a series of identical or almost identical genes all of which are transcribed so as to generate identical or almost identical RNA products and, in the case of protein coding genes, identical proteins
5. a type of DNA with a presumed structural role that makes up much of the chromosome in the centromere region
6. an extragenic DNA sequence that is repeated more than a million times

---

⬤ THOUGHT QUESTION

The bacteriophage PBS2 is very unusual in that its DNA uses uracil in place of thymine and therefore runs the risk of being "corrected" by the deamination enzymes of the bacterium that it infects. It therefore injects an inhibitor of bacterial uracil DNA glycosidase to prevent this from happening. Consider a section of the bacterial chromosome with the structure

5'TGAA 3'

3'ACTT 5'

and suppose that at around the same time that it is infected by PBS2 a deamination event converts the cytosine to a uracil. Given that with uracil DNA glycosidase deactivated the bacterium cannot repair the error by base excision repair, can it repair the error using mismatch repair enzymes? If the error is not corrected, suggest what will happen to the DNA sequence (i) when the bacterium replicates its DNA and generates two daughter cells (ii) when these cells replicate their DNA to generate four second generation cells?

5

TRANSCRIPTION AND THE CONTROL OF GENE EXPRESSION

Transcription (RNA synthesis) is the process whereby the information held in the nucleotide sequence of DNA is transferred to RNA. Until recently, RNAs were thought to belong to one of three major classes – **ribosomal RNA (rRNA), transfer RNA (tRNA),** and **messenger RNA (mRNA)** – all of which play key roles in protein synthesis. However current research is revealing new roles for a wide range of **noncoding RNAs** which do not fit into these three classes. We will describe these later.

Genes encoding mRNAs are known as protein-coding genes. A gene is said to be **expressed** when its genetic information is transferred to mRNA and then to protein. Two important questions are addressed in this chapter: how is RNA synthesized, and what factors control how much mRNA is made?

STRUCTURE OF RNA

Ribonucleic acid is a polymer made up of monomeric nucleotide units. RNA has a chemical structure similar to that of DNA but there are two major differences. First, the sugar in RNA is ribose instead of deoxyribose (Figure 5.1). Second, although RNA contains the two purine bases, adenine and guanine, and the pyrimidine cytosine, the fourth base is different. The pyrimidine uracil (U) replaces thymine (Figure 5.1). The building blocks of RNA are therefore the four ribonucleoside triphosphates guanosine 5′-triphosphate,

adenosine 5′-triphosphate, uridine 5′-triphosphate, and cytidine 5′-triphosphate. These four nucleotides are joined together by phosphodiester bonds (Figure 5.2). Like DNA, the RNA chain has direction. In the first nucleotide in the chain, the 5′ carbon of the ribose is phosphorylated. This is the 5′ end of the RNA chain. At the other end is a ribose with a free hydroxyl group on its 3′ carbon. This is the 3′ end. As in DNA, phosphodiester bonds are made between the α phosphate group on the 5′ carbon of the ribose of the incoming nucleotide and the free 3′ hydroxyl group of the last nucleotide in the chain. RNA molecules are single-stranded along much of their length, although they often contain regions that are double-stranded due to intramolecular base pairing (see for example Figure 5.5b).

RNA POLYMERASE

In any gene only one DNA strand acts as the template for transcription. The sequence of nucleotides in RNA depends on their sequence in the DNA template. The bases G, A, T, and C in the DNA template will specify the bases C, U, A, and G, respectively, in RNA. DNA is transcribed into RNA by the enzyme **RNA polymerase.** Transcription requires that this enzyme recognize the beginning of the gene to be transcribed and catalyze the formation of phosphodiester bonds between nucleotides that have been selected according to the sequence within the DNA template.

Cell Biology: A Short Course, Fourth Edition. Stephen Bolsover, Andrea Townsend-Nicholson, Greg FitzHarris, Elizabeth Shephard, Jeremy Hyams and Sandip Patel.
© 2022 John Wiley & Sons Ltd. Published 2022 by John Wiley & Sons Ltd.
Companion website: www.wiley.com/go/bolsover/cellbiology4

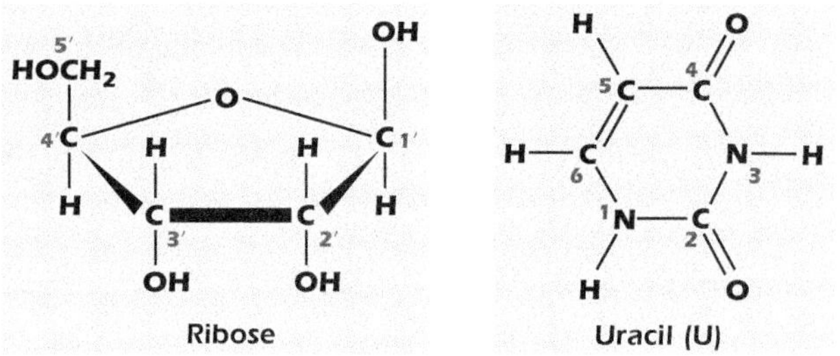

Figure 5.1. RNA contains the sugar ribose and the base uracil in place of deoxyribose and thymine.

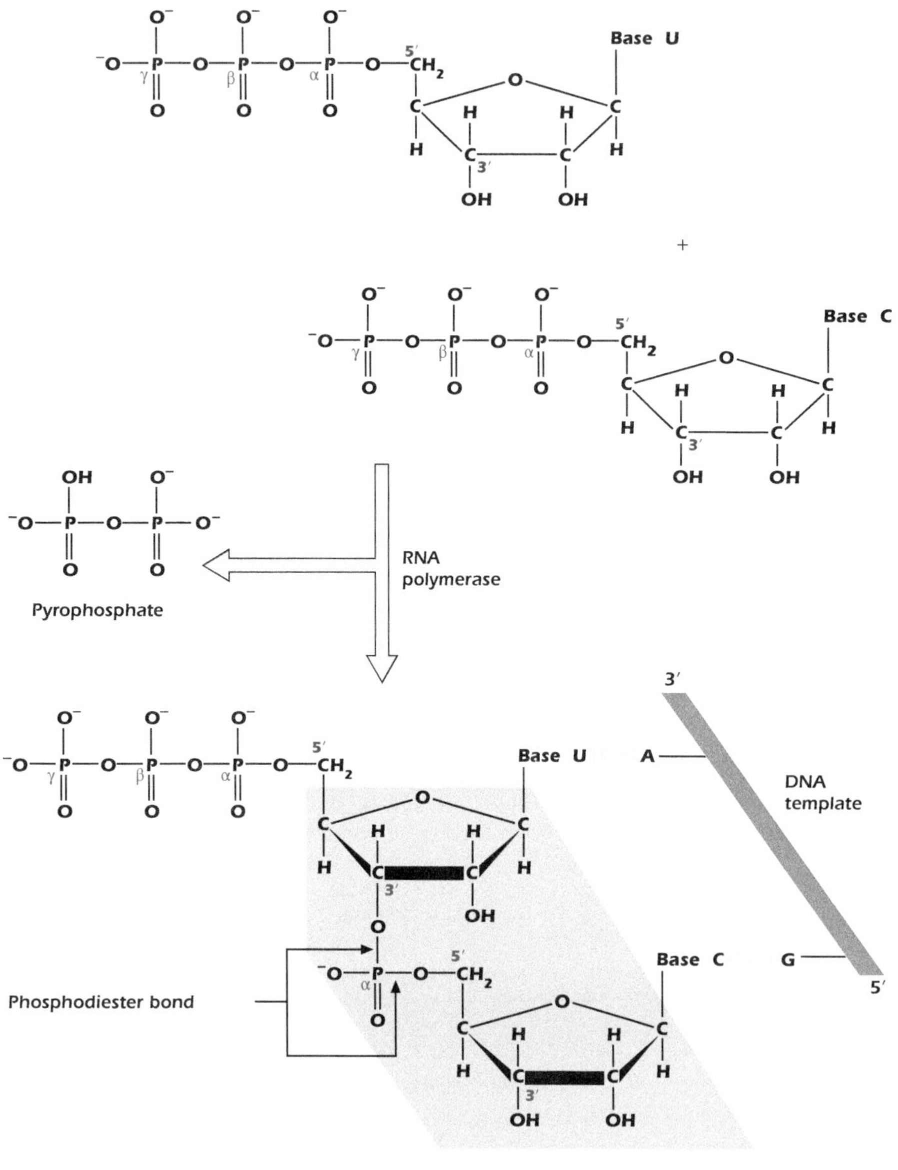

Figure 5.2. Synthesis of an RNA strand.

GENE NOTATION

Figure 5.3 shows the notation used in describing the positions of nucleotides within and adjacent to a gene. The nucleotide in the template strand at which transcription begins is designated with the number +1. Transcription proceeds in the **downstream** direction, and nucleotides that will be transcribed into RNA are given successive positive numbers. Downstream sequences are drawn, by convention, to the right of the transcription start site. Nucleotides that lie to the left of this site are the **upstream** sequences and are identified by negative numbers. Because a nucleotide will either be upstream or downstream of the start of transcription, there is no 0.

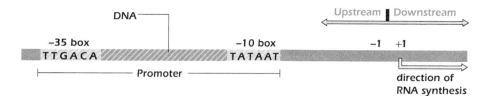

Figure 5.3. Numbering on a DNA sequence.

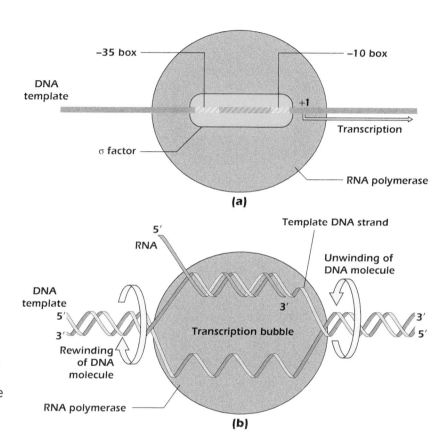

Figure 5.4. (a) RNA polymerase binds to the promoter to form the closed promoter complex. (b) The open promoter complex: The DNA helix unwinds and RNA polymerase synthesizes an RNA molecule.

BACTERIAL RNA SYNTHESIS

Escherichia coli genes are all transcribed by the same RNA polymerase. This enzyme is made up of five subunits (polypeptide chains). The subunits are named α (there are two of these), β, β′, and σ. Each of the subunits has its own job to do in transcription. The role of the sigma (σ) factor is to recognize a specific DNA sequence called the **promoter,** which lies just upstream of the gene to be transcribed (Figure 5.3). *E. coli* promoters contain two important regions. One centered around nucleotide –10, usually has the sequence TATAAT. This sequence is called the –10 box (or the Pribnow box). The second, centered near nucleotide –35 often has the sequence TTGACA. This is the –35 box.

On binding to the promoter sequence (Figure 5.4a), the σ factor brings the other subunits (two of α plus one each of β and β′) of RNA polymerase into contact with the DNA to be transcribed. This forms the **closed promoter complex.** For transcription to begin, the two strands of DNA must separate, enabling one strand to act as the template for the

synthesis of an RNA molecule. This formation is the **open promoter complex.** The separation of the two DNA strands is helped by the AT-rich sequence of the –10 box. As mentioned in Chapter 3, there are only two hydrogen bonds between the bases adenine and thymine; thus it is relatively easy to separate the two strands at this point. DNA unwinds and rewinds as RNA polymerase advances along the double helix, synthesizing an RNA chain as it goes. This produces a **transcription bubble** (Figure 5.4b). The RNA chain grows in the 5′ to 3′ direction, and the template strand is read in the 3′ to 5′ direction.

When the RNA chain is about 10 bases long, the σ factor is released from RNA polymerase and plays no further role in transcription. The β subunit of RNA polymerase binds ribonucleotides and joins them together by catalyzing the formation of phosphodiester links as it moves along the DNA template. The β′ subunit helps to keep the RNA polymerase attached to the DNA. The two α subunits are important as they help RNA polymerase to assemble on the promoter.

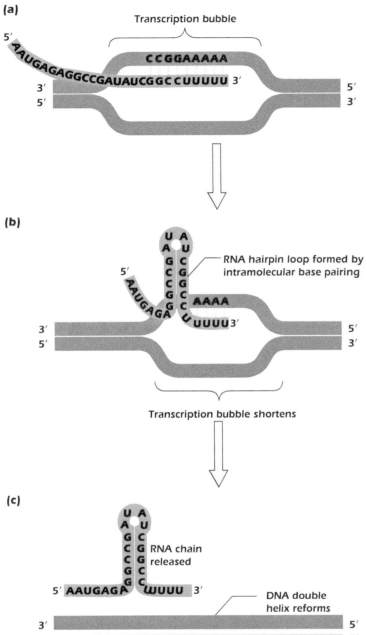

Figure 5.5. Rho-independent transcription termination in *Escherichia coli.*

RNA polymerase must recognize when it has reached the end of a gene. *E. coli* has specific sequences, called **terminators,** at the ends of its genes that cause RNA polymerase to stop transcribing DNA. A terminator sequence consists of two regions rich in the bases G and C that are separated by about 4 bp. This sequence is followed by a stretch of A bases. Figure 5.5 shows how the terminator halts transcription. When the GC-rich regions are transcribed, a hairpin loop forms in the RNA with the first and second GC-rich regions aligning and pairing up. Formation of this structure within the RNA molecule causes the transcription bubble to shrink because where the template DNA strand can no longer bind to the RNA molecule it

reconnects to its sister DNA strand. The remaining interactions between the adenines in the DNA template and the uracils in the RNA chain have only two hydrogen bonds per base pair and are therefore too weak to maintain the transcription bubble. The RNA molecule is then released, transcription terminates, and the double helix reforms. This type of transcription termination is known as rho-independent termination.

Some *E. coli* genes contain a different type of terminator site. These are recognized by a protein, known as rho, which frees the RNA from the DNA. In this case transcription is terminated by a process known as rho-dependent termination.

CONTROL OF BACTERIAL GENE EXPRESSION

Many bacterial proteins are present in the cell in a constant amount. However, the amount of other proteins is regulated by the presence or absence of a particular nutrient. To grow and divide and make the most efficient use of the available nutrients, bacteria have to adjust quickly to changes in their environment. They do this by regulating the production of proteins required for either breakdown or synthesis of a particular compound. Gene expression in bacteria is controlled mainly at the level of transcription. This is because bacterial cells have no nuclear envelope, and RNA synthesis and protein synthesis are not physically separated but occur simultaneously. This is one reason why bacteria lack the more sophisticated control mechanisms that regulate gene expression in eukaryotes.

Each bacterial promoter usually controls the transcription of a cluster of genes coding for proteins that work together on a particular task. This collection of related genes is called an **operon** and is transcribed as a single mRNA molecule called a **polycistronic mRNA.** As shown in Figure 5.6, translation of this mRNA produces the required proteins because there are several start and stop codons for protein synthesis along its length. Each start and stop codon (page 45) specifies a region of RNA that will be translated into one particular protein. The organization of genes into operons ensures that all the proteins necessary to metabolize a particular compound are made at the same time and hence helps bacteria to respond quickly to environmental changes.

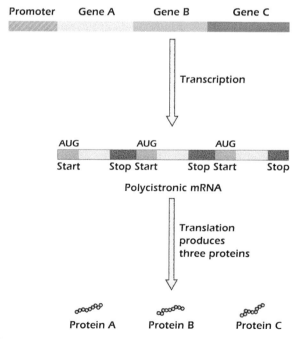

Figure 5.6. A bacterial operon is transcribed into a polycistronic mRNA.

The three major factors involved in regulating how much RNA is made are (i) nucleotide sequences within or flanking a gene, (ii) proteins that bind to these sequences, and (iii) the environment. The human intestine contains many millions of *E. coli* cells that must respond very quickly to the sudden appearance of a particular nutrient. For instance, while most foods do not contain the sugar lactose, milk contains large amounts. Within minutes of our drinking a glass of milk, *E. coli* in our intestines start to produce the enzyme β-galactosidase that cleaves lactose to glucose and galactose (Figure 5.7). In general, the substrates of β-galactosidase are compounds like lactose that contain a β-galactoside linkage and are therefore called β-galactosides.

Lac, an Inducible Operon

β-Galactosidase is encoded by one of the genes that make up the lactose (*lac*) operon, which is shown in Figure 5.8. The operon contains three protein-coding genes: *lacZ, lacY,* and *lacA.* β-Galactosidase is encoded by the *lacZ* gene. As noted before, gene names are always italicized, while the protein product is always in standard type. *LacY* encodes β-galactoside permease, a carrier (page 147) that helps lactose get into the cell. The *lacA* gene codes for transacetylase. This protein is thought to remove compounds that have a structure similar to lactose but that are not useful to the cell.

In the absence of β-galactoside compounds like lactose, there is no need for *E. coli* to produce β-galactosidase or β-galactoside permease, and the cell contains only a few molecules of these proteins. The *lac* operon is said to be inducible because its rate of transcription increases greatly when a β-galactoside is present. How is the transcription of the *lacA, lacY,* and *lacA* genes switched on and off? A repressor protein (the product of the *lacI* gene) binds to a sequence in the *lac* operon known as the operator. The operator lies next to the promoter so that, when the repressor is bound, RNA polymerase is unable to bind to the promoter. In the absence of a β-galactoside, the *lac* operon spends most of its time in the state shown in Figure 5.8a. The repressor is bound to the operator, RNA polymerase cannot bind, and no transcription occurs. Only for the small fraction of time that the operator is unoccupied by the repressor can RNA polymerase bind and generate mRNA. Thus, in the absence of a β-galactoside, only very small amounts of β-galactosidase, β-galactoside permease, and transacetylase are synthesized.

If lactose appears, it is converted to an isomer called allolactose. This conversion is carried out by β-galactosidase (Figure 5.7); as we have seen, a small amount of β-galactosidase is made even when β-galactoside is absent. The repressor protein has a binding site for allolactose and undergoes a conformational change when bound to this compound (Figure 5.8b). This means that the repressor is no longer able to bind to the operator. The way is then clear for RNA polymerase to bind to the promoter and to transcribe the operon. Thus, in a short time the bacteria produce the proteins

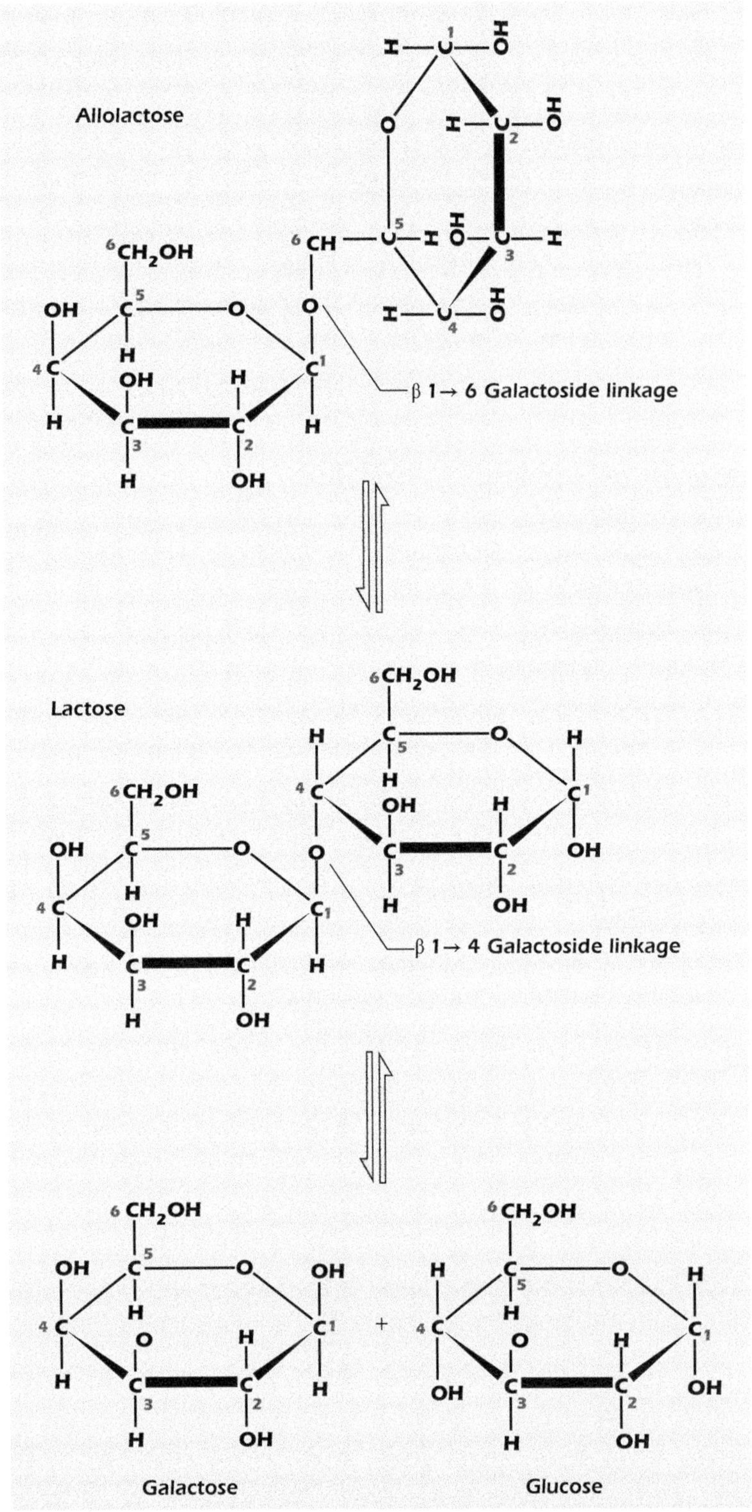

Figure 5.7. Reactions catalyzed by β-galactosidase.

necessary for utilizing the new food source. The concentration of the substrate (lactose in this case) determines whether or not mRNA is synthesized. The *lac* operon is said to be under negative regulation by the repressor protein.

The transcription of the *lac* operon is controlled not only by the repressor protein but also by another protein, the **catabolite activator protein (CAP).** If both glucose and lactose are present, it is more efficient for the cell to use glucose as the carbon source because the utilization of glucose requires no new RNA and protein synthesis, all the proteins necessary being already present in the cell. Only in the absence of glucose, therefore, does *E. coli* transcribe the

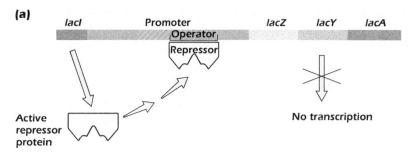

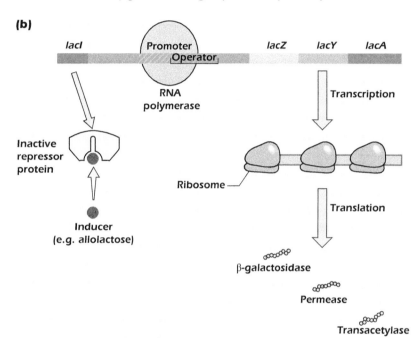

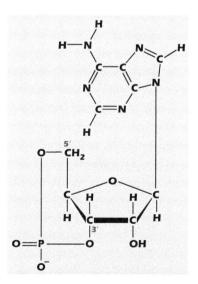

○ Figure 5.8. Transcription of the *lac* operon requires the presence of an inducer.

○ Figure 5.9. Cyclic adenosine monophosphate, also called cyclic AMP or just cAMP.

lac operon at a high rate. This control operates through the intracellular messenger molecule **cyclic adenosine monophosphate (cyclic AMP or cAMP)** (Figure 5.9). (It is worth mentioning that cAMP is always pronounced "cy-clic A-M-P," with five syllables.) When glucose concentrations are low, the concentration of cAMP increases. cAMP binds to CAP, and the complex then binds to a sequence upstream of the *lac* operon promoter (Figure 5.10) where it deforms the DNA strand, causing the DNA surrounding CAP to bend by about 90°. This allows both α subunits of RNA polymerase to make contact with CAP at the same time so that the affinity of RNA polymerase for the *lac* promoter is increased. The result is that now the *lac*Z, *lac*Y, and *lac*A genes can be transcribed very efficiently. The *lac* operon is said to be under **positive regulation** by the CAP–cAMP complex.

To recap, the control of the *lac* operon is not simple. Several requirements need to be met before it can be transcribed. The repressor must not be bound to the operator, and the CAP–cAMP complex and RNA polymerase must be bound to their respective DNA binding sites. These

Example 5.1 Quorum Sensing: Squid That Glow in the Dark

The bacterium *Vibrio fischeri* lives free in seawater but is also found at high densities in the light-emitting organs of the nocturnal squid *Euprymna scolopes,* where it synthesizes an enzyme called luciferase that generates light. This phenomenon is called bioluminescence. When living free in seawater *V. fischeri* synthesizes almost no luciferase; indeed, there would be little point in doing so because the light emitted by a single bacterium would be too dim for anything to see. *V. fischeri* only begins making lots of luciferase when the density of bacteria is high – just as it is in the squid light organs. The word *quorum* is defined in Webster's Dictionary as "the number of . . . members of a body that when duly assembled is legally competent to transact business." Thus *quorum sensing* is a good description of *V. fischeri's* behavior. How does it work?

Luciferase is a product of the *luxA* and *luxB* genes in a bacterial operon called the *lux* operon. A region in the *lux* operon promoter binds a transcription factor, LuxR, which is only active when it has bound the small, uncharged molecule N-acyl-HSL (also called VAI for *V. fischeri* autoinducer). N-acyl-HSL in turn is made by the enzyme N-acyl-HSL synthase (or VAI synthase)

that is encoded by the gene *luxI,* which is part of the *lux* operon. In free-living *V. fischeri* the *lux* operon is transcribed at a low level. Small amounts of N-acyl-HSL are made, which immediately leak out of the cell into the open sea without binding to LuxR. When the bacteria are concentrated in the squid's light organs, then some N-acyl-HSL binds to LuxR, increasing transcription of the *lux* operon. This makes more luciferase – but it also makes more N-acyl-HSL synthase. The concentration of N-acyl-HSL therefore rises – so transcription of the *lux* operon increases further. This means that the genes for N-acyl-HSL synthase, luciferase, and the enzymes that produce the substrate for luciferase are now transcribed at a high rate. This **autoinduction** of the *lux* operon by N-acyl-HSL is the very opposite of the kind of negative feedback that we see with the *trp* operon: the product of a process is acting to accelerate that process, so this is **positive feedback.** When luciferase carries out its reaction, the squid will luminesce at the intensity of moonlight so that as it glides at night over the coral reefs of Hawaii it does not cast a dark shadow; thus, its prey are less likely to notice its presence.

requirements are only met when glucose is absent and a β-galactoside, such as the sugar lactose, is present.

Other compounds such as isopropylthio-ß-D-galactoside (**IPTG**) (Figure 5.11) can bind to the repressor but are not metabolized. These **gratuitous inducers** are very useful in DNA research and in biotechnology.

Trp, a Repressible Operon

Operons that code for proteins that synthesize amino acids are regulated in a different way from the *lac* operon. These operons are only transcribed if the amino acid is not present,

and transcription is switched off if there is already enough of the amino acid around. In this way the cell carefully controls the concentration of free amino acids. The tryptophan (*trp*) operon is made up of five structural genes encoding enzymes that synthesize the amino acid tryptophan (Figure 5.12). This is a **repressible operon.** The cell regulates the amount of tryptophan produced by preventing transcription of the *trp* operon mRNA when there is sufficient tryptophan about. As with the *lac* operon, the transcription of the *trp* operon is controlled by a regulatory protein. The gene *trpR* encodes an inactive repressor protein that is called an **aporepressor.** Tryptophan binds to this to produce an active repressor

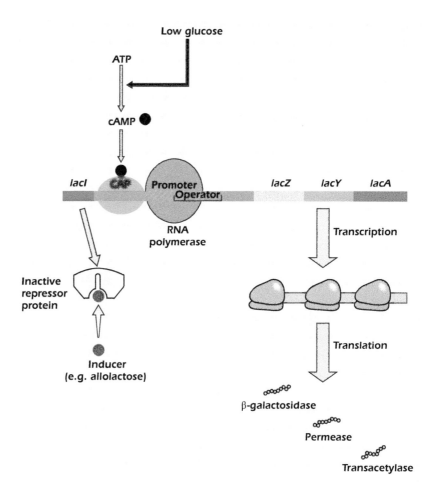

Figure 5.10. For efficient transcription of the *lac* operon, both cAMP and a β-galactoside sugar must be present.

complex. The active repressor complex binds to the operator sequence of the *trp* operon and prevents the attachment of RNA polymerase to the *trp* promoter sequence. Therefore, when the concentration of tryptophan in the cell is high, the active repressor complex will form, and transcription of the *trp* operon is prevented. However, when the amount of tryptophan in the cell decreases, the active repressor complex cannot be formed. RNA polymerase binds to the promoter, transcription of the *trp* operon proceeds, and the enzymes needed to synthesize tryptophan are produced.

Negative feedback is an important concept in biology and engineering. It is said to occur when the end product of a process inhibits that process. The control of the *trp* operon is an example of negative feedback since tryptophan, produced by the enzymes generated from expression of the operon, inhibits expression when the concentration of tryptophan is sufficient for the needs of the cell.

Many other operons are regulated by similar mechanisms in which specific regulatory proteins interact with specific small molecules.

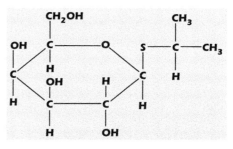

IPTG (isopropylthio-β-D-galactoside)

Figure 5.11. Isopropylthio-β-D-galactoside (IPTG), which can bind to the *lac* repressor protein but which is not metabolized.

EUKARYOTIC RNA SYNTHESIS

Eukaryotes have three types of RNA polymerase. **RNA polymerase I** transcribes the genes that code for most of the ribosomal RNAs. All messenger RNAs are synthesized

(a)

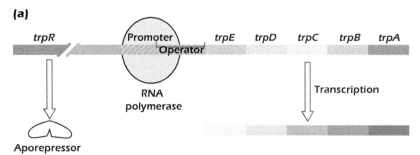

No tryptophan; operon transcribed

(b)

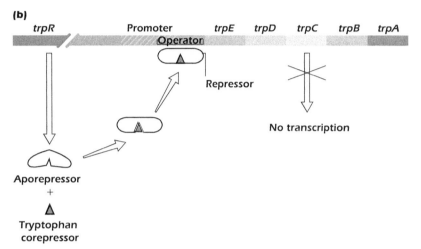

Tryptophan present; operon repressed

Figure 5.12. Transcription of the *trp* operon is controlled by the concentration of the amino acid tryptophan.

Messenger RNA Processing in Eukaryotes

using **RNA polymerase II.** Transfer RNA genes are transcribed by **RNA polymerase III.** The chemical reaction catalyzed by these three RNA polymerases, the formation of phosphodiester bonds between nucleotides, is the same as in archaea and bacteria.

A newly synthesized eukaryotic mRNA undergoes several modifications before it leaves the nucleus (Figure 5.13). The first is known as capping. Very early in transcription the 5′-terminal triphosphate group is modified by the addition of a guanosine via a 5′-5′-phosphodiester bond. The guanosine is subsequently methylated to form the **7-methyl guanosine cap.** Next, the 3′ ends of nearly all eukaryotic mRNAs are modified by the addition of a long stretch of adenosine residues, the **poly(A) tail.** A sequence 5′ AAUAAA 3′ is found in most eukaryotic mRNAs about 20 bases from where the poly(A) tail is added and is probably a signal for the enzyme poly(A) polymerase to bind and to begin the polyadenylation process. The length of the poly(A) tail varies; it can be as long as 250 nucleotides. Unlike DNA, RNA is an unstable molecule, and the capping of eukaryotic mRNAs at their 5′ ends and the addition of a poly(A) tail to their 3′ end increases the lifetime of mRNA molecules by protecting them from digestion by nucleases.

Many eukaryotic protein-coding genes are split into exon and intron sequences. Both the exons and introns are transcribed into the initial mRNA transcript. The introns have to be removed and the exons joined together by a process known as RNA splicing before the mRNA can be used to make protein. Removal of introns takes place within the nucleus. Splicing is complex and not yet fully understood. It has, however, certain rules. Within an mRNA the first two bases following an exon are always GU and the last two bases of the intron are AG. Several **small nuclear RNAs (snRNAs)** are involved in splicing. These are complexed with a number of proteins to form a structure known as the **spliceosome.** One of the snRNAs is complementary in sequence to either end of the intron sequence. It is thought that binding of this snRNA to the intron, by complementary base pairing, brings the two exon sequences together, which causes the intron to loop out (Figure 5.13). The proteins in the spliceosome remove the intron and join the exons together. Splicing is the final modification made to the mRNA in the nucleus. The mature mRNA is now transported to the cytoplasm for protein synthesis.

As well as removing introns, splicing can sometimes remove exons in the process of **alternative splicing.** This allows the same gene to give rise to different proteins at different times or in different cells. Alternative splicing is a powerful mechanism that allows the ~19 116 protein-coding

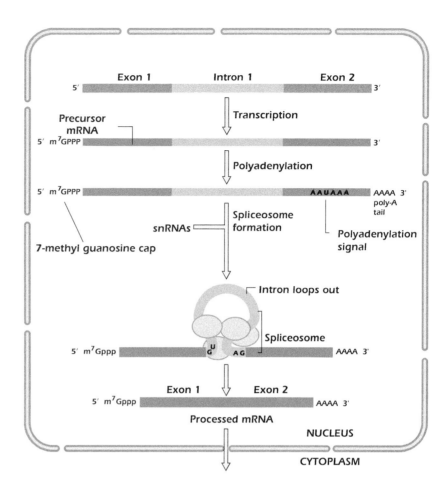

Figure 5.13. mRNA processing in eukaryotes.

genes in the human genome to code for 100 000s of proteins (In Depth 4.1 on page 61). For example, alternative splicing of the gene for the molecular motor dynein produces motors that transport different types of cargo (page 223).

CONTROL OF EUKARYOTIC GENE EXPRESSION

Since most eukaryote species are multicellular with many cell types, gene expression must be controlled so that different cell lineages develop differently and remain different. A brain cell is quite different from a liver cell because it contains different proteins, even though the DNA in the two cell types is identical. During development and differentiation, different sets of genes are switched on and off. Hemoglobin, for example, is only expressed in developing red blood cells even though the globin genes are present in all types of cell. Genetic engineering technology (Chapter 8) has made the isolation and manipulation of eukaryotic genes possible. This has given us some insight into the extraordinarily complex processes that regulate the transcription of eukaryotic genes and allow a fertilized egg to develop into a multicellular, multi-tissue adult.

Unlike the situation in bacteria, the eukaryotic cell is divided by the nuclear envelope into nucleus and cytoplasm.

Transcription and translation are therefore separated in space and in time. This means that the expression of eukaryotic genes can be regulated at more than one place in the cell. Although gene expression in eukaryotes is controlled primarily by regulating transcription in the nucleus, there are many instances in which expression is controlled at the level of translation in the cytoplasm or by altering the way in which the primary mRNA transcript is processed.

The interaction of RNA polymerase with its promoter is far more complex in eukaryotes than it is in bacteria (Figure 5.14). RNA polymerase II is responsible for transcribing protein-coding genes, but in contrast to bacterial RNA polymerase, RNA polymerase II cannot recognize a promoter sequence. Instead, other proteins known as **transcription factors** bind to the promoter and guide RNA polymerase II to the beginning of the gene to be transcribed.

The promoter sequence of most eukaryotic genes encoding mRNAs contains an AT-rich region about 25 bp upstream of the transcription start site. This sequence is called the **TATA box,** not to be confused with the −10 box of prokaryotes, although similarly rich in A and T. The TATA box binds a protein called transcription factor IID (**TFIID**), one of whose subunits is the TATA-binding protein, or TBP (Figure 5.14a). Several other transcription factors (TFIIA, TFIIB, TFIIE, TFIIF, and TFIIH) then bind to TFIID and to the promoter region (Figure 5.14b). TFIIF is the protein that

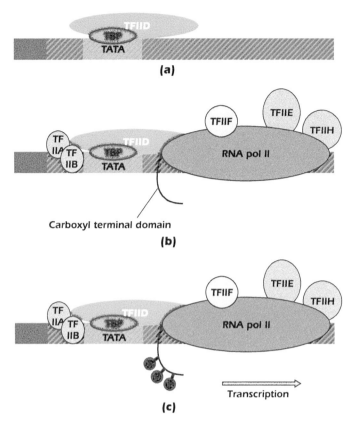

Figure 5.14. In eukaryotes, RNA polymerase II is guided to the promoter by TFII accessory proteins. (a) TBP binds to the TATA box. (b) The complete transcription preinitiation complex. (c) Phosphorylated RNA polymerase is active.

guides RNA polymerase II to the beginning of the gene to be transcribed. The complex formed between the TATA box, TFIID, the other transcription factors, and RNA polymerase is known as the transcription preinitiation complex. Note that the proteins with the prefix TFII are so named because they are transcription factors that help RNA polymerase II to bind to promoter sequences.

Although many gene promoters contain a TATA box, some do not. These TATA-less genes usually encode proteins that are needed in every cell and are hence called **housekeeping genes.** The promoters of these genes contain the sequence 5′ GGGGCGGGGC 3′, called the GC box. A protein called Sp1 binds to the GC box and is then able to recruit TATA-binding protein to the DNA even though there is no TATA box for the latter to bind to. The TATA-binding protein then recruits the rest of the transcription preinitiation complex so that transcription can proceed.

In either case, transcription begins when the carboxy-terminal domain of RNA polymerase II is phosphorylated. This region is rich in the amino acids serine and threonine which contain OH groups in their side chains. When these OH groups are phosphorylated (page 107), RNA polymerase II breaks away from the preinitiation complex and proceeds to transcribe DNA into mRNA (Figure 5.14c).

Although the formation of the transcription preinitiation complex is sometimes enough to produce a few molecules of RNA, the binding of other proteins to sequences next to the gene greatly increases the rate of transcription, producing much more mRNA. These proteins are also called

transcription factors, and the DNA sequences to which they bind are **enhancers,** so named because their presence enhances transcription. Enhancer sequences often lie upstream of a promoter, but they have also been found downstream. Enhancer sequences and the proteins that bind to them play an important role in determining whether a particular gene is to be transcribed. Some transcription factors bind to a gene to ensure that it is transcribed at the right stage of development or in the right tissue. Figure 5.15 shows how one gene can be transcribed in skeletal muscle but not in the liver simply because of the presence or absence of proteins that bind to enhancer sequences.

Glucocorticoids Cross the Plasma Membrane to Activate Transcription

Glucocorticoids are steroid hormones produced by the adrenal cortex that increase the transcription of several genes important in carbohydrate and protein metabolism. Because they are uncharged and relatively hydrophobic, steroid hormones can pass through the plasma membrane by simple diffusion to enter the cytosol. Here they encounter a class of transcription factors that have a binding site for steroid hormones and are therefore called **steroid hormone receptors** (Figure 5.16). In the absence of glucocorticoid hormone, its receptor remains in the cytosol and is inactive because it is complexed to two molecules of an inhibitor protein known as Hsp90. However, when the glucocorticoid hormone enters the cell and binds to its receptor, the Hsp90 protein is

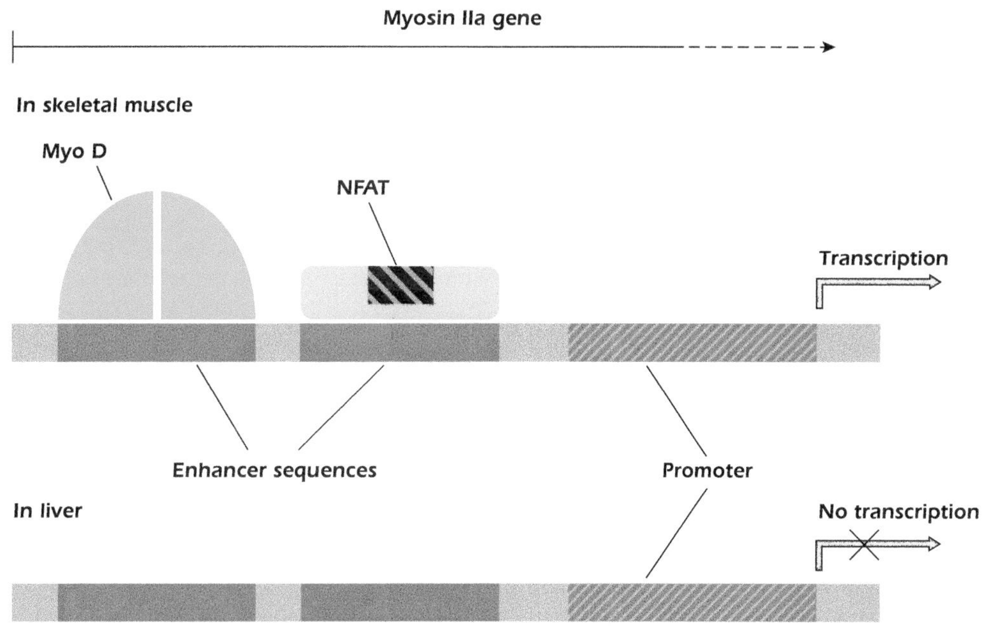

Myosin IIa gene

In skeletal muscle

Myo D

NFAT

Transcription

Enhancer sequences

Promoter

Figure 5.15. Tissue-specific transcription. The myosin IIa gene is not transcribed in liver cells, which do not contain the transcription factors MyoD and NFAT.

In liver

No transcription

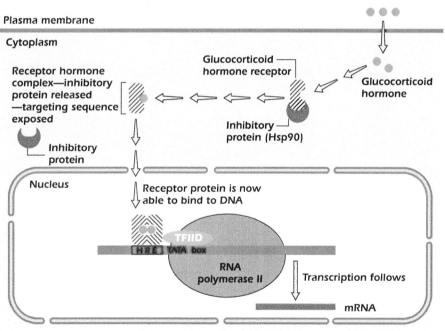

Plasma membrane

Cytoplasm

Glucocorticoid hormone receptor

Receptor hormone complex—inhibitory protein released —targeting sequence exposed

Glucocorticoid hormone

Inhibitory protein

Inhibitory protein (Hsp90)

Nucleus

Receptor protein is now able to bind to DNA

TFIID

HRE TATA box

RNA polymerase II

Transcription follows

mRNA

Figure 5.16. The glucocorticoid hormone receptor acts to increase gene transcription in the presence of hormone.

displaced. The targeting sequence (page 204) that targets the receptor to the nucleus is uncovered, and the glucocorticoid receptor–hormone complex can now move into the nucleus. Here, two molecules of the complex bind to a 15-bp sequence known as the **hormone response element** (**HRE**) that lies upstream of the TATA box. The HRE is an enhancer sequence. The glucocorticoid receptor–hormone complex interacts with the preinitiation complex bound to the TATA

Medical Relevance 5.1 Aldosterone in the Kidney

The kidney uses a channel called ENaC to recover sodium that would otherwise be lost in the urine. When body sodium is low, the adrenal gland produces the steroid hormone aldosterone. This binds to glucocorticoid hormone receptors in kidney cells. The receptor proteins then dimerize and bind in turn to a hormone response element in the enhancer region of the *ENaCα* gene. The result is that more ENaC protein is made and more sodium is reabsorbed.

Medical Relevance 5.2   The Glucocorticoid Receptor, Inflammation, and COVID-19

We have described how the dimerized glucocorticoid receptor binds to the hormone response element and activates gene transcription. However glucocorticoid receptors have a multitude of effects, in part because rather than forming a dimer with a second glucocorticoid receptor, a monomeric receptor can associate with a wide range of other transcription factors and either activate or repress transcription of target genes.

Among these target genes are pro- and anti-inflammatory proteins, and it has long been known that glucocorticoid analogues such as dexamethasone reduce inflammation and are therefore used to treat, for example, asthma and arthritis. Recently dexamethasone has been found highly effective in reducing lung inflammation and therefore the risk of death in severe COVID-19.

---

box, and the rate at which RNA polymerase transcribes genes containing the HRE is increased.

Figure 5.17 shows why the glucocorticoid hormone receptor binds to DNA as a dimer. The HRE is a palindrome – the sequence on both the top and bottom strands is the same when read in the 5′ to 3′ direction. Each strand of the HRE has the 6-bp sequence 5′ AGAACA 3′ that is known as the core recognition motif. This is the sequence to which a single receptor molecule binds. Because the HRE contains two recognition motifs, it binds two molecules of glucocorticoid receptor. The two 6-bp sequences are separated by three base pairs shown as X, which presumably are there to provide sufficient space for the receptor homodimer to fit snugly on the double helix. The binding of the glucocorticoid receptor to the HRE is unaffected by the identity of these three base pairs.

Chemicals that are released from one cell and that alter the behavior of other cells are in general called **transmitters.** Glucocorticoids are an example of transmitters that alter gene transcription. The cells of multicellular organisms turn the transcription of genes on and off in response to many extracellular chemicals. Unlike steroid hormones, most of these transmitters cannot enter the cell and must activate **intracellular messenger** systems inside the cell that in turn carry the signal onward from the plasma membrane to the nucleus (Chapter 11).

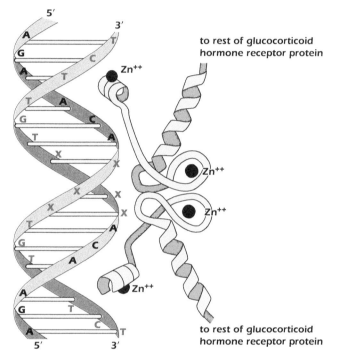

Figure 5.17. The dimerized glucocorticoid hormone receptor binds to a palindromic HRE.

NONCODING RNAs AND THE CONTROL OF EUKARYOTIC GENE EXPRESSION

The number of noncoding RNAs (**ncRNAs**) in the human genome is at least as large as the number of protein-coding genes. The genes encoding two of three major classes of RNA, rRNA, and tRNA, are not protein-coding and other noncoding RNAs include **micro RNA (miRNA)**, **long noncoding RNA (lncRNA)**, and **circular RNA (circRNA)**. While rRNA and tRNA are directly involved in the translation of mRNA into protein (Chapter 6), miRNA, lncRNA, and circRNA are involved in transcription regulation, controlling gene expression by influencing chromatin structure and RNA splicing and by negatively regulating mRNA expression, leading to a reduction in the amount of protein

produced. The entire suite of RNAs expressed in a cell is called the **transcriptome** and is increasingly the subject of research and documentation (Chapter 8).

Micro RNAs

There are more than 2300 mature human **micro RNAs (miRNAs).** miRNAs are encoded in the genome and transcribed by RNA polymerase II. The single-stranded primary miRNA (pri-miRNA) transcript is matured through a series of enzymatic cleavages that give rise to a short double-stranded RNA that associates with an **Argonaute protein** (AGO) to form a **ribonucleoprotein** complex. This complex unwinds and releases the "passenger" strand of the miRNA. The "guide" strand remains, forming part of the mature RNA-induced silencing complex (**RISC**), where it recognizes and base pairs with target mRNAs (Figure 5.18). mRNAs bound to a

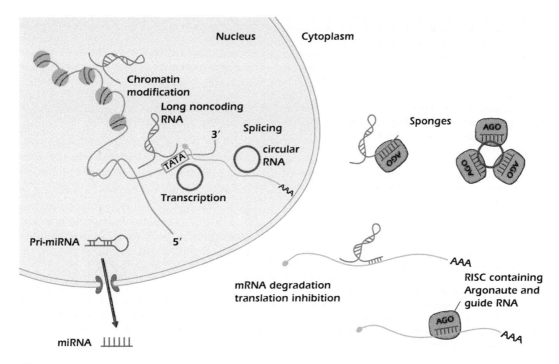

Figure 5.18. Noncoding RNAs such as microRNA (miRNA), long noncoding RNA (lncRNA), and circular RNA (circRNA) play an important role in the control of gene expression from influencing the chromatin structure of the gene to controlling the amount of protein that is produced.

miRNA are either degraded or their translation into protein is inhibited. For example, we will later describe how the transcription factor p53 suppresses cell division. One way it does this is by causing expression of human microRNA 34a which then prevents translation of the cyclin-dependent protein kinases Cdk4 and Cdk6 and the protein phosphatase Cdc25, proteins involved in control of the cell cycle (page 238). Production of miRNAs is a normal function of a cell. However, when increased amounts of miRNAs are produced this can lead to disease because important genes are silenced.

Long Noncoding RNAs

lncRNAs are defined as RNA transcripts greater than 200 nucleotides in length that are not translated into protein. Although much remains to be discovered about lncRNA, it is known to be involved in the biological activities of the cell and is associated with a large number of diseases, including cancer and cardiovascular and neurological diseases. lncRNA sequences are involved in influencing the gene expression and chromatin structure of nearby genes. The best-known example of this in humans and other placental mammals is the lncRNA *Xist* (X-inactive specific transcript). In females, *Xist* RNA promotes the modification of histones (page 39), initiating a cascade of events that culminates in one of the two X chromosomes of the cell being inactivated. This X chromosome-wide transcription silencing results in the 3D remodeling of the inactivated chromosome into the characteristic structure known as the **Barr body.** Like miRNA, lncRNAs can also regulate gene expression by interacting with target mRNAs to degrade them or to influence the translation of the mRNA into protein (Figure 5.18).

Circular RNAs

circRNAs are single-stranded RNA molecules. Their free 5′ and 3′ ends are joined together by a phosphodiester bond. These closed-loop structures are formed from precursor RNAs by the spliceosome during the process of RNA splicing. The biological role of circRNAs has not yet been fully established. CircRNAs are found in the nucleus, where they influence transcription and splicing. In the cytoplasm they act as miRNA sponges, mopping up specific miRNAs and affecting the ability of these miRNAs to interact with their mRNA targets (Figure 5.18). While this "sponge effect" may be an important method of regulating gene expression by preventing the silencing of important genes, it may also lead to disease by preventing protein levels from being appropriately controlled. For example, we will later describe the role of the cyclin CDK6 in regulating the cell cycle (page 238). A circRNA, circTCF25, has been shown to function as a miRNA sponge in cancerous bladder tissue, leading to the increased expression of the Cdk6 protein, which is associated with the development of bladder cancer. lncRNAs and miRNAs can also regulate each other through a similar "sponge effect."

SUMMARY

1. DNA is transcribed into RNA by the enzyme RNA polymerase. Guanine, adenine, uracil, and cytosine are the four bases in RNA.

2. In bacteria, RNA polymerase binds to the promoter sequence just upstream of the start site of transcription. The enzyme moves down the DNA template and synthesizes an RNA molecule. RNA synthesis stops once the enzyme has transcribed a terminator sequence.

3. Three types of RNA are directly involved in protein synthesis: ribosomal RNA (rRNA), transfer RNA (tRNA), and messenger RNA (mRNA).

4. Bacterial genes encoding proteins for the same metabolic pathway are often clustered into operons. Some operons are induced in the presence of the substrate of the pathway, for example, the *lac* operon. Others are repressed in the presence of the product of the pathway, for example, the *trp* operon.

5. Eukaryotic mRNAs are modified by the addition of a 7-methyl-guanosine cap at their 5′ end. A poly(A) tail is added to their 3′ end. Intron sequences are removed, and the exon sequences joined together by the process known as splicing. The fully processed, mature mRNA is then ready for transport to the cytoplasm and protein synthesis.

6. In eukaryotes, there are three RNA polymerases – RNA polymerases I, II, and III. RNA polymerase II needs the help of the TATA-binding protein and other transcription factors to become bound to a promoter. This group of proteins is called the transcription pre-initiation complex, and this is sufficient to make a small number of RNA molecules. However, to make a lot of RNA in response to a signal, such as a hormone, other proteins bind to sequences called enhancers. These proteins interact with the preinitiation complex and increase the rate of RNA synthesis.

7. Noncoding RNAs such as miRNA, lncRNA, and circRNA play an important role in the control of gene expression in eukaryotes. Noncoding RNAs have been shown to change the activity of a gene through chromatin modification, influence the transcription of DNA into mRNA, modulate mRNA splicing, and decrease protein levels by degrading target mRNAs or by inhibiting their translation into protein. The ability of these noncoding RNAs to interact with each other ("sponge effects") may result in abnormal gene expression and lead to disease.

Answer to thought question: Exons are the parts of the newly synthesized mRNA that are spliced together to produce the mRNA that will be translated into protein. Exons can contain sequences that code for proteins (protein-coding exons), but exons also contain noncoding sequence (i.e. sequences found at the 5′ and 3′ ends of the mature mRNA). It is quite possible therefore for alternative RNA splicing to produce two mRNAs that have different 5′ leader sequences, or different so-called 3′ untranslated regions (3′UTRs), but for the protein product to be identical because all the protein-coding exons are common to both mature mRNAs. This phenomenon is often cell-specific, so one mature mRNA is produced in one cell type and an alternative mRNA product is produced in another cell type. Different 5′ noncoding sequences may help to regulate how efficiently the mRNA is translated in different tissues. So, two cells may still produce the same protein, but alternative splicing, by modifying the 5′ sequence of the mRNA, may help to determine how much of a protein is made in each cell. 3′UTRs are thought to contain sequences that allow the mRNA to be picked up by intracellular transport systems (page 222) so that one form of the mRNA might remain in the cell body and generate protein there, while the other might be carried out to distant cell processes so that the same protein is now preferentially made at the remote location.

FURTHER READING

Chen, L.L. (2020 Aug). The expanding regulatory mechanisms and cellular functions of circular RNAs. *Nature Reviews. Molecular Cell Biology* 21 (8): 475–490. doi: https://doi.org/10.1038/s41580-020-0243-y. Epub 2020 May 4. PMID: 32366901.

Du, W. W., Zhang, C., Yang, W., Yong, T., Awan, F. M., & Yang, B. B. (2017 Sep 26). Identifying and characterizing circRNA-protein interaction. Theranostics 7 (17): 4183–4191. doi: https://doi.org/10.7150/thno.21299. PMID: 29158818; PMCID: PMC5695005.

Giono, L.E. and Kornblihtt, A.R. (2020 Aug 28). Linking transcription, RNA polymerase II elongation and alternative splicing. *The Biochemical Journal* 477 (16): 3091–3104. https://doi.org/10.1042/BCJ20200475. PMID: 32857854.

Lloyd, G., Landini, P., and Busby, S. (2001). Activation and repression of transcription initiation in bacteria. *Essays in Biochemistry* 37: 17–31.

Panni, S., Lovering, R.C., Porras, P., and Orchard, S. (2020 Jun). Non-coding RNA regulatory networks. *Biochimica et Biophysica Acta – Gene Regulatory Mechanisms* 1863 (6): 194417. doi: https://doi.org/10.1016/j.bbagrm.2019.194417. Epub 2019 Sep 4. PMID: 31493559.

Roberts, G.C. and Smith, C.W. (2000). Alternative splicing: combinatorial output from the genome. *Current Opinion in Chemical Biology* 6: 375–383.

Statello, L., Guo, C.J., Chen, L.L., and Huarte, M. (2021 Feb). Author correction: gene regulation by long non-coding RNAs and its biological functions. *Nature Reviews. Molecular Cell Biology* 22 (2): 159. https://doi.org/10.1038/s41580-021-00330-4. Erratum for: Nat Rev Mol Cell Biol. 2021 Feb;22(2):96-118. PMID: 33420484.

Tjian, R. (1995). Molecular machines that control genes. *Scientific American* 272: 38–45.

⬤ REVIEW QUESTIONS

5.1 Theme: Codes within the base sequence

A bending of the DNA double helix through a right angle
B formation of a 7-methyl guanosine cap on eukaryotic mRNA
C polyadenylation of eukaryotic mRNA
D removal of introns in eukaryotic mRNA
E transcription initiation in eukaryotes
F transcription initiation in prokaryotes
G transcription termination in eukaryotes
H transcription termination in prokaryotes

From the above list of processes, identify the process triggered by or associated with each of the base sequences below.

1. a stretch of DNA rich in adenine and thymine, called the TATA box
2. a stretch of DNA rich in guanine and cytosine, followed by a string of adenines
3. the DNA sequence 5′ GGGGCGGGGC 3′, called the GC box
4. the DNA sequence 5′ TATAAT 3′, called the −10 or Pribnow box
5. the RNA motif 5′ GU. . ..AG 3′, where the . . . indicates a long sequence of bases
6. the RNA sequence 5′ AAUAAA 3′

5.2 Theme: The control of transcription

A A protein that binds to a regulatory site on DNA only in the absence of a small soluble molecule. Binding of the protein to the DNA increases transcription of the relevant gene. If the small soluble molecule appears, the protein adopts a conformation that cannot bind to the regulatory site, so the transcription promoting effect disappears. (Increases transcription when small soluble molecule absent)

B A protein that binds to a regulatory site on DNA only in the absence of a small soluble molecule. Binding of the protein to the DNA reduces transcription of the relevant gene. If the small soluble molecule appears, the protein adopts a conformation that cannot bind to the regulatory site, so the blocking effect on transcription disappears. (Inhibits transcription when small soluble molecule absent)

C A protein that binds to a regulatory site on DNA only in the presence of a small soluble molecule. Binding of the protein to the DNA increases transcription of the relevant gene. If the small soluble molecule disappears, the protein adopts a conformation that cannot bind to the regulatory site, so the transcription promoting effect disappears. (Increases transcription when small soluble molecule present)

D A protein that binds to a regulatory site on DNA only in the presence of a small soluble molecule. Binding of the protein to the DNA reduces transcription of the relevant gene. If the small soluble molecule disappears, the protein adopts a conformation that cannot bind to the regulatory site, so the blocking effect on transcription disappears. (Inhibits transcription when small soluble molecule present)

From the above list of descriptions, select the description applicable to each of the regulatory proteins listed below.

1. catabolite activator protein
2. glucocorticoid hormone receptor
3. *lac* repressor protein
4. *trp* aporepressor protein

5.3 Theme: Events that occur after transcription in eukaryotes

A capping: the addition of a methylated guanosine to the end of the RNA chain
B digestion by nucleases
C export from the nucleus
D polyadenylation
E RNA splicing

From the above list of processes, select the one described by each of the descriptions below.

1. a chemical modification of the 3′ end of the RNA molecule
2. a chemical modification of the 5′ end of the RNA molecule
3. a process that allows two or more polypeptide chains of different amino acid sequence to be generated from the same mRNA transcript
4. a process that reduces, often dramatically, the length of the RNA molecule prior to its subsequent translation into protein

---

● THOUGHT QUESTION

How is it that alternative splicing of a primary mRNA transcript sometimes generates two different mature mRNA molecules, both of which are translated, yet the proteins produced when the two mature mRNAs are translated are identical?

6

MANUFACTURING PROTEIN

The genetic code (page 44) dictates the sequence of amino acids in a protein molecule. The synthesis of proteins is quite complex, requiring three types of RNA. Messenger RNA (**mRNA**) contains the code and is the template for protein synthesis. Transfer RNAs (**tRNAs**) are adapter molecules that carry amino acids to the mRNA. Ribosomal RNAs (**rRNAs**) form part of the ribosome that brings together all the components necessary for protein synthesis. Several enzymes also help in the construction of new protein molecules. This chapter describes how the nucleotide sequence of an mRNA molecule is translated into the amino acid sequence of a protein.

Figure 6.1 shows the basic mechanism of protein synthesis, also called translation. In the first step, free amino acids are attached to tRNA molecules. In the second step, a ribosome assembles on the mRNA strand to initiate synthesis. In the third step, the ribosome travels along the mRNA. At each codon on the mRNA a tRNA binds, bringing the amino acid defined by that codon to be added to the growing polypeptide chain. In the last, fourth, step the ribosome encounters a stop codon and protein synthesis is terminated.

ATTACHMENT OF AN AMINO ACID TO ITS tRNA

Amino acids are not directly incorporated into protein on a messenger RNA template. An amino acid is carried to the mRNA chain by a tRNA molecule. tRNAs are small, about 7000 nucleotides in length, and are folded into precise three-dimensional structures because of hydrogen bonding between bases in particular stretches of the molecule. This gives rise to four double-stranded regions, and it is these that give tRNA its characteristic cloverleaf structure when drawn in two dimensions as in Figure 6.2.

Each tRNA molecule has an amino acid attachment site at its 3′ end and an **anticodon,** three bases that are complementary in sequence to a codon on the mRNA. The tRNA binds to the mRNA molecule because hydrogen bonds form between the anticodon and codon. For example, the codon for methionine is 5′ AUG 3′, which will base pair with the anticodon 5′ CAU 3′ (Figure 6.3a).

Transfer RNA, the Anticodon, and Wobble

Although 61 codons specify the 20 different amino acids, there are not 61 tRNAs; instead the cell economizes. The codons for some amino acids differ only in the third position of the codon. Figure 3.9 on page 44 shows that when an amino acid is encoded by only two different triplets the third bases will be either U and C, or A and G. For example aspartate is coded by GAU and GAC and glutamine by CAA and CAG. The wobble phenomenon allows a single tRNA to recognize more than one codon for a given amino acid. The wobble position is the first base (at the 5′ end) of the

Cell Biology: A Short Course, Fourth Edition. Stephen Bolsover, Andrea Townsend-Nicholson, Greg FitzHarris, Elizabeth Shephard, Jeremy Hyams and Sandip Patel.
© 2022 John Wiley & Sons Ltd. Published 2022 by John Wiley & Sons Ltd.
Companion website: www.wiley.com/go/bolsover/cellbiology4

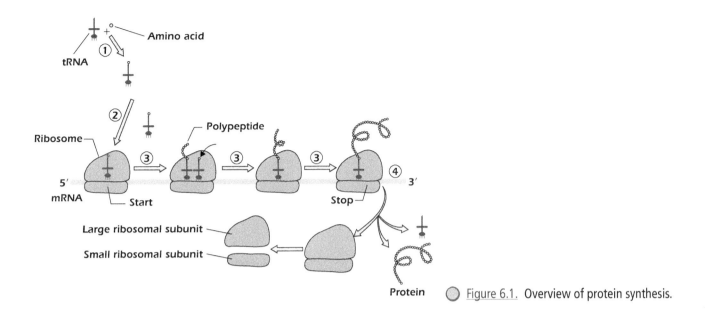

Figure 6.1. Overview of protein synthesis.

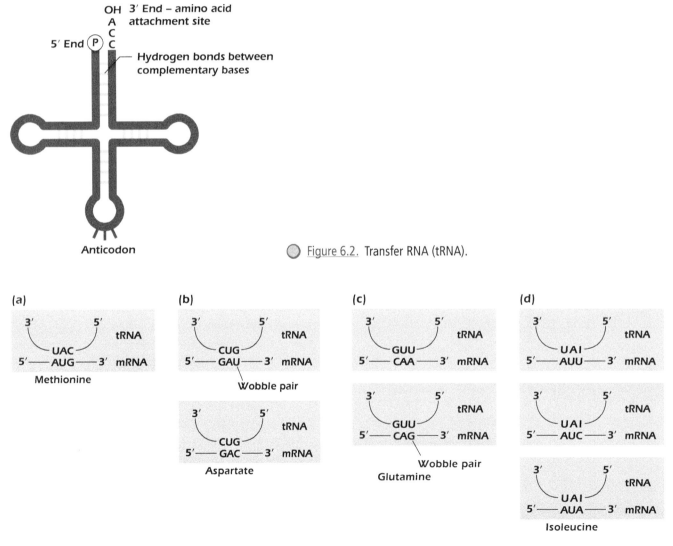

Figure 6.2. Transfer RNA (tRNA).

Figure 6.3. Wobble at the third position of the codon allows one tRNA to pair with two or even three triplet combinations.

anticodon. When either U or G are the first base of the anticodon, the tRNA can "wobble" and form a base pair with one of two different bases at the third position of the mRNA codon (see Table 6.1) according to the following rules. Any pyrimidine (that is, C and U) in the third position of the codon can base pair with the tRNA when G is the wobble base in the anticodon (Figure 6.3b). Any purine (that is, A and G) in the third position of the codon can base pair with the tRNA when U is the wobble base in the anticodon (Figure 6.3c). The anticodon of some tRNAs contains the unusual base inosine (I). When I is the wobble base in the anticodon, it can base pair with A or C or U in the third position of the codon (Figure 6.3d).

The attachment of an amino acid to its correct tRNA molecule is illustrated in Figure 6.4. This process occurs in two stages, both catalyzed by the enzyme **aminoacyl tRNA synthase.** During the first reaction (A in figure), the amino acid is joined, via its carboxyl group, to an adenosine monophosphate (**AMP**) and remains bound to the enzyme. All tRNA molecules have at their 3′ end the nucleotide sequence CCA. In the second reaction (B in figure) aminoacyl tRNA synthase transfers the amino acid from AMP to the tRNA, forming an ester bond between its carboxyl group and either the 2′- or 3′-hydroxyl group of the ribose of the terminal adenosine (A) on the tRNA to form an **aminoacyl tRNA.** This step is often referred to as amino acid activation because the energy of the ester bond can be used in the formation of a lower energy peptide bond between two amino acids. A tRNA that is attached to an amino acid is known as a **charged tRNA.** There are at least 20 aminoacyl tRNA synthases, one for each amino acid and its specific tRNA.

TABLE 6.1. Allowed combinations at the wobble base.

Base-pairing combinations			
3rd base of mRNA codon	Wobble base (1st base) of tRNA anticodon	Base pair	Type of base pair
G	C	G–C	Watson-Crick
U	A	U–A	Watson-Crick
A	U	A–U	Watson-Crick
G	U	G–U	Wobble
C	G	C–G	Watson-Crick
U	G	U–G	Wobble
A	I	A–I	Wobble
C	I	C–I	Wobble
U	I	U–I	Wobble

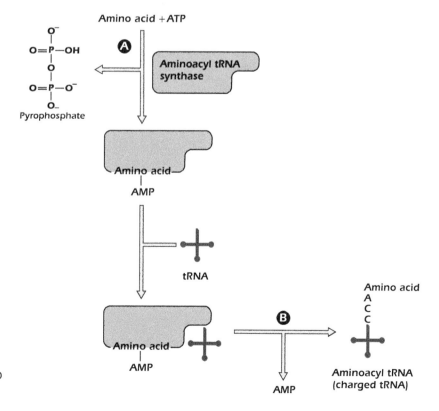

Figure 6.4. Attachment of an amino acid to its tRNA.

IN DEPTH 6.1   HOW WE SEPARATE PROTEINS IN ONE DIMENSION

The technique known as Sodium dodecyl sulfate (**SDS**) – **PAGE** is widely used to analyze the spectrum of proteins made by a particular tissue, cell type, or organelle. It is also invaluable for assessing the purity of isolated proteins. SDS stands for sodium dodecyl sulfate and PAGE for polyacrylamide gel electrophoresis.

The aim of the technique is to denature the proteins to be analyzed and then to separate them according to their size in an electrical field. To do this we first add the chemical 2-mercaptoethanol to the protein sample. This will break any **disulfide bonds** (page 104) within a protein or between protein subunits. Next SDS, which is an anionic detergent, is added and the protein sample is

boiled. Each individual polypeptide in the sample becomes covered with an overall net negative charge. The sample is now placed at the top of a slab of poly-acrylamide gel which provides an open matrix through which the coated proteins can move. A voltage is applied with the +ve terminal at the bottom. Because any inherent charge on different proteins is overwhelmed by the negative charge on the SDS, all the proteins are pulled toward the anode and separate with the smallest proteins moving the quickest.

When electrophoresis is complete, the proteins are stained by incubating the gel in a solution of Coomassie brilliant blue. Each protein band stains blue and is

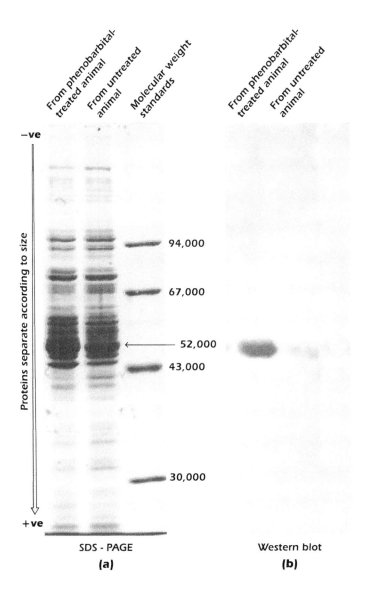

SDS - PAGE
(a)

Western blot
(b)

detectable by eye. However, if the amount of protein is very low, a more sensitive detection system is needed, such as a silver stain. Proteins of known molecular mass are also electrophoresed on the gel. By comparison with the standard proteins, the mass of an unknown protein can be determined.

If we want to follow the fate of a single protein in a complex mixture of proteins, we combine SDS – PAGE with a technique called **western blotting.** The name western blotting is, like northern blotting, a play on the name of Dr. Ed Southern who devised the technique of Southern blotting to analyze DNA (Table 8.2 on page 131).

The protein mixture is separated by SDS – PAGE. A nylon membrane is then placed up against the polyacrylamide gel and picks up the proteins, so that the pattern of protein bands on the original polyacrylamide gel is preserved on the nylon membrane. The nylon membrane is then incubated with an antibody specific for the protein of interest. This antibody, the primary antibody, will seek out and bind to its target protein on the nylon membrane. A second antibody is added that will bind to the primary antibody. To be able to detect the specific protein of interest on the membrane, the secondary antibody is attached to an enzyme or is fluorescently tagged. In the figure shown, the enzyme used was horseradish peroxidase.

A substrate is added and is converted by the enzyme into a colored product. The protein of interest is seen as a colored band on the nylon membrane. The same enzyme-linked secondary antibody can be used in other laboratories or at other times for western blotting of many different proteins because the specificity is determined by the unlabeled primary antibody.

Part A of the figure shows the Coomassie brilliant-blue-stained pattern of proteins isolated from the endoplasmic reticulum of liver. The leftmost lane is from an animal treated with the barbiturate phenobarbital, while the middle lane is from an untreated control animal. The dark bands indicate the presence of protein. The spectrum of proteins is very similar in the two samples, except that a band with a relative molecular mass (Mr) of about 52 000 is darker in the sample from the treated animal. This tells us that drug treatment has caused an increase in the production of a protein with this relative molecular mass. Western blotting (part B) using an antibody against a detoxification enzyme called CYP2B1 reveals that the induced protein is CYP2B1, showing that the *CYP2B1* gene is activated by phenobarbital. The upregulated CYP2B1 helps destroy the phenobarbital.

SDS is a major constituent of hair shampoo, where it is usually called by its alternative name of sodium lauryl sulfate.

THE RIBOSOME

The ribosome is the cell's factory for protein synthesis. Each ribosome consists of two subunits, one large and one small, each of which is made up of rRNA plus a large number of proteins. The ribosomal subunits and their RNAs are named using a parameter, called the **S value.** The S value, or **Svedberg unit,** provides an estimate of a particle's size based on its sedimentation rate. It is a measure of how fast a molecule moves to the bottom of a solution. The bigger a ribosomal subunit, the quicker it will sediment and the larger the S value. Because the S value refers to the sedimentation rate rather than the molecular mass, S values are not additive. Prokaryotic ribosomes, and those found inside mitochondria and chloroplasts comprise a larger, 50S subunit and a smaller 30S one and assemble to form a 70S ribosome (Figure 6.5a). Eukaryotic ribosomes are 80S when fully assembled and comprise a larger, 60S subunit and a smaller 40S one (Figure 6.5b). The formation of a peptide bond (page 42) between two amino acids takes place on the ribosome. The ribosome has binding sites for the mRNA template and for two charged tRNAs. An incoming tRNA with its linked amino acid occupies the **aminoacyl site (A site),** and the tRNA attached to the growing polypeptide chain occupies the **peptidyl site (P site).** The ribosome has a third

binding site for tRNA, the **exit site (E site).** This is the site to which a tRNA moves before it leaves the ribosome.

BACTERIAL PROTEIN SYNTHESIS

Ribosome-Binding Site

For protein synthesis to take place, a ribosome must first attach to the mRNA template. AUG is not only the start codon for protein synthesis; it is used to code for all the other methionines in the protein. How does the ribosome recognize the correct AUG at which to begin protein synthesis? All bacterial mRNAs have at their 5′ end a stretch of nucleotides called the **untranslated** (or leader) **sequence.** These nucleotides do not code for the protein but are nevertheless essential for the correct placing of the ribosome on the mRNA. A nucleotide sequence 5′ GGAGG 3′ (or similar) is usually found with its center about 8–13 nucleotides upstream of (5′ of) the AUG start codon (Figure 6.6). This sequence is complementary to a short stretch of sequence, 3′ CCUCC 5′, found at the 3′ end of the rRNA molecule within the 30S ribosomal subunit. The mRNA and the rRNA interact by complementary base pairing to place the 30S ribosomal subunit in the correct position to start protein synthesis. The sequence on the mRNA molecule is called the ribosome-binding site.

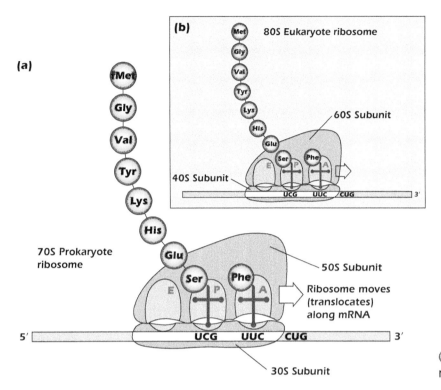

Figure 6.5. Prokaryote and eukaryote ribosomes.

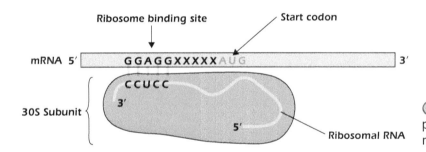

Figure 6.6. The initial binding of the prokaryote 30S subunit to mRNA. X indicates "any nucleotide."

This is sometimes referred to as the **Shine–Dalgarno sequence** after the two scientists who found it.

Because the genetic code is read in triplets of three bases, there are three possible reading frames (Figure 3.10 on page 45). The reading frame that is actually used by the cell is defined by the first AUG that the ribosome encounters downstream of the ribosome-binding site.

Chain Initiation

The first amino acid incorporated into a new bacterial polypeptide is always a modified methionine, formyl methionine (fMet) (Figure 6.7). Methionine first attaches to a specific tRNA molecule, tRNA^{fMet}, and is then modified by the addition of a formyl group that attaches to its amino group. tRNA^{fMet} has the anticodon sequence 5′ CAU 3′ that binds to its complementary codon, the universal start codon AUG.

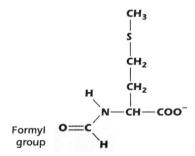

Figure 6.7. Formyl methionine.

Initiation Factor 2 Is a GTPase

Assembly of the ribosome involves a number of proteins called **initiation factors,** IFs. Of these IF2 is a **GTPase,** one of an enormous family of proteins that we will meet many

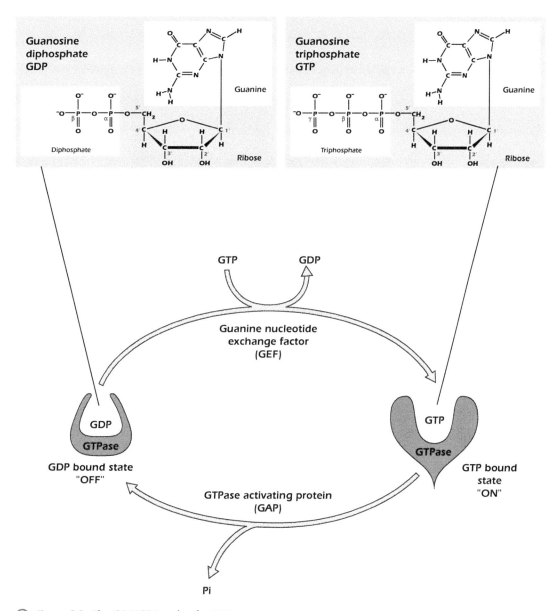

<u>Figure 6.8.</u> The GDP/GTP cycle of a GTPase.

times in this book. GTPases are often called molecular switches, because they toggle between on and off states, controlling specific processes in the cell. However, the analogy is not particularly good because many GTPases have the effect of driving a process in a particular direction. Figure 6.8 shows the basic cycle of operation of a GTPase. On the left the protein is in its off state. A nucleotide-binding domain holds the nucleotide GDP. The GTPase switches to its on state when it ejects the GDP allowing GTP, which is at a much higher concentration in the cytosol, to occupy the nucleotide binding site. For most GTPases this does not happen spontaneously but requires the assistance of a **guanine nucleotide exchange factor** or **GEF** specific to that GTPase. Once the GTPase has GTP bound it is active and able to bind to its target. The nature of the target will be

different for all GTPases, and the processes driven are extremely diverse. However, all GTPases turn off in the same way: by hydrolysis of the GTP to GDP and inorganic phosphate. Some GTPases can do this by themselves, if rather slowly, but most require assistance from another protein, their specific **GTPase-activating protein** or **GAP.**

The 70S Initiation Complex

The initiation phase of protein synthesis begins with the assembly of the 30S initiation complex on the mRNA (Figure 6.9). IF3 binds first, and prevents the 50S subunit binding to the 30S subunit. This gives time for tRNA^{fMet} to bind to the mRNA at the AUG start codon. The complex is a GEF for IF2, which releases GDP, binds GTP and becomes active, forming together

BrainBox 6.1   John Shine

In 1973, as a PhD student in Lynn Dalgarno's lab at the Australian National University, John Shine identified the RNA sequence responsible for recruiting the ribosome to the mRNA, leading to the initiation of protein synthesis in bacteria and archaea. The discovery of the Shine–Dalgarno sequence took place at the time when recombinant DNA techniques were being developed and genetic engineering technology established. In 1975 John Shine realised the importance of the Shine-Dalgarno sequence for the successful expression of recombinant proteins in bacteria. The key was to insert into a plasmid the ribosome binding site sequence in front of the cloned insulin sequence. Shine was therefore a central figure in the success of recombinant human insulin production, which was one of the first biotechnology products to be produced commercially. Since its introduction as a clinical therapy in 1982, it has become the standard of care for the treatment of diabetes.

John Shine. Source: Fairfax Media Archives/Getty Images.

Example 6.1   The Irritating Formyl Methionine

White blood cells are strongly attracted by any peptide that begins with formyl methionine: they assume an amoeboid shape and begin to crawl toward the source of the peptide. To a white blood cell, the presence of a peptide beginning with formyl methionine means that there is an infection nearby that needs to be fought. This is because the body's own proteins do not contain formyl methionine: only prokaryotes begin protein synthesis with this modified amino acid.

In mice, scent-sensitive neurons in a specialized region of the nose respond strongly to formyl methionine peptides. The hypothesis, as yet unproven, is that this allows the mice to smell the presence of bacteria and therefore avoid spoiled food.

with IF1 the complete 30S initiation complex (Figure 6.9a). IF3 now leaves and the 50S subunit attaches (Figure 6.9b). The 50S subunit is the GAP for IF2, which hydrolyzes GTP to GDP and Pi, switches to the inactive state, and leaves accompanied by IF1. The ribosome is now complete (Figure 6.9c), and the first tRNA and its amino acid are in place in the P site of the ribosome. A 70S initiation complex has been formed, and protein synthesis can begin. The ribosome is orientated so that it will move along the mRNA in the 5′ to 3′ direction, the direction in which the information encoded in the mRNA molecule is read. IF2 has used the energy released by GTP hydrolysis to drive assembly of the complex.

Elongation of the Protein Chain in Bacteria

The synthesis of a protein begins when an aminoacyl tRNA enters the A site of the ribosome (Figure 6.10). The identity of the incoming aminoacyl tRNA is determined by the codon on the mRNA. If, for example, the second codon is 5′ AAA 3′, then lysyl tRNA^{Lys}, whose anticodon is 5′ UUU 3′, will occupy the A site. The P site has, of course, already been occupied by tRNA^{fMet} during the formation of the initiation complex.

Elongation of a polypeptide chain needs the help of proteins called **elongation factors** (**EFs**). The aminoacyl tRNA cannot bind on its own to the A site. Instead, the aminoacyl tRNA must first form a complex with the GTPase EF-Tu in its activated state (Figure 6.10a). The aminoacyl tRNA is now able to enter the A site (Figure 6.10b). The presence of EF-Tu is important because it both guides and helps to correctly place the aminoacyl tRNA in the A site. However, for the peptide bond to form between two amino acids, EF-Tu must first be released from the ribosome. This happens when GTP is hydrolyzed to GDP plus an inorganic phosphate ion (Pi). Now that both the A and P sites are occupied, the enzyme **peptidyl**

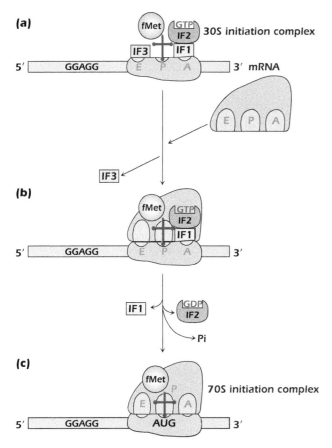

Figure 6.9. Formation of the prokaryote 70S initiation complex.

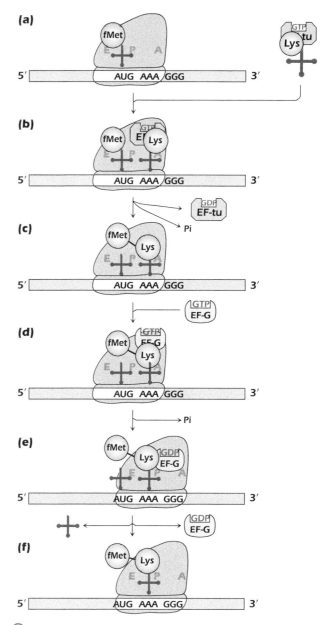

Figure 6.10. Elongation of the protein chain.

transferase catalyzes the formation of a peptide bond between the two amino acids (fMet and Lys in this example) (Figure 6.10c). The dipeptide is attached to the tRNA occupying the A site.

For the polypeptide chain to grow, the ribosome must move down the mRNA; this is known as **translocation.** For this to happen the GTPase EF-G enters the A site (Figure 6.10d). When the GTP molecule is hydrolyzed to GDP and inorganic phosphate (Pi) the ribosome moves (Figure 6.10e). This causes the tRNA in the P site, which is no longer attached to an amino acid, to move to the E site from where it is released from the ribosome (Figure 6.10f). The tRNA and attached peptide chain move to occupy the P

site. The movement of the ribosome releases EF-G from the A site (Figure 6.10f) which is once again available for an incoming aminoacyl tRNA. The process of peptide bond formation, followed by translocation, is repeated until the

---

IN DEPTH 6.2 PEPTIDYL TRANSFERASE IS A RIBOZYME

Peptidyl transferase, which catalyzes the formation of the peptide bond between two amino acids, is not a protein. In *Escherichia coli* it is the rRNA molecule within the large ribosomal subunit that is responsible for peptide bond formation and this is therefore called a **ribozyme.** This came as a surprise to scientists who discovered that the large subunit of the ribosome had peptidyl transferase activity even when all its protein had been removed.

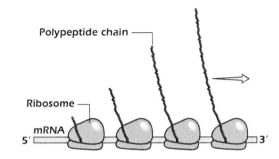

Figure 6.11. The polyribosome.

ribosome reaches a stop signal and protein synthesis termi-nates. Proteins are synthesized beginning at their **amino** or **N terminus** (page 104). The first amino acid hence has a free (although formylated) amino group. The last amino acid in the chain has a free carboxyl group and is known as the **carboxyl** or **C terminus** (page 104).

The Polyribosome

More than one polypeptide chain is synthesized from an mRNA molecule at any given time. Once a ribosome has begun translocating along the mRNA, the start AUG codon is free, and another ribosome can bind. A second 70S initia-tion complex forms. Once this ribosome has moved away, a third ribosome can attach to the start codon. This process is repeated until the mRNA is covered with ribosomes. Each of these spans about 80 nucleotides. The resultant structure, the **polyribosome** or **polysome** (Figure 6.11), is visible under the electron microscope. This mechanism allows many protein molecules to be made at the same time on one mRNA.

Termination of Protein Synthesis

There are three codons, UAG, UAA, and UGA, that have no corresponding tRNA molecule. These are stop codons. Instead of interacting with tRNAs, the A site occupied by one of these codons is filled by proteins known as **release factors** (**RF**) (Figure 6.12). In the presence of these factors the newly synthesized polypeptide chain is freed from the

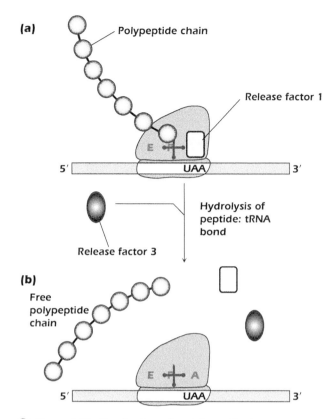

Figure 6.12. Termination of protein synthesis.

ribosome. RF1 causes polypeptide chain release from UAA and UAG, and RF2 terminates chains with UAA and UGA. The RF proteins mimic the structure of tRNA. Despite being made up of amino acids instead of nucleotides, RF1 and RF2 have a very similar three-dimensional structure to tRNA. When the A site is occupied by a release factor (Figure 6.12a), the enzyme peptidyl transferase is unable to add an amino acid to the growing polypeptide chain and instead catalyzes the hydrolysis of the bond joining the poly-peptide chain to the tRNA. The carboxyl (COOH) end of the protein is therefore freed from the tRNA (Figure 6.12b), and the protein is released. The release factors themselves must be removed from the ribosome. A GTPase called RF3 helps remove RF1 from the ribosome.

Medical Relevance 6.1   A Premature STOP

The largest known human gene, called *DMD*, encodes the protein dystrophin, part of a protein complex in muscle cells that links the cell's cytoskeleton with the extracellular matrix and which is needed for muscles to function properly (see In Depth 13.2 on page 230). More than 700 nonsense mutations in *DMD* have been identified that give rise to Duchenne and Becker muscular dystrophies. The position of these nonsense mutations within the gene can predict the severity of the disease. Ataluren (Translarna™) is a therapeutic that allows ribosomes to read through premature stop codons and is approved for use in Europe for the treat-ment of nonsense mutation Duchenne muscular dystrophy.

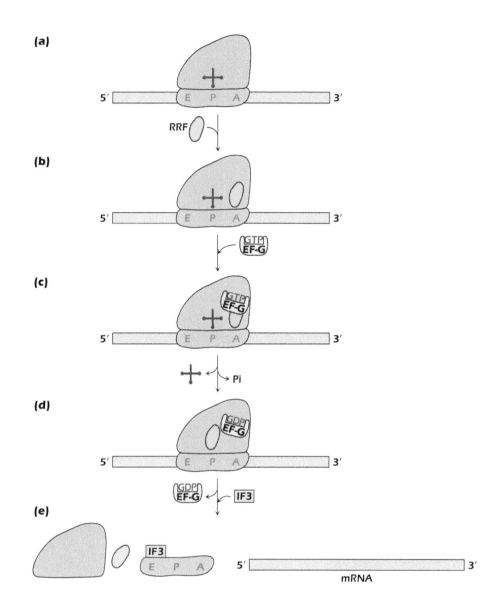

Figure 6.13. The ribosome is recycled.

The Ribosome Is Recycled

The end of protein synthesis sees the polypeptide and release factor proteins dissociate from the ribosome. At this stage the ribosome is still bound to the mRNA and the P site is still occupied by a tRNA. Bacterial cells contain a protein called **ribosome recycling factor (RRF)**. RRF, like the release factors RF1 and RF2, mimics the structure of a tRNA and binds to the empty A site (Figure 6.13). The RRF protein is then able to recruit the EF-G protein, the same GTPase used in ribosome translocation. It is thought that EF-G causes a movement of the RRF from the A site to the P site, with the loss of the tRNA in the P site, in a process similar to its role during chain elongation (compare panels D-F of Figure 6.10). The two ribosomal subunits then dissociate as initiation factor 3 binds to the 30S subunit. The 30S subunit has been prepared to begin the assembly of the 30S initiation complex and a new round of protein synthesis.

EUKARYOTIC PROTEIN SYNTHESIS IS A LITTLE MORE COMPLEX

Elongation of the polypeptide chain and the termination of protein synthesis in eukaryotes does not differ very much from that described for bacteria. The elongation factors and release factors found in eukaryotes have different names to those found in bacteria, but the equivalent proteins do a very similar job. However, the initiation of protein synthesis is more complex in eukaryotes. Their proteins always start with methionine instead of the formyl methionine used in bacterial protein synthesis. A special transfer RNA, $tRNA_i^{met}$, is used to initiate protein synthesis from the AUG start codon. The methionine is often removed from the protein after synthesis. Eukaryotic mRNAs do not contain the bacterial Shine–Dalgarno sequence for ribosome binding. Eukaryotic mRNAs have at their 5′ end a 7-methyl

BrainBox 6.2   Venkatraman Ramakrishnan, Thomas Steitz, and Ada Yonath

Venkatraman Ramakrishnan, Thomas Steitz, and Ada Yonath. Source: The Nobel Foundation. Photo: U. Montan.

Ribosomes are the protein-making machines in our cells and the only way our bodies can make proteins. Ribosomes are large complex molecular machines that read the information contained within the mRNA and produce the protein encoded by the gene. Venkatraman Ramakrishnan, Thomas Steitz, and Ada Yonath were the three key people who made the determination of the structure of ribosomes possible and uncovered important functional aspects of how the ribosome works. From the structures they obtained of the ribosome in different functional states we now understand how the ribosome inserts amino acids with accuracy and how the amino acids are joined together in the growing protein chain. In 2009 they were awarded the Nobel Prize in Chemistry for this achievement. About half of all known antibiotics target the bacterial ribosome, including penicillin. The ribosomal structures obtained by the three Laureates have shown the way in which different antibiotics affect the ribosome and have provided insight for the design of new antibiotics. With the award of the Nobel Prize, Ada Yonath became the fourth woman Laureate in Chemistry, 45 years after Dorothy Crowfoot Hodgson was awarded the prize in 1964 for her determinations by X-ray techniques of the structures of important biochemical substances, including the structure of penicillin.

guanosine cap (page 76). The cap is a key feature in the assembly of the ribosome on mRNA. The initiator $tRNA_i^{met}$ binds directly to the P site on the small ribosome subunit (the 40S subunit) along with several proteins called **eukaryotic initiation factors (eIF).** One of the eIF proteins binds directly to the cap and so positions the small subunit at the 5′ end of the mRNA. The small subunit then has to move forward to find the AUG start codon. It is helped to do this by other eIF proteins, which bind ATP that provides the power for the subunit to move. One of the eIFs is a helicase that unwraps any kinks in the mRNA produced by intramolecular hydrogen bonds so that the ribosome is able to slide along the mRNA.

All eukaryotic mRNAs have a sequence very similar to

$$5'GCC\frac{A}{G}CC\underline{AUG}G3'$$ that contains the AUG initiation

codon (underlined). The base at position 4 of the sequence is a purine and can be an A or G. This sequence (known as the Kozak sequence after Marilyn Kozak, the scientist who noted it) tells the ribosome that it has reached the start AUG codon. The recognition of the AUG codon that specifies the start site for translation also requires the help of at least nine eIF proteins. Once the large 60S ribosomal subunit has attached to form the 80S initiation complex protein synthesis can begin.

Example 6.2   The Diptheria Bacterium Inhibits Protein Synthesis

Some bacteria cause disease because they inhibit eukaryotic protein synthesis. Diphtheria was once a widespread and often fatal disease caused by infection with the bacterium *Corynebacterium diphtheriae*. This organism produces an enzyme (diphtheria toxin) that inactivates eukaryotic elongation factor 2 (the equivalent of the bacterial elongation factor G). Diphtheria toxin attaches an ADP-ribose molecule to elongation factor 2 in a process known as ADP ribosylation. The protein is now inactive and is unable to assist in the movement of the ribosome along the mRNA template. Protein synthesis therefore stops in the affected human cells. All the amino acids that the host was using to make its own protein are now available for the bacterium's use.

ANTIBIOTICS AND PROTEIN SYNTHESIS

Many antibiotics work by blocking protein synthesis, a property that is extensively exploited in research and medicine. Many antibiotics only inhibit protein synthesis in bacteria and not in eukaryotes. They are therefore extremely useful in the treatment of infections because the invading bacteria will die but protein synthesis in the host organism remains unaffected. Examples are chloramphenicol, which blocks the peptidyl transferase reaction, and tetracycline, which inhibits the binding of an aminoacyl tRNA to the A site of the ribosome. Both of these antibiotics therefore block chain elongation. Streptomycin, on the other hand, inhibits the formation of the 70S initiation complex because it prevents $tRNA^{fmet}$ from binding to the P site of the ribosome.

Puromycin causes the premature release of polypeptide chains from the ribosome and acts on both bacterial and eukaryotic cells. This antibiotic has been widely used in the study of protein synthesis. Puromycin can occupy the A site of the ribosome because its structure resembles the 3′ end of an aminoacyl-tRNA (Figure 6.14). However, puromycin does not bind to the mRNA. Puromycin blocks protein synthesis because peptidyl transferase uses it as a substrate and forms a peptide bond between the growing polypeptide and the antibiotic. Once translocation has occurred, the growing polypeptide has no strong attachment to the mRNA and is therefore released from the ribosome.

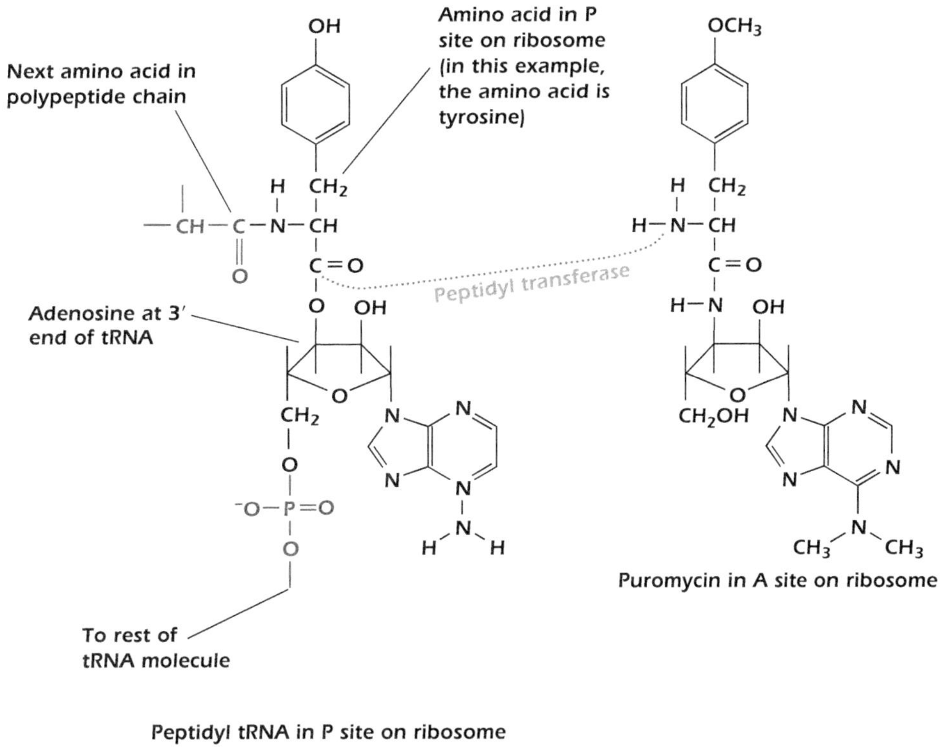

Figure 6.14. Puromycin can occupy the ribosome A site.

PROTEIN DESTRUCTION

Different proteins have enormously different lifetimes. The **keratin** (page 229) in hair lasts months until it is cut or falls out, while **calmodulin** (page 115) has a half life of seven hours, and many other proteins live even shorter lives. Cytosolic proteins are marked for destruction by being **ubiquitinated.** Ubiquitin is a small soluble protein found in most tissues of most eukaryotes, which is why it is called ubiquitin. An enzyme called **ubiquitin ligase** attaches ubiquitin to a protein by a covalent bond. The doomed protein is then sent to the cell's equivalent of an office shredder, the **proteasome.** This is a barrel-shaped proteolytic machine that chops proteins up into short lengths of peptide that are then in turn hydrolyzed by cytosolic peptidases into their constituent amino acids, ready to be used again for protein synthesis.

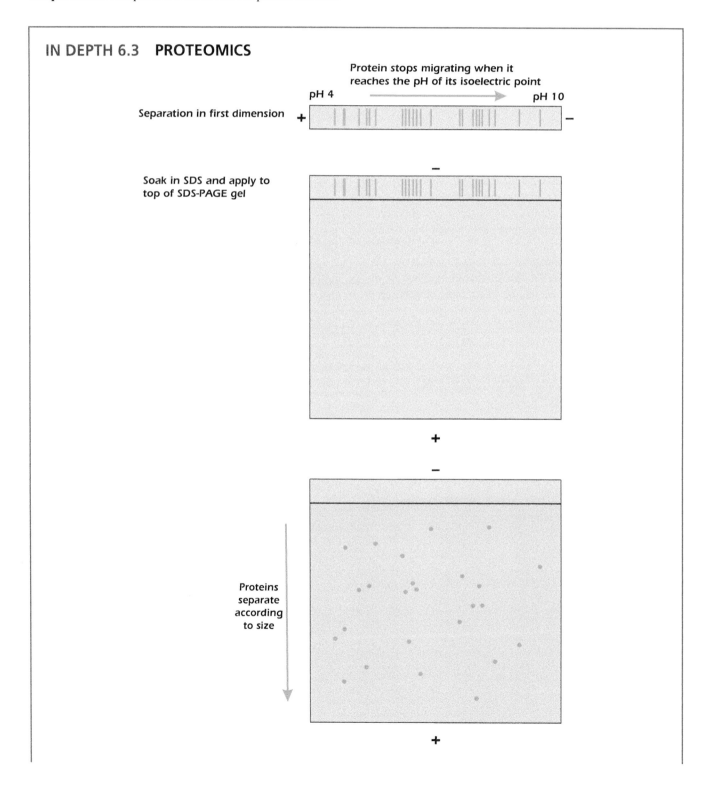

IN DEPTH 6.3 PROTEOMICS

Protein stops migrating when it reaches the pH of its isoelectric point

pH 4 → pH 10

Separation in first dimension + −

Soak in SDS and apply to top of SDS-PAGE gel

−

+

−

Proteins separate according to size

+

Proteomics is the study of the **proteome** – the complete protein content of a cell. It is the proteins a cell makes that allow it to carry out its specialized functions. Although, for example, a liver and a kidney cell have many proteins in common, they also each possess a unique subset of proteins, which gives to them their own characteristics. Similarly, a cell will need to make different proteins, and proteins in different amounts, according to its metabolic state. The goal of proteomics is to identify all of the proteins produced by different cells and how a particular disease changes a cell's protein profile.

One of the ways of tackling the separation of a cell's protein mixture into individual components is by using a technique known as two-dimensional polyacrylamide gel electrophoresis. This produces a pattern of protein spots. These patterns are recorded and serve as templates for the comparison of the proteomes of different cells. Spots that change during a cell's development, or changes that occur in disease, can be easily identified. A protein spot of interest is excised from the polyacrylamide gel and the protein broken into small, overlapping peptide fragments by proteolytic enzymes. The fragments are fed into a mass spectrometer, and their peptide mass fingerprint determined. The fingerprint identifies the protein.

The dream of many scientists, now that we know the base sequence of the human genome, is to define the human proteome and to determine all of its variations. However, the genome project used a single automated method to sequence the order of the four bases in DNA, and data was obtained at a relatively rapid pace. In contrast it is very labor intensive to identify proteins and much effort is being made to devise ways of speeding things up.

SUMMARY

1. Amino acids are prepared for use in protein synthesis by being attached to the 3′ end of a tRNA to form an aminoacyl tRNA.

2. The genetic code in an mRNA is translated into a sequence of amino acids on the ribosome, which comprises a small and large subunit and has two aminoacyl tRNA binding sites, the P site and the A site, plus an E site from which tRNAs are lost.

3. Initiation of protein synthesis involves the binding of the small ribosomal subunit to the mRNA. A special tRNA (tRNA^{fMet} in prokaryotes, tRNA$_i^{Met}$ in eukaryotes) binds to the initiation codon, and then the large ribosomal subunit attaches and the initiation complex is formed.

4. Protein synthesis begins when a second aminoacyl tRNA occupies the A site. Each incoming amino acid is specified by the codon on the mRNA. The anticodon on the tRNA hydrogen bonds to the codon, thus positioning the amino acid on the ribosome.

5. A peptide bond is formed, by peptidyl transferase, between the amino acids in the P and A sites. The newly synthesized peptide occupies the P site, and another amino acid is brought into the A site. This process of elongation requires a number of proteins (elongation factors); as it continues, the peptide chain grows.

6. When a stop codon is reached, the polypeptide chain is released with the help of proteins known as release factors.

7. More than one ribosome can attach to an mRNA. This forms a polyribosome, and many protein molecules can be made simultaneously from the same mRNA.

8. Many antibiotics fight disease because they inhibit particular steps in protein synthesis.

Answer to thought question: SDS-PAGE separates multi-subunit proteins into their component polypeptides. *E. coli* RNA polymerase is made of five subunits, one each of β, β′, and σ plus two identical α subunits. Four bands will therefore be visible on the stained polyacrylamide gel. The two α subunits are identical and will run together down the gel to form one band.

FURTHER READING

Arnez, J.G. and Moras, D. (1997). Structural and functional considerations of the aminoacylation reaction. *Trends in Biochemical Sciences* 22: 211–216.

Moore, P.B. and Steitz, T.A. (2002). The involvement of RNA in ribosome function. *Nature* 418: 229–235.

Müller, J. B., Geyer, P. E., Colaço, A. R. et al. (2020). The proteome landscape of the kingdoms of life. *Nature* 582: 592–596.

Ribas de Pouplana, L. and Schimmel, P. (2001). Aminoacyl-tRNA synthetases: potential markers of genetic code development. *Trends in Biochemical Sciences* 26: 591–596.

⬤ REVIEW QUESTIONS

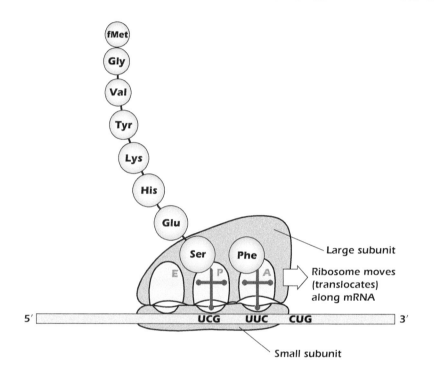

The figure above shows a ribosome during translation to provide a reminder of the basic structure.

6.1 Theme: Translation initiation

A   A site
B   E site
C   P site
D   5′ AUG 3′

E   5′GCC$\frac{A}{G}$CCAUGG3′

F   5′AGC$\frac{A}{G}$GUCACC3′

G   5′ CUG 3′
H   5′ GGAGG 3′
I   5′ UGCUUC 3′
J   formyl methionine

K   methionine
L   poly(A) tail
M   7-methyl guanosine cap

From the above list of base sequences and other components, select the one relevant to each of the steps in translation initiation which are described below in sequence.

1. In the early stages of translation initiation in prokaryotes, the small ribosomal subunit attaches by complimentary base pairing to this sequence at the 5′ end of the mRNA, sometimes called the Shine–Dalgano sequence.
2. In contrast in the early stages of translation initiation in eukaryotes, the small ribosomal subunit attaches to this group at the extreme 5′ end of the mRNA molecule.
3. The eukaryotic small subunit then slides along the mRNA until it encounters this sequence, known as the Kozak sequence.

4. The subsequent steps are similar in prokaryotes and eukaryotes. The small subunit slides a few more bases along until it encounters this sequence, the start codon for translation.

5. Initiation factors then act to catalyze the assembly of the complete ribosome. The first tRNA, with its formyl methionine (in prokaryotes) or methionine (in eukaryotes) attached, locates in one of the three tRNA binding sites on the ribosome: state which one.

6.2 Theme: Translation elongation and termination

A   A site
B   E site
C   P site
D   EF-G
E   EF-Tu
F   IF-3
G   IF-Tu
H   puromycin
I   release factor 1 or 2
J   tRNA^{fMet} in prokaryotes, tRNA$_i^{Met}$ in eukaryotes

From the above list of proteins, compounds, and components, select the one relevant to each of the steps in translation initiation which are described below in sequence.

1. Following translocation of the ribosome three bases along the mRNA, this site on the ribosome is empty and can be occupied by a charged tRNA whose anticodon is complimentary to the corresponding codon on the mRNA.

2. Peptidyl transferase now catalyzes the formation of a peptide bond between the new amino acid and the existing polypeptide chain. Immediately following this, the polypeptide chain is only attached to the mRNA via the tRNA in which of the three sites on the ribosome?

3. The next step is translocation, the physical movement of the ribosome three bases along the mRNA. Energy from

this is provided by the hydrolysis of GTP by which enzyme, which occupies the A site on the ribosome?

4. As a result of translocation the uncharged tRNA, which gave up its amino acid during the formation of the peptide bond, moves to this site on the ribosome, from where it is released.

5. When translocation brings a stop codon UGA, UAA, or UAG into the position facing the A site on the ribosome, the A site becomes occupied not by a normal charged tRNA molecule but rather by this molecule.

6. Finally the ribosome splits into two subunits as a result of energy released in GTP hydrolysis by EF-G. The released small ribosomal subunit already has one initiation factor attached, ready to accept the other initiation factors and the first charged amino acid to initiate synthesis of a new polypeptide. Name this pre-attached initiation factor.

6.3 Theme: Wobble

You should refer to Figure 3.9 on page 44 and Table 6.1 on page 87 while answering this question.

A   5′ CAU 3′
B   5′ GAA 3′
C   5′ GAI 3′
D   5′ GUU 3′
E   5′ IAU 3′
F   5′ ICC 3′
G   5′ UAC 3′
H   5′ UUG 3′

From the above list of base sequences, select the one that is likely to be the anticodon on a tRNA specific for each of the amino acids below.

1. methionine
2. asparagine
3. phenylalanine
4. glycine

⬤ THOUGHT QUESTION

Predict the appearance of an SDS-PAGE gel used to study a purified sample of *E. coli* RNA polymerase (see page 69).

7

PROTEIN STRUCTURE

Virtually everything cells do depends upon proteins. We are all, regardless of build, made up of water plus more or less equal amounts of fat and protein. Although the DNA in our cells contains the information necessary to make our bodies, DNA itself is not a significant part of our body mass. Nor is it a chemically interesting molecule, in the sense that one length of DNA is much the same as another in terms of shape and chemical reactivity. The simplicity of DNA arises because it is a polymer made up of only four fairly similar monomers, and this is appropriate because the function of DNA is simply to remain as a record and to be read during transcription. In contrast, proteins made using the instructions in DNA vary enormously in physical characteristics and function. Silk, hair, the lens of an eye, an immunoassay (such as found in a pregnancy-test kit or a COVID-19 lateral flow test), and cottage cheese are all just protein plus more or less water, but they have very different properties because the proteins they contain are different. Proteins carry out almost all of the functions of the living cell including, of course, the synthesis of new DNA. The entire set of proteins expressed by the cell is called the **proteome** (In Depth 6.3 on page 98).

Most proteins have functions that depend on their ability to recognize, and bind to, other molecules and especially other proteins. This recognition depends on specific three-dimensional **binding sites** that make multiple interactions with the **ligand,** the molecule being bound. To do this a protein must itself have a specific three-dimensional structure. Each of the huge number of protein functions demands its own protein structure. Evolution has produced this diversity by using a palette of 20 amino acid monomers, each with its own unique shape and chemical properties, as the building blocks of proteins. Huge numbers of very different structures are therefore possible.

NAMING PROTEINS

To be able to discuss proteins, we need to give them names. Naming conventions vary between different areas of biology. We have already seen how enzymes such as DNA polymerase are named for the reaction they catalyze, and their names usually end in *ase*. Many proteins have names that describe their structure or their role in cells. A large number of proteins have not been given proper names but are simply referred to by their size: p53 (page 240) and p75 (page 248) have relative molecular masses of about 53 000 and 75 000 respectively. Clearly this could cause confusion, so if the name of the gene is different we sometimes add it as a superscript: the cyclin-dependent kinase inhibitor p16^{INK4a} (Chapter 14) is a protein of relative molecular mass (Mr) of about 16 000 that is the product of the *INK4a* gene.

Cell Biology: A Short Course, Fourth Edition. Stephen Bolsover, Andrea Townsend-Nicholson,
Greg FitzHarris, Elizabeth Shephard, Jeremy Hyams and Sandip Patel.
© 2022 John Wiley & Sons Ltd. Published 2022 by John Wiley & Sons Ltd.
Companion website: www.wiley.com/go/bolsover/cellbiology4

POLYMERS OF AMINO ACIDS

Translation produces linear polymers of α-amino acids. If there are fewer than around 50 amino acids in a polymer, we tend to call it a **peptide.** More and it is a **polypeptide.** Proteins are polypeptides, and most have dimensions of a few nanometers (nm), although structural proteins like keratin in hair are much bigger. The relative molecular masses of proteins can range from 5000 to hundreds of thousands. The dystrophin protein (Medical Relevance 6.1, page 94) has a relative molecular mass of over 425 000.

The Amino Acid Building Blocks

The general structure of α-**amino acids,** the building blocks of polypeptides and proteins, is shown in Figure 7.1a. R is the side chain. It is the side chain that gives each amino acid its unique properties.

During the process of translation (page 85), peptidyl transferase joins the amino group of one amino acid to the carboxyl of the next to generate a peptide bond. A generalized polypeptide is shown in Figure 7.1b. The backbone, a series of peptide bonds separated by the α carbons, is shown in red. At the left-hand end is a free amino group; this is known as the N terminus or amino terminus (in a freshly synthesized prokaryote polypeptide, the amino group would be masked by a formyl group, page 90). At the right-hand end is a free carboxyl group; this is known as the C terminus or carboxy terminus. Peptides are normally written this way, with the N terminus to the left and the C terminus to the right.

The properties of individual polypeptides are conferred by the side chains of their constituent amino acids. Many different properties are important – size, electrical charge, the ability to participate in particular reactions – but the most important is the affinity of the side chain for water. Side chains that interact strongly with water are hydrophilic. Those that do not are hydrophobic. We have already encountered the 20 amino acids coded for by the genetic code (page 44), but we will now describe each in turn, beginning with the most hydrophilic and ending with the most hydrophobic.

Each amino acid has both a three-letter abbreviation and a one-letter code (Figure 7.2). In the following section we refer to each amino acid by its full name and give the three- and one-letter abbreviations. This will help you to familiarize yourself with the amino acid abbreviations, which are used in other sections of the book.

Several amino acids have hydrophilic side chains. Of these, four amino acids have side chains that are either acid groups or derived from them. Aspartate (Asp, D) and glutamate (Glu, E) have acidic carboxyl side chains. At the pH of the cytoplasm, these side chains bear negative charges, which interact strongly with water molecules. As they are usually charged, we generally name them as their ionized forms: aspartate and glutamate rather than aspartic acid and glutamic acid. A polypeptide made entirely of these amino acids is very soluble in water. Asparagine (Asn, N) and glutamine (Gln, Q) are the amides of aspartate and glutamate. They are hydrophilic but not charged.

Lysine (Lys, K) and arginine (Arg, R) have basic side chains. At the pH of the cytoplasm, these side chains bear positive charges, which again interact strongly with water molecules.

Figure 7.3 shows histidine (His, H), whose side chain is weakly basic with a **pKa** of 7. At neutral pH about half the histidine side chains bear a positive charge. The fact that histidine is equally balanced between these forms gives it important roles in enzyme catalysis. At neutral pH a polypeptide made entirely of histidine will be very soluble in water.

Figure 7.3 also shows cysteine (Cys, C) whose thiol (–SH) group is weakly acidic with a pKa of about 8. At neutral pH most (about 90% of) cysteine side chains will have their hydrogen attached. Even so the charge on the remaining 10% means that a polypeptide made entirely of cysteine will be soluble in water. Cysteine's thiol group is chemically reactive and has important roles in some enzyme active sites. Under oxidizing conditions the thiol groups of two cysteine residues can link together to form a **disulfide bond** (or **disulfide bridge**) (Figure 7.4). Proteins made for use outside the cell often have disulfide bonds that confer additional rigidity on the protein molecule. If all the peptide

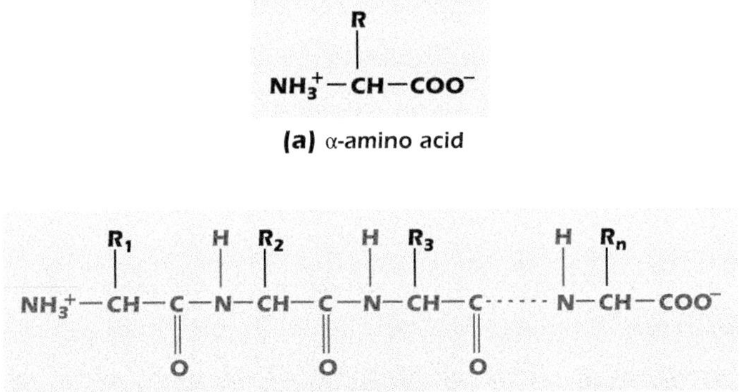

$$NH_3^+ - CH - COO^-$$

(a) α-amino acid

$$NH_3^+ - CH - C - N - CH - C - N - CH - C \cdots N - CH - COO^-$$

(b) Polypeptide

Figure 7.1. α-amino acids and the peptide bond.

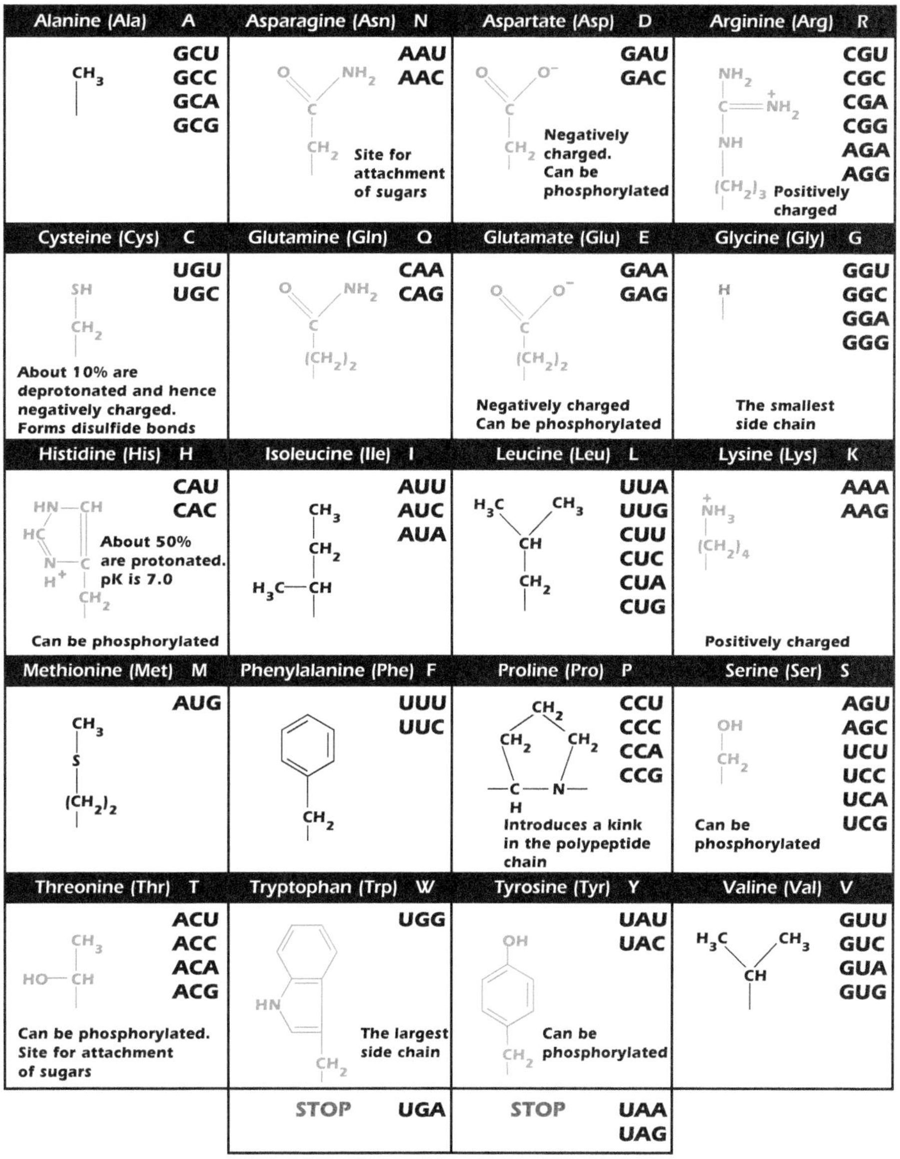

Figure 7.2. The genetic code. Amino acid side chains are shown in alphabetical order together with the three- and one-letter amino acid abbreviations. Hydrophilic side chains are shown in green, hydrophobic side chains in black. Other important characteristics of each side chain are noted. To the right of each amino acid we show the corresponding mRNA codons.

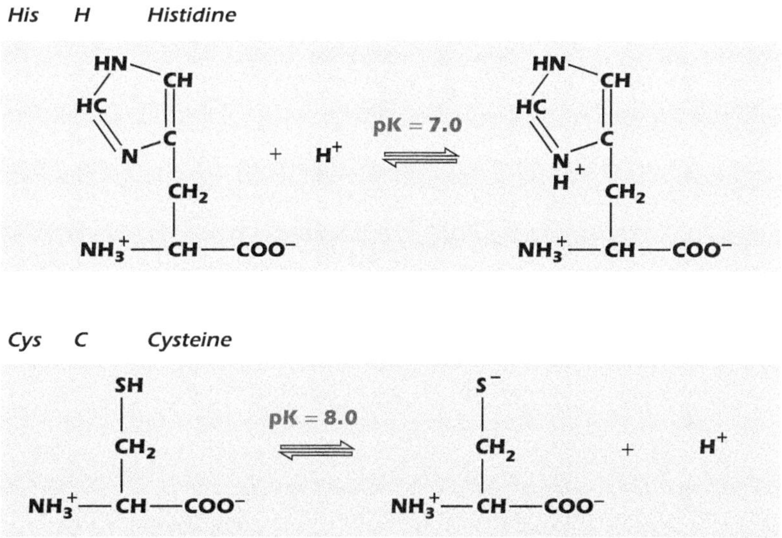

Figure 7.3. Histidine and cysteine have pKa values in the physiological range.

Figure 7.4. Oxidation of adjoining cysteine residues produces a disulfide bond. Proteolysis of the polypeptide releases cystine.

bonds in a polypeptide are hydrolyzed, any cysteines that were linked by disulfide bonds remain connected in dimers called – very confusingly – cystine.

The three amino acids serine (Ser, S), threonine (Thr, T), and tyrosine (Tyr, Y) are hydrophilic as their side chains have hydroxyl (– OH) groups that can hydrogen bond with water molecules. A polypeptide composed of these amino acids is soluble in water.

Tryptophan (Trp, W) is the largest of the amino acids. The NH group can hydrogen bond with water, making the side chain weakly hydrophilic.

Glycine (Gly, G) has nothing but a hydrogen atom for its side chain. It is relatively indifferent to its surroundings.

Five amino acids have hydrophobic side chains of carbon and hydrogen only. These are alanine (Ala, A), valine, (Val, V), leucine (Leu, L), isoleucine (Ile, I), and phenylalanine (Phe, F). The side chains cannot interact with water so they are hydrophobic. A polypeptide composed entirely of these amino acids does not dissolve in water but will dissolve in olive oil.

Methionine (Met, M) is also hydrophobic: its sulfur atom is in the middle of the chain so cannot interact with water. Last comes proline (Pro, P). Proline is not really an amino acid at all – it is an **imino acid,** but biologists give it honorary amino acid status. Because the side chain (which is hydrophobic) is connected to the nitrogen atom and therefore to the peptide bond, including proline in the polypeptide chain introduces a kink.

IN DEPTH 7.1   HYDROPATHY PLOTTING – THE PDGF RECEPTOR

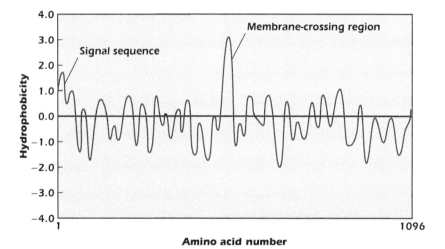

One of the central problems in structural biology has been the prediction of the structure of a protein from the primary structure or indeed from the DNA sequence once the gene has been found. One of the simpler things that can be looked for is regions of hydrophobic amino acids: if these are 21–22 amino acids long, they are likely to be membrane-spanning α helices. The figure shows a hydropathy plot for the platelet-derived growth factor (PDGF) receptor described in Chapter 11 (page 183). Each amino acid in the protein has a hydrophobicity allotted to it, so that ionized groups like those on aspartate and arginine get a big negative score, while groups like those on phenylalanine and leucine get a big positive score. The plot is a running average of the hydrophobicity along the polypeptide chain. The protein begins with a somewhat hydrophobic region: this is the signal sequence (page 204) that directs the growing protein to the endoplasmic reticulum (ER). The rest of the protein is neutral or somewhat hydrophilic except for a prominent short hydrophobic region in the center. We can therefore predict that this protein will cross the membrane once at this location. Figure 7.6 shows the full predicted structure of this protein.

The Unique Properties of Each Amino Acid

Although we have classified side chains on the basis of their affinity for water, their other properties are important.

Charge. If a side chain is charged it will be hydrophilic. But charge has other effects too. A positively charged residue such as a lysine will attract a negatively charged residue such as glutamate. If the two residues are buried deep within a folded polypeptide, where neither can interact with water, then it will be very difficult to pull them apart. We call such an electrostatic bond inside a protein a **salt bridge.** Negatively charged residues will attract positively charged ions out of solution, so a pocket on the surface of a protein lined with negatively charged residues will, if it is the right size, form a binding site for a particular positively charged ion like sodium or calcium.

Example 7.1 The Salt Bridges in ROMK Hold the Channel Open

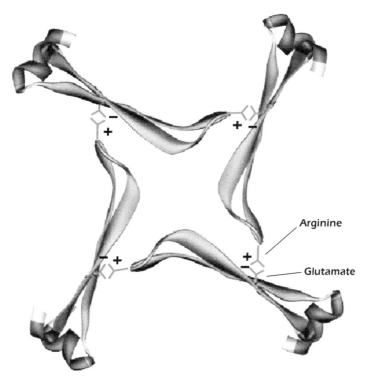

A potassium channel called ROMK plays a critical role in allowing the kidney to recover sodium that would otherwise be lost in the urine. The complete channel is formed from four identical protein subunits. The pore through which potassium ions pass lies at the center where all four subunits meet.

The illustration shows this critical pore region. The quaternary structure is stabilized by salt bridges that form between positively charged arginine residues and negatively charged glutamate residues on the neighboring subunit. A rare genetic disease called Bartter's syndrome is caused by the absence of one or other of the arginine or glutamate partners. In the absence of the salt bridge the four subunits slump inwards, occluding the central pore. Without a functional ROMK, patients lose considerable sodium in the urine and must take salt supplements.

Post-translational modifications. After a polypeptide has been synthesized some of the amino acid side chains may be modified. A good example of this is glycosylation, in which chains of sugars may be added to asparagine or threonine side chains. A very specific post-translational modification occurs in the connective tissue protein collagen. After the polypeptide chain of collagen has been synthesized, specific prolines and lysines are hydroxylated to hydroxyproline and hydroxylysine. Because hydroxyl groups can hydrogen bond with water, this helps the extracellular matrix to form a hydrated gel.

Not all post-translational modifications are so permanent. One very important modification is **phosphorylation,** the attachment of a phosphate group. The side chains of six amino acids: serine, threonine, tyrosine, aspartate, gluta-mate, and histidine can all be phosphorylated (Figure 7.5). Usually the phosphate group comes from ATP, in which case the enzyme that does the job is called a **kinase.** Phosphate groups carry two negative charges and can therefore mark-edly alter the balance of electrical forces within a protein. Another group of enzymes called **phosphatases** remove the phosphate groups. Since phosphorylation is readily

reversible, it is often used to turn protein activity off and on. For example, the transcription factor NFAT (page 214) is only active in the dephosphorylated state, while eukaryote RNA polymerase II (page 79) is activated by phosphoryla-tion. In this book the majority of the protein kinases we meet will fall into two classes: **serine–threonine kinases,** which can phosphorylate on either of these two side chains, and **tyrosine kinases.**

Phosphoesters

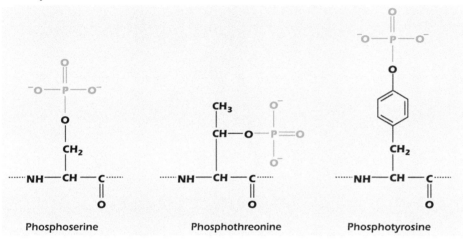

Phosphoanhydrides

Phosphoimides

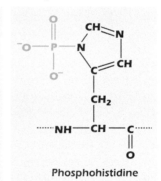

Figure 7.5. Six amino acid side chains can be phosphorylated.

Medical Relevance 7.1   Adding a Hydrophobic Group to Ras

Ras is an important protein involved in control pathways. It is a GTPase that turns on an intracellular cascade of phosphorylation that will be described in Chapter 11. Ras as made at the ribosome is a small, soluble protein. It is post-translationally modified by the attachment of a long hydrophobic group. This targets it to the inner surface of

the plasma membrane where it participates in processes leading to cell survival and cell division. Mutations in Ras can disrupt these pathways and lead to some types of cancer. The anti-cancer drug Zarnestra (also called Tipifarnib) inhibits the enzyme that carries out the addition of the hydrophobic group.

OTHER AMINO ACIDS ARE FOUND IN NATURE

Ornithine and citrulline are α-amino acids that play a vital role in the body – they are used as part of a biochemical pathway called the urea cycle to get rid of ammonium ions, which would otherwise poison us. However, they are not used as building blocks in the synthesis of polypeptides.

There are many other nonprotein amino acids. Very early in the evolution of life, the palette of amino acids used to make polypeptides became fixed at the 20 that we have described. Almost the entire substance of all living things on earth is either polypeptide, composed of these 20 monomers, or other molecules synthesized by enzymes that are themselves proteins made of these 20 monomers. Natural selection has directed evolution within the constraints imposed by the palette.

THE THREE-DIMENSIONAL STRUCTURES OF PROTEINS

Proteins are polypeptides folded into specific, complex three-dimensional shapes. Generally, hydrophobic amino acids pack into the interior of the protein while hydrophilic amino acid side chains end up on the surface where they can interact with water. The amino acid side chains pack together tightly in the interior: there are no spaces. It is the three-dimensional structure of a particular protein that allows it to carry out its role in the cell. The shape can be fully defined by stating the position and orientation of each amino acid, and such knowledge lets us produce representations of the protein.

Protein structures are held together by a large number of relatively weak interactions between the amino acid side chains interacting with one another, with the peptide bonds of the polypeptide backbone, and with other bound molecules; and between the peptide bonds themselves. These interactions are often between amino acids far apart in the linear sequence because the chain is folded in three dimensions. Although individually weak, collectively these interactions are sufficient to produce protein molecules that are stable in their environments.

The forces that stabilize the three-dimensional structures are hydrogen bonds, electrostatic interactions, van der Waals forces, hydrophobic interactions and, in some proteins only, the covalent disulphide bond.

Hydrogen Bonds

As we will see, hydrogen bonds (page 37) are important in all of the higher levels of protein structure.

Electrostatic Interactions

If positive and negatively charged amino acid residues are buried in the hydrophobic interior of a protein, where neither can interact with water, then they will attract each other and the force between them will be stronger than it would be in water. Such an electrostatic bond inside a protein is called a salt bridge.

Groups such as hydroxyls and amides are dipoles, which means that they have an excess of electrons at one atom and a corresponding deficiency at another. The partial charges of dipoles will be attracted to other dipoles and to fully charged groups.

Van der Waals Forces

These are relatively weak close-range interactions between atoms. Imagine two atoms sitting close together. At a given instant more of the electrons on one atom may be on one side, and this exposes the positive charge on the nucleus. This positive charge attracts the electrons of the adjacent atom thus exposing its nuclear charge, which would attract the electrons of another atom and so on. The next instant the electrons will have moved, so we have a situation of fluctuating attractions between atoms. These forces are important in the close-packed interiors of proteins and membranes and in the specific binding of a ligand to its binding site.

Hydrophobic Interactions

Molecules that do not interact with water (and are therefore classed as hydrophobic) force the water around them to become more organized: hydrogen-bonded "cages" form. This organization is thermodynamically undesirable and is minimized by the clustering together of such molecules. This is called the **hydrophobic effect.** A polypeptide with hydrophilic and hydrophobic residues will spontaneously adopt a configuration in which the hydrophobic residues are not exposed to water. They can achieve this either by sitting in a lipid bilayer (Figure 7.6) or by adopting a globular shape in which the hydrophobic residues are clustered in the center of the protein.

Disulfide Bonds

Extracellular proteins often have disulfide bonds between specific cysteine residues. These are strong covalent bonds, and they tend to lock the molecule into its conformation. Although relatively few proteins contain disulfide bonds, those that do are more stable and are therefore easy to purify and study. For this reason many of the first proteins studied in detail, such as the digestive enzymes chymotrypsin and ribonuclease, and the bacterial-cell-wall-degrading enzyme lysozyme, have disulfide bonds.

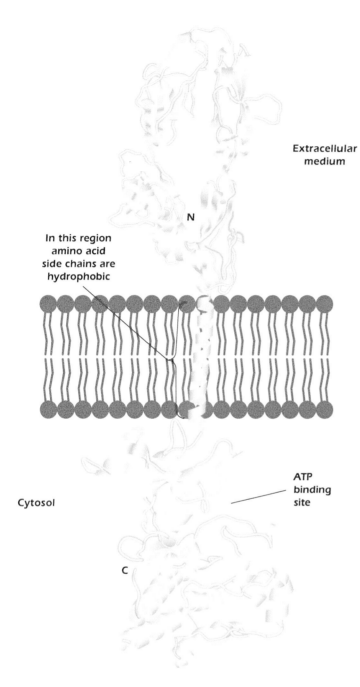

Extracellular
medium

N

In this region
amino acid
side chains are
hydrophobic

Cytosol

ATP
binding
site

C

 Figure 7.6   Structure of the platelet-derived growth
factor receptor.

LEVELS OF COMPLEXITY

When discussing protein structure, it is helpful to think in terms of a series of different levels of complexity. These levels are (unimaginatively) called primary, secondary, and tertiary structures. Some proteins are made up of individually folded, tertiary-structured subunits and such proteins are said to have quaternary structures.

The Primary Structure

The **primary structure** of a protein is the sequence of amino acids. For example, lysozyme is an enzyme that attacks

bacterial cell walls. It is found in secretions such as tears and in the white of eggs. Lysozyme has the following primary structure:

(NH2) KVFGRCELAAAMKRHGLDNYRGYSLGN
WVCAAKFESNFNTQATNRNTDGSTDYGILQIN
SRWWCDNGRTPGSRNLCNIPCSALLSSDITASV
NCAKKIVSDGDGMNAWVAWRNRCKGTDVQAWI
RGCRL(COOH).

Numbering is always from the amino terminal end where synthesis of the protein begins on the ribosome. Lysozyme has four disulfide bonds between four pairs of cysteines. The 129 amino acids of hen-egg-white lysozyme

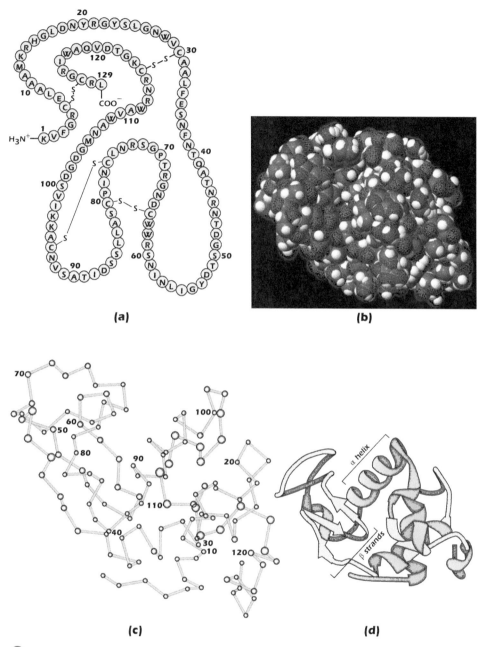

Figure 7.7 Lysozyme: (a) Linear map. (b) Space-filling model in which carbon atoms appear gray, oxygen red, nitrogen blue, and hydrogen white. None of the sulfur atoms are visible on the surface. (c) Backbone representation. (d) Cartoon. Source: Panel b uses data from Wang et al. (2007) *Acta Crystallogr. Sect. D* 63, p. 1254 viewed via the Research Collaboratory for Structural Bioinformatics (www.rcsb.org) database.

are shown in linear order in Figure 7.7a with the disulfide bonds indicated. Lysozyme was the first enzyme to have its three-dimensional structure fully determined (in 1965). Figure 7.7b shows that structure, with all of the atoms that form the molecule displayed. We see little except an irregular surface. However, if the amino acid side chains are stripped away and the path of the peptide-bonded backbone drawn (Figure 7.7c), we see that some regions of the protein backbone are ordered in a repeating pattern.

The Secondary Structure

Two types of protein backbone organization are common to many proteins. These are named the α **helix** and the β **sheet.** Figure 7.7d redraws the peptide backbone of lysozyme to emphasize these patterns, with the lengths of peptide participating in β sheets represented as arrows. These repeating patterns are known as **secondary structures.** There are other regions of the protein that do not have any such ordered pattern.

In an α helix the polypeptide chain twists around in a spiral, each turn of the helix taking 3.6 amino acid residues. This allows the nitrogen atom in each peptide bond to form a hydrogen bond with the oxygen four residues ahead of it in the polypeptide chain (Figure 7.8a and b). All the peptide bonds in the helix are able to form such hydrogen bonds (Figure 7.8c), producing a rod in which the amino acid side chains point outward (Figure 7.8d). Because it introduces a kink into the polypeptide chain and has no hydrogen to donate to a hydrogen bond, proline cannot be within an α helix.

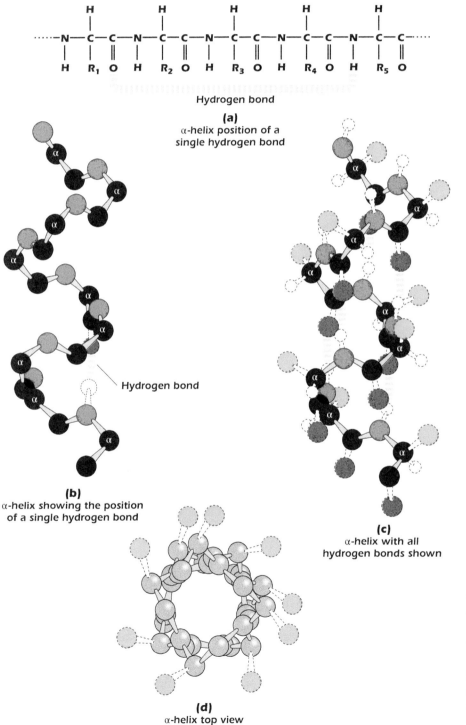

(a)
α-helix position of a
single hydrogen bond

(b)
α-helix showing the position
of a single hydrogen bond

(c)
α-helix with all
hydrogen bonds shown

(d)
α-helix top view

Figure 7.8. The α helix. In b and c the carbon atoms of the backbone are black, with α carbons so marked. Peptide bond nitrogens are blue and their attached hydrogens are white. Peptide bond oxygens are red. Green spheres represent the amino acid side chains. d shows only the backbone with all its atoms shown gray, plus the side chains in green.

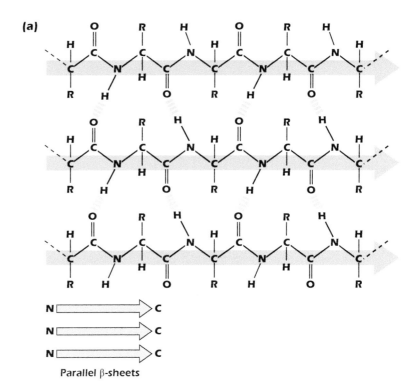

(a)

Parallel β-sheets

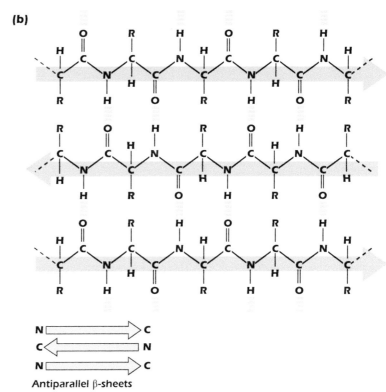

(b)

Antiparallel β-sheets

Figure 7.9. β sheets.

In a β sheet lengths of polypeptide run alongside each other, and hydrogen bonds form between the peptide bonds of the strands. This generates a sheet that has the side chains protruding above and below it (Figure 7.9). Along a single strand the side chains alternate up then down, up then down. Because the actual geometry prevents them from being completely flat, they are sometimes called β pleated sheets. A polypeptide chain can form two types of β sheet: Either all of the strands in the sheet are running in the same direction (Figure 7.9a) forming a **parallel β sheet** or they can alternate in direction (Figure 7.9b), making an **antiparallel β sheet.** The polypeptide chains in sheets are fully extended, unlike the chain in an α helix. Again, proline is unlikely to be found within a β sheet strand.

IN DEPTH 7.2   CHIRALITY AND AMINO ACIDS

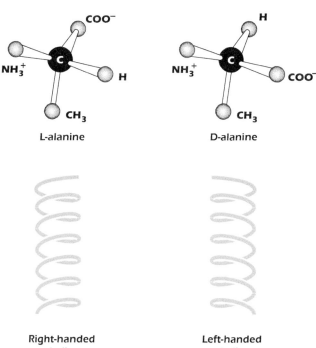

A chiral structure is one in which the mirror image of the structure cannot be superimposed on the structure. The shape of your whole body is not chiral. Your mirror image could be rotated through 180°, so that it faces into the mirror, then step back and be superimposed on you. Your right hand, however, is chiral: Its mirror image is not a right hand anymore but a left hand. Molecules are often chiral – this arises if a carbon atom has four different groups attached to it and is therefore asymmetric. The α carbon of α-amino acids is asymmetric for all except glycine. It is possible for there to be more than one asymmetric carbon in a molecule. If there is an asymmetric carbon in a molecule, there will be two different ways that the groups on that carbon can be arranged and so there will be two **optical isomers.** These isomers interact with polarized light differently. Although a different system is used in chemistry, optical isomers of amino acids

(and sugars) are denoted L and D. L amino acids are exclusively used in proteins and predominate in the metabolism of amino acids generally. However D amino acids are found in nature. D alanine occurs in bacterial cell walls while some antibiotics such as valinomycin and gramicidin A contain D amino acids. These molecules are synthesized in entirely different ways from proteins. Among sugars it is the D forms that are mainly used by cells.

Helices are chiral too. The α helix found in proteins is right-handed, like a regular screw thread. Reflect a length of right-handed α helix in a mirror, and you will get a left-handed α helix composed of D amino acids. Because of the actual structures, it turns out that L amino acids fit nicely in a right-handed α helix, while D amino acids fit nicely in a left-handed α helix.

In structural proteins like keratin in hair or fibroin in silk, the whole polypeptide chain is ordered into one of these secondary structures. Such fibrous proteins have relatively simple repeating shapes and do not have binding sites for other molecules. In contrast most proteins have regions without secondary structures, the precise folding and packing of the amino acids being unique to the protein, side by side with regions of secondary structure.

Tertiary Structure: Domains and Motifs

The three-dimensional protein structure often has protrusions, clefts, or grooves on the surface where particular amino acids are positioned to form sites that bind ligands and, in the case of enzymes, catalyze reactions within or between ligands. The whole three-dimensional arrangement of the amino acids in the protein is called the tertiary

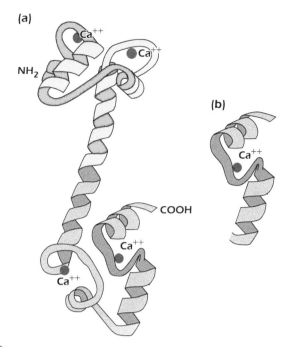

Figure 7.10. (a) The complete calmodulin molecule, composed of two very similar domains. (b) An EF hand.

structure. The tertiary structure is stabilized by all the interactions we have listed above. In particular, hydrogen bonds, which we have already seen to stabilize the α helix and β sheet, also mold the tertiary structure by linking pairs of amino acid side chains, or one side chain to a physically adjacent section of the peptide backbone.

The tertiary structure is often seen to divide into discrete regions. The calcium-binding protein **calmodulin** shows this clearly (Figure 7.10a). Its single chain is organized into two **domains,** one shown in green, the other in pink, joined by only one strand of the polypeptide chain. In calmodulin the two domains are very similar, and the modern gene probably arose through duplication of an ancestral gene that was half as big. Domains are easier to see than to define. "A separately folded region of a single polypeptide chain" is as good a definition as any.

Motifs are distinctive structures, built from standard amino acid sequences, that are found in more than one protein. For example, each domain of calmodulin contains two EF hand motifs, each of which can bind a calcium ion (Figure 7.10b). EF hands can be recognized structurally as formed of two α helices arranged at right angles with the calcium binding loop at the join. But they can also be recognized from the amino acid sequence: each EF hand in calmodulin is formed of an amino acid sequence that fits the pattern **ENXXNNXXNDXDXXGXNXXXENXXNN** where N is a hydrophobic amino acid and X is any amino acid. The same or a very similar pattern can be recognized in many other proteins and is the **consensus sequence** of an EF hand. We now know, for example, that the side chains of the

two aspartates (D) create a negatively charged cradle for the positive calcium ion while the glycine (G), whose side chain takes up very little space, allows the chain to make a tight bend.

Domains may be similar or different in both structure and function. The catabolite activating protein (CAP) of *Escherichia coli* (page 72) binds to a specific sequence of bases on DNA, assisting RNA polymerase to bind to its promoter and initiate transcription of the *lac* operon. It does this only when it has bound cAMP (which in turn only happens when the intracellular glucose concentration is low). One of the domains of CAP has the job of binding cAMP. Another recognizes DNA sequences using a **helix-turn-helix** motif (Figure 7.11). One of these helices fits into the major groove of the DNA where it can make specific interactions with the exposed edges of the bases.

Proteins that interact with DNA often do so via **zinc finger** motifs. The DNA binding domain of the glucocorticoid receptor (page 79) contains two zinc fingers (Figure 7.12a). Each finger is generated by the binding of four cysteine residues with a Zn^{2+} ion. The domain contains two α helices (Figure 7.12b), one in each finger. The first helix, also known as the recognition helix, fits into the major groove of the DNA and makes contacts with specific bases. The second helix is bent at right angles to the first and helps to stabilize the receptor homodimer on DNA by promoting dimerization between the two receptor monomers. This finger can alternatively interact with other proteins that help to activate or repress transcription.

Domains usually correspond to exons (page 59) in the gene. It is therefore relatively easy for evolution to create new proteins by mixing and matching domains from existing proteins. The calcium-dependent transcription factor NFAT (Medical Relevance 12.2 on page 214) is thought to have arisen in the ancestor of the vertebrates when a mutation occurred that spliced the DNA-binding domain from the transcription factor NFκB or a related protein into a preexisting protein that was a target for the calcium-activated phosphatase calcineurin.

Two domains that can often be recognized in proteins are the **β barrel** and the **transmembrane helix.** Many proteins with different functions show a β-barrel structure where β sheet is rolled up to form a tube. Green fluorescent protein (page 15) is one example (Figure 7.13). Integral membrane proteins, which must cross the hydrophobic interior of the membrane, often do so in an α helix comprising 22 amino acids with hydrophobic side chains, and this can be recognized in the amino acid sequence of the protein. Figure 7.6 shows an integral membrane protein called the platelet-derived growth-factor receptor. Hydrophobic side chains on the transmembrane α helix interact with the hydrophobic interior of the membrane. The protein is held tightly in the membrane because for it to leave would expose these hydrophobic side chains to water.

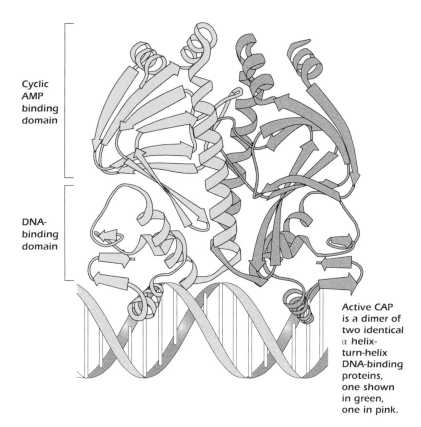

Cyclic AMP binding domain

DNA-binding domain

Active CAP is a dimer of two identical α helix-turn-helix DNA-binding proteins, one shown in green, one in pink.

Figure 7.11. Active catabolite activator protein is a dimer.

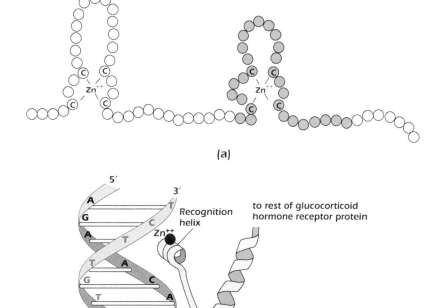

(a)

5′

3′

Recognition helix

to rest of glucocorticoid hormone receptor protein

Zn++

Zn++

Zn++

Zn++

5′

3′

to rest of glucocorticoid hormone receptor protein

(b)

Figure 7.12. (a) Two zinc finger motifs in the glucocorticoid receptor. Each circle is one amino acid residue. Four cysteine residues, indicated c, bind each zinc ion. (b) Drawing of the zinc finger domains of a dimerized pair of glucocorticoid hormone receptors interacting with DNA. One zinc finger is indicated in green and corresponds to the amino acids colored green in (a).

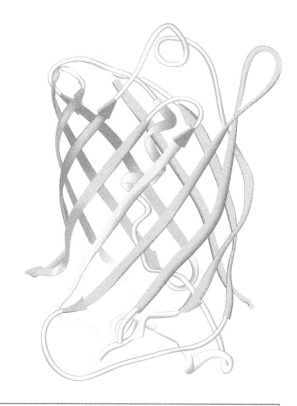

 Figure 7.13. The green fluorescent protein molecule comprises a β barrel (shown gray) and a central α helix.

 BrainBox 7.1   Christine Orengo and Janet Thornton

Christine Orengo and Janet Thornton. Sources: ismb, https://commons.wikimedia.org/wiki/File:108-ISMBECCB15-AGM_(20849977588).jpg. Licensed under CC BY 2.0; PLOS, https://commons.wikimedia.org/wiki/File:Plos_thornton.jpg. Licensed under CC BY 2.5.

Protein domains are easier to see than define, making the use of computational tools that enable protein domains to be visualized a very helpful resource for researchers. One such computational tool, the CATH Protein Structure Classification Database, was developed by Christine Orengo and Janet Thornton. Experimentally determined protein structures are analyzed computationally to identify protein domains which are then classified according to the CATH hierarchy (Class Architecture Topology Homologous super-family). At the Class (C) level, domains are assigned according to their secondary structure. Increasing levels of complexity are added as one proceeds through to the Class (H) level, where assignment a Homologous super-family level is made when there is evidence that domains are related by evolution. GFP (Chapter 1 page 15, Figure 9.10) is in CATH Superfamily 2.40.155.10. The December 2020 release of the CATH Protein Structure Classification Database contains more than 500 000 classified protein domains.

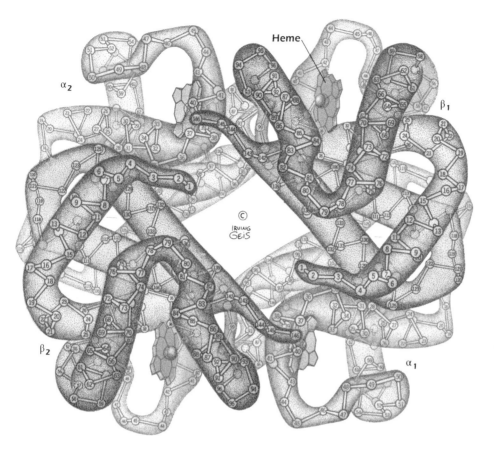

 <u>Figure 7.14.</u> Hemoglobin, a tetrameric protein with heme prosthetic groups. Illustration by Irving Geis. Rights owned by Howard Hughes Medical Institute. Source: Reproduced by permission.

Quaternary Structure: Assemblies of Protein Subunits

Many proteins associate to form multiple molecular structures that are held together tightly but not covalently by the same interactions that stabilize tertiary structures. For example, the collagen helix (Medical Relevance 3.2, page 47) is formed from two identical α1 chains and an α2 chain. CAP is only active when it dimerizes (Figure 7.11). Hemoglobin, the protein that carries oxygen in our red blood cells, is formed from four individual polypeptide chains, two α chains and two β chains (Figure 7.14). In all these cases, we call the three-dimensional arrangement of the protein subunits as they fit together to form the multimolecular structure the **quaternary structure** of the protein.

⬤ PROSTHETIC GROUPS

Even with the enormous variety of structures available, there are some functions that proteins need help with because the 20-protein amino acid side chains do not cover the properties required. Moving electrons in oxidation and reduction reactions and binding oxygen are good examples. Proteins therefore associate with other chemical species that have the required chemical properties. Hemoglobin uses iron-containing heme groups to carry oxygen molecules (Figure 7.15). It is worth noting that the iron remains as Fe^{2+}

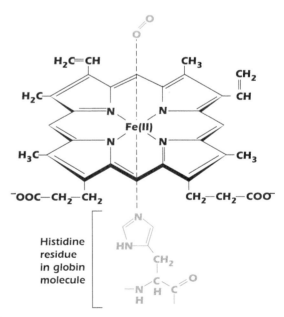

<u>Figure 7.15.</u> The iron-containing prosthetic group heme in the form in which it is found in oxygenated hemoglobin. The oxygen molecule is at the top. Source: From D. Voet and J. D. Voet, *Biochemistry*, 2nd ed., p. 216. © 1995 John Wiley & Sons, Inc. Used by permission of John Wiley & Sons, Inc.

throughout: although it binds the oxygen molecule it is not oxidized to Fe^{3+} when hemoglobin is working normally. The general name for a nonprotein species that is tightly bound to a protein and helps it perform its function is **prosthetic group.**

Some proteins have tightly bound metal ions that are essential to their function. We have already met the zinc fingers of the glucocorticoid receptor and the iron in hemoglobin. Other proteins use molybdenum, manganese, or copper.

THE PRIMARY STRUCTURE CONTAINS ALL THE INFORMATION NECESSARY TO SPECIFY HIGHER-LEVEL STRUCTURES

Protein structures are stable and functional over a small range of environmental conditions. Outside this range the pattern of interactions that stabilizes the tertiary structure is disrupted and the molecule **denatures** – activity disappears as the molecule loses its structure. Denaturation may be caused by many factors, which include excessive temperature, change of pH, and detergents. Concentrated solutions

of urea (8 mol per liter) have long been used by biochemists to denature proteins. Unlike heat and pH, urea does not cause the protein to precipitate. Physicochemical techniques have shown that in urea solution all of the higher levels of structure are lost and that polypeptide chains adopt random, changing conformations. Reagents such as urea that do this are called **chaotropic reagents.** If the urea is removed (by dialysis or simply by dilution), the protein refolds, regaining its structure and biological activity. This shows that the sequence of amino acids contains all of the information necessary to specify the final structure. The refolding of a urea-denatured protein cannot be random. It would take longer than the life of the universe to calculate the 10^{300} different possible configurations for an average protein, even if this sampling were to take place at rates much faster than the timescales on which the proteins fold inside our cells. Even a small protein with 100 amino acids would take some 10^{50} years to try out all of the structural conformations available. The fact that refolding does happen and happens on a time scale of seconds tells us that there must be a folding pathway, and the process is not random. Secondary structures may form first and act as folding units. In the cell folding is assisted by proteins called **chaperones** (page 213).

Medical Relevance 7.2   Protein Folding Gone Awry: Congenital Hypothyroidism

Congenital hypothyroidism (CH) is one of the most common endocrine diseases in children. In the UK, one in 3500 newborn babies has CH. The disease is caused by a failure of the thyroid gland to work properly. This can be due to problems in thyroid gland development or to inadequate production of the thyroid hormone thyroxine.

Many of the genetic mutations that cause CH affect protein folding. These misfolded proteins are retained in the ER and targeted for degradation by the proteasome. The disease is therefore caused by a protein deficiency and is successfully treated by giving patients the thyroxine that the body cannot produce.

Medical Relevance 7.3   Protein Folding Gone Awry: Alzheimer's Disease

Misfolded proteins are degraded in the proteasome. Many neurodegenerative diseases, including Alzheimer's Disease (AD), are characterized by the accumulation of misfolded proteins that form protein aggregates which become toxic and lead to cell death. Two different proteins, amyloid β and tau, form the aggregates that cause AD. Proteasomes can only degrade single polypeptide

chains and, once formed, aggregates are very difficult to remove. How can you break down protein aggregates? In 2020 there were 121 unique therapies in clinical trials for AD. Five drugs, all antibodies that bind to amyloid plaques and oligomers, are in Phase 3 trials. If successful, these therapies will be approved for use in the general population.

PROTEIN–PROTEIN INTERACTIONS UNDERLIE ALL OF CELL BIOLOGY

The vast web of interactions and reactions that operate in the living cell are mediated by recognition and temporary

binding between macromolecules. We have already met many examples of nucleic acid: nucleic acid recognition mediated by complimentary base pairing. We have also met protein: nucleic acid recognition, such as the association of ribosome recycling factor with the ribosome (page 95).

The vast majority of macromolecules in cells are proteins, and recognition and binding between proteins are critical to their function. Because recognition involves protein motifs, we can recognize likely interactions by simply interrogating the protein sequence database. For example, the FFAT motif, standing for FF in Acidic Tract, is found in many proteins that associate with the ER. The motif binds to a cleft in VAPs, which are integral membrane proteins of the ER (page 26). The consensus sequence is EFFDAXE with the two phenylalanines bracketed by the acidic

residues glutamate and aspartate; X can be any amino acid. As noted in Chapter 2 (page 26), VAPs are often responsible for forming membrane contact sites, and they do so by binding FFAT motif proteins. For example, the protein OSBP is a peripheral membrane protein of the Golgi complex whose FFAT motif EFFDAPE binds to VAP on the ER to bring the two membranes close and facilitate the transfer of cholesterol from the ER, where it is made, to the Golgi. We will meet many more examples of protein–protein interactions in this book.

IN DEPTH 7.3   LEVINTHAL'S PARADOX

Newly synthesized proteins fold spontaneously and rapidly – a process that can start while the protein is still being translated by the ribosome and one that is often complete within milliseconds, sometimes microseconds, after translation has finished. However, a completely random sampling of the enormous number of configurations that a protein could theoretically fold into before finding and adopting its final 3D structure would take infinitely longer. This is Levinthal's Paradox. The organization called CASP (Critical Assessment of Structure Prediction) was founded to improve methods for predicting a protein

structure from an amino acid sequence. Since 1994, they have held a biannual competition to identify "state of the art" in protein structure prediction. In 2020, "group 427" provided a prediction that was the most accurate to date, with errors comparable to the width of an atom. Group 427 was revealed to be DeepMind, whose program AlphaFold2 uses deep learning, a form of Artificial Intelligence. Computational methods provide a complementary strand to the experimental methods used to address biological questions and will revolutionize protein structure prediction.

SUMMARY

1. Polypeptides are linear polymers of α-amino acids linked by peptide bonds. There are 20 amino acids coded for by the genetic code. They differ in the properties of their side chains, which range from hydrophobic groups to charged and uncharged hydrophilic groups.

2. In addition to their hydrophilicity, side chains of individual amino acids have specific properties of which the most important are charge, the ability to form disulfide bonds, and the ability to undergo post-translational modification.

3. Phosphorylation (on S, T, Y, D, E, or H) is a post-translational modification that allows the balance of electrical charges on a protein to be dramatically and reversibly altered.

4. Proteins are polypeptides that have a complex three-dimensional shape.

5. It is convenient to consider protein structure as having three levels. The primary structure is the linear sequence of the amino acid monomers.

6. In some parts of a protein regular, repeated foldings of the polypeptide chain can be seen: these are secondary structures. Secondary structures are held together by hydrogen bonds between the carboxyl oxygens and the hydrogens on the nitrogens of the peptide bonds.

7. There are two common types of secondary structure. In the α helix the chain coils upon itself making a spiral with hydrogen bonds running parallel to the length of the spiral. In β sheets the hydrogen bonds are between extended strands of polypeptide that run alongside one another. The hydrogen bonds are at right angles to the strands and in the plane of the sheet. There are two types – parallel and antiparallel.

8. The final, complex folding of a protein is its tertiary structure. Interactions between side chains stabilize the tertiary structure.

9. Some proteins have a quaternary structure. This is an association of subunits, each of which has a tertiary structure.

Answer to thought question: In addition to any charges on the side chain all free amino acids bear two charges, a positive charge on the amino group and a negative charge on the carboxyl group, at cellular pH values. Therefore, like all

small ions they are soluble in water. Formation of the peptide bond removes these charges, leaving only any charges present on the side chain.

FURTHER READING

Branden, C. and Tooze, J. (1999). *Introduction to Protein Structure*, 2e. New York: Garland.

Creighton, T. (1992). *Proteins, Structures and Molecular Properties*, 2e. New York: W. H. Freeman.

Fersht, A. (2017). *Structure and Mechanism in Protein Science: A Guild to Enzyme Catalysis and Protein Folding*. World Scientific.

Gershenson, A., Gosavi, S., Faccioli, P., and Wintrode, P.L. (2020). Successes and challenges in simulating the folding of large proteins. *The Journal of Biological Chemistry* 295 (1): 15–33.

Levinthal, C. (1969) How to fold graciously. https://web.archive.org/web/20110523080407/http://www-miller.ch.cam.ac.uk/levinthal/levinthal.html.

McGee, H. (2017). *On Food and Cooking*. London: Unwin.

Perutz, M. (1992). *Protein Structure: New Approaches to Disease and Therapy*. New York: W. H. Freeman.

Tanford, C. and Reynolds, J. (2001). *Nature's Robots – A History of Proteins*. Oxford: Oxford University Press.

Voet, D. and Voet, J.D. (2011). *Biochemistry*, 4e. Wiley: Hoboken.

REVIEW QUESTIONS

7.1 Theme: Amino acids

A alanine

C cysteine

E glutamate

F phenylalanine

G glycine

M methionine

N asparagine

P proline

R arginine

V valine

W tryptophan

From the list of amino acids above, choose the amino acid that best matches each of the descriptions below.

1. can be phosphorylated
2. has a strongly acidic side chain
3. has a strongly basic side chain
4. is an imino acid, not an amino acid, and therefore imposes greater constraints on the shape of the polypeptide chain
5. two can form a disulphide bond

7.2 Theme: Terms used to describe proteins

A α helix

B β sheet

C denaturation

D disulfide bond

E domain

F hydrophobic effect

G motif

H phosphorylation

I post-translational modification

J primary structure

K salt bridge

L subunit

M van der Waals

From the above list of compounds, select the one described by each of the descriptions below.

1. a covalent bond between the side chains of two cysteine residues
2. a protein secondary structure where the backbone coils in a right-handed helix with hydrogen bonds between the nitrogen and oxygen atoms of the peptide bonds with the hydrogen bonds running parallel with the direction of the helix
3. a separately folded region of a single polypeptide chain
4. loss of all of the higher levels of structure with accompanying loss of biological activity of a protein
5. the tendency for hydrophobic molecules to cluster together away from water
6. the general term for a structural element that can be found in a number of different proteins

7.3 Theme: Specific binding partners

A a specific base sequence on DNA

B β-galactoside sugars

C calcium ions

D Connexin 43

E glucose

F NFAT

G Valium

Many of the functions of proteins depend on their ability to bind very specifically to other ions or molecules. From the list above choose a ligand to which each of the proteins below binds with high specificity.

1. calmodulin
2. catabolite activator protein (CAP)
3. Connexin 43
4. glucocorticoid receptor

THOUGHT QUESTION

We have stated that hydrophobic solutes will dissolve in nonpolar solvents but will not dissolve in water. How is it that amino acids with hydrophobic side chains are able to dissolve in the aqueous cytosol, ready to be picked up by their tRNAs?

8

RECOMBINANT DNA TECHNOLOGY AND GENETIC ENGINEERING

Deoxyribonucleic acid (**DNA**) is the cell's database. Within the base sequence is all the information necessary to encode ribonucleic acid (**RNA**) and protein. A number of biological and chemical methods now give us the ability to isolate DNA molecules and to determine their base sequence. Once we have the DNA and know the sequence, many possibilities open up. We can identify mutations that cause disease, make a human therapeutic protein in a bacterial or eukaryotic cell, or alter a sequence and hence the protein it encodes. The knowledge of the entire base sequence (In Depth 4.2 on page 62) of the human genome, and of the genomes of many other organisms, such as bacteria that cause disease, is revolutionizing medicine and biology. Our increasing knowledge about the influence of noncoding RNAs has added to the DNA-based techniques we have available to us, giving us the means to manipulate genomes, transcriptomes, and proteomes in individual cells and whole organisms. In future years the power of these different forms of genetic engineering is likely to impact ever more strongly on industry and on the way we live. This chapter begins by describing some of the important methods involved in recombinant DNA technology at the heart of which is DNA cloning.

DNA CLONING

DNA cloning has had an enormous impact on our understanding of the information stored within cells. This is because the technology used in DNA cloning allows us to fragment large DNA molecules (e.g. a chromosome) into smaller ones and to separate these from each other. Cloning is a way to make many copies of selected DNA molecules and to store particular DNA sequences for later copying. It is very difficult to work with just one copy of a molecule of DNA: cloning provides the investigator with many copies of an identical DNA sequence that are then amenable to analysis.

Since all DNA molecules have very similar chemical properties, it is extremely difficult to purify individual species of DNA by classical biochemical techniques similar to those used successfully for the purification of proteins. However, we can use DNA cloning to help us to separate DNA molecules from each other. A **clone** is a population of cells that arose from one original cell and, in the absence of mutation, all members of a clone will be genetically identical. If a foreign gene or gene fragment is introduced into a

Cell Biology: A Short Course, Fourth Edition. Stephen Bolsover, Andrea Townsend-Nicholson,
Greg FitzHarris, Elizabeth Shephard, Jeremy Hyams and Sandip Patel.
© 2022 John Wiley & Sons Ltd. Published 2022 by John Wiley & Sons Ltd.
Companion website: www.wiley.com/go/bolsover/cellbiology4

cell and the cell then grows and divides repeatedly, many copies of the foreign gene can be produced, and the gene is then said to have been cloned. A DNA fragment can be cloned from any organism. The basic approach to cloning a gene is to take the genetic material from the cell of interest, which in the examples we will describe is a human cell, and to introduce this DNA into bacterial cells. Clones of bacteria are then generated, each of which contains and replicates one fragment of the human genetic material. The clones that contain the gene we are interested in are then identified and grown separately. We therefore use a biological approach to isolate DNA molecules rather than physical or chemical techniques.

● CREATING THE CLONE

How do we clone a human DNA sequence? The human genome has 3×10^9 base pairs of DNA, and with the exception of gametes (Chapter 14) the DNA content of each cell is identical. However, each cell expresses only a fraction of its genes. Different types of cells express different sets of genes and thus their mRNA content is not the same. In addition, processed mRNA is shorter than its parent DNA sequence and contains no introns (page 59). Consequently, it is much easier to isolate a DNA sequence by starting with mRNA. We therefore start the cloning process by isolating mRNA from the cells of interest. The mRNA is then copied into DNA by an enzyme called **reverse transcriptase** that is found in some viruses (Medical Relevance 3.1 on page 39). As the newly synthesized DNA is complementary in sequence to the mRNA template, it is known as **complementary DNA**, or **cDNA.** The sample of cDNAs, produced from the mRNA, will represent the products of many different genes.

The way in which a cDNA molecule is synthesized from mRNA is shown in Figure 8.1. Most eukaryotic mRNA molecules have a string of As at their 3′ end, the poly(A) tail (page 76). A short run of T residues can therefore be used to prime the synthesis of DNA from an mRNA template using reverse transcriptase. The resulting double-stranded molecule is a hybrid containing one strand of DNA and one of RNA. The RNA strand is removed by digestion with the enzyme ribonuclease H (page 55). This enzyme cleaves phosphodiester bonds in the RNA strand of the paired RNA–DNA complex, making a series of nicks down the length of the RNA. DNA polymerase (page 52) is then added. This homes in on the nicks and then moves along replacing ribonucleotides with deoxyribonucleotides. Lastly, DNA ligase is used to reform any missing phosphodiester bonds. In this way a double-stranded DNA molecule is generated by the replacement of the RNA strand with a DNA strand. If the starting point had been mRNA isolated from liver cells, then a collection of cDNA molecules representative of all the mRNA molecules within the liver will

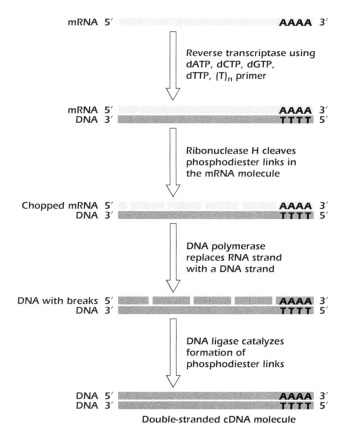

● Figure 8.1. Synthesis of a double-stranded cDNA molecule.

have been produced. These DNA molecules now have to be introduced into bacteria.

Introduction of Foreign DNA Molecules into Bacteria

Cloning Vectors. To ensure the survival and propagation of foreign DNAs, they must be inserted into a vector that can replicate inside bacterial cells and be passed on to subsequent generations of the bacteria. The vectors used for cloning are derived from naturally occurring bacterial plasmids or bacteriophages. Plasmids (page 41) are small circular DNA molecules found within bacteria. Each contains an origin of replication (page 51) and thus can replicate independently of the bacterial chromosome and produce many copies of itself. Plasmids often carry genes that confer antibiotic resistance on the host bacterium. The advantage of this to the scientist is that bacteria containing the plasmid can be selected for in a population of other bacteria simply by applying the antibiotic. Those bacteria with the antibiotic resistance gene will survive, whereas those without it will die. Figure 8.2 shows the basic components of a typical plasmid cloning vector: an antibiotic resistance gene, a restriction endonuclease site (see next section) at which foreign DNA can be inserted, and an origin of replication so the plasmid can copy itself many times inside the bacterial cell.

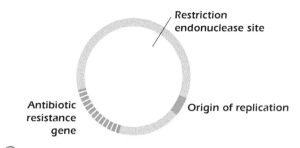

Figure 8.2. A plasmid cloning vector.

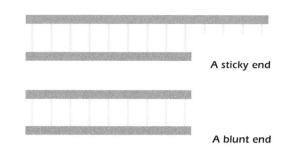

Figure 8.4. Restriction endonucleases generate two types of cut ends in double-stranded DNA.

Joining Foreign DNAs to a Cloning Vector

Enzymes known as **restriction endonucleases** are used to insert foreign DNA into a cloning vector. Each restriction endonuclease recognizes a particular DNA sequence of (usually) 4 or 6 bp. The enzyme binds to this sequence and then cuts both strands of the double helix. Many restriction endonucleases have been isolated from bacteria. The names and recognition sequences of a few of the common ones are shown in Figure 8.3. Restriction endonuclease names are conventionally written in italics because they are derived from the Latin name for the bacterium in which the protein occurs.

Some enzymes such as *Bam* H1, *Eco* R1, and *Pst* 1 make staggered cuts on each strand. The resultant DNA molecules are said to have **sticky ends** (Figure 8.4) because such fragments can associate by complementary base-pairing to any other fragment of DNA generated by the same enzyme. Other enzymes, such as *Eco* RV, cleave the DNA smoothly to produce **blunt ends** (Figure 8.4). DNA

fragments produced in this way can be joined to any other blunt-ended fragment.

Figure 8.5 illustrates how human DNA is inserted into a plasmid that contains a *Bam* H1 restriction endonuclease site. A short length of synthetic DNA (an **oligonucleotide**) that includes a *Bam* H1 recognition site is added to each end of the human DNA fragment. Both the human DNA and the cloning vector are cut with *Bam* H1. The cut ends are now complementary and will anneal together by hydrogen bonding. DNA ligase then catalyzes the formation of phosphodiester bonds between the vector and the human DNA. The resultant molecule is known as a **recombinant plasmid.** If our starting material was mRNA from a sample of liver, we would now have a collection of plasmids each carrying a cDNA from one of the genes that was being transcribed in this organ.

A new method of creating recombinant DNA molecules without using restriction enzymes has recently been developed. Gibson Assembly joins multiple DNA fragments in a

Bacterial species/strain	Enzyme name	Recognition sequences and cleavage sites
Bacillus amyloliquefaciens **H**	*Bam* H1	5′ GGATCC 3′ 3′ CCTAGG 5′
Escherichia coli Ry13	*Eco* R1	5′ GAATTC 3′ 3′ CTTAAG 5′
Eco RV (*Escherichia coli* J62 pLG74)	Eco RV	5′ GATATC 3′ 3′ CTATAG 5′
Nocardia otitidis-caviarum	Not 1	5′ GCGGCCGC 3′ 3′ CGCCGGCG 5′
Providencia stuartii 164	Pst 1	5′ CTGCAG 3′ 3′ GACGTC 5′
Thermus aquaticus YT-1	Taq I	5′ TCGA 3′ 3′ AGCT 5′

Figure 8.3. Recognition sites of some common restriction endonucleases.

Medical Relevance 8.1   mRNA Therapeutics

Perhaps the best-known mRNA therapeutic in the world, the mRNA-based vaccines directed against the severe acute respiratory syndrome coronavirus 2 (**SARS-CoV-2**) virus have captured the vivid attention of both scientists and the general public. These mRNA vaccines have been developed for the prevention of COVID-19, a disease that in 2020 unleashed a global pandemic infecting over 100 million people and causing millions of deaths. The means by which SARS-CoV-2 enters human cells is through an interaction between the viral spike protein and the host's ACE2 receptor. The COVID-19 mRNA vaccines encode a partial, and harmless, part of the spike protein. After immunization, these mRNA molecules are translated into protein, creating a small piece of the spike protein that is used by our immune cells to display the protein piece on their surface. Our immune systems recognize this piece as foreign and build an immune response, producing antibodies to protect us should we be infected with the SARS-CoV-2 virus at some future point. The mRNA used for immunization is rapidly degraded and does not affect or interact with our DNA in any way. One of the advantages of mRNA vaccines is that they can be modified quickly if a variant emerges that looks like it would escape immunity. Humans are not the only species to be infected by SARS-CoV-2. Computational analysis of ACE2 receptor sequences has identified susceptible species. SARS-CoV-2 is very infectious and despite the vigilance of their keepers, zoo animals have tested positive for COVID-19. Reports of various species contracting COVID-19 have prompted further investigation and additional studies have suggested that cats may be particularly susceptible. We may see mRNA vaccines being used to protect both humans and their domesticated animal companions.

single reaction using three enzymes: an exonuclease, DNA polymerase, and DNA ligase. This method has the advantage of requiring fewer steps and fewer reagents and it allows multiple fragments to be combined simultaneously. It can also be used to incorporate site-specific mutations to investigate the biological activity of DNA, RNA, and protein molecules, and for protein engineering.

Introduction of Recombinant Plasmids into Bacteria

Figure 8.6 summarizes how recombinant plasmids are introduced into bacteria such as *Escherichia coli*. Bacteria are first treated, either chemically or electrically, to make the cell wall more permeable to DNA. DNA can now enter these cells, which are said to have been made **competent.** Cells that take up DNA in this way are said to be **transformed.** The transformation process is very inefficient, and only a small percentage of cells actually take up the recombinant molecules. This means that it is extremely unlikely that any one bacterium has taken up two plasmids. The presence of an antibiotic resistance gene in the cloning vector makes it possible to select those bacteria that have taken up a molecule of foreign DNA, since only the transformed cells can survive in the presence of the antibiotic. The collection of bacterial colonies produced after this selection process is a **clone library.** All the cells of a single colony harbor identical recombinant molecules that began as one mRNA molecule in the original cell sample. Other colonies in the same clone library contain plasmids carrying different DNA inserts. Isolating individual bacterial colonies will produce different clones of foreign DNA. In the example we have described, where the starting DNA material used to produce these clones was a population of cDNA molecules, the collection of clones is called a **cDNA library.** Although it is possible to screen a cDNA library to obtain a specific cDNA of interest there is a much faster way to accomplish this goal if the nucleotide sequence of the cDNA is known. Later in this chapter, we will see that the polymerase chain reaction (**PCR**) technique can be used to amplify a specific DNA sequence in sufficient quantity to be used for cloning into a vector. Using this method, where the starting material was cDNA produced from a specific mRNA, individual bacterial colonies will each contain the same recombinant plasmid and produce the same cDNA clone.

Genomic DNA Clones

The approach described in the previous section permits the isolation of cDNA clones. Complementary DNA clones have many important uses, some of which are described below. However, as a cDNA is a copy of mRNA only, when we want to isolate a gene to investigate its structure and function, we need to create genomic DNA clones. Genes contain exons and introns and have regulatory sequences at their 5′ and 3′ ends and are therefore much larger than cDNAs. The vectors used to clone genes must therefore be able to hold long stretches of DNA. Plasmids used for cDNA cloning cannot do this. A selection of vectors used to clone genes is shown in Table 8.1. Vectors such as the P1 artificial chromosomes (**PACs**) – based on the bacteriophage P1 – can hold about 150 000 bp of DNA. Vectors called **BACs**

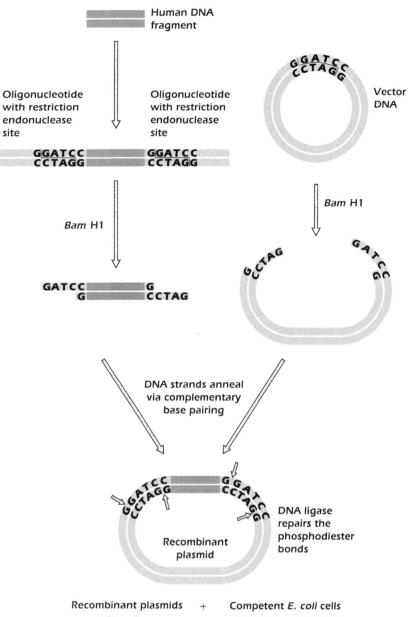

Figure 8.5. Generation of a recombinant plasmid.

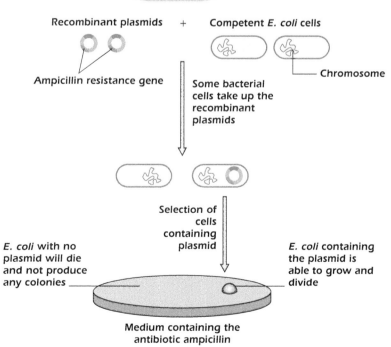

Figure 8.6. Introduction of recombinant plasmids into bacteria.

Example 8.1   Cloning a Receptor Protein cDNA

1. mRNA from brain used to make cDNA

2. ⎯ cDNA clones

Subdivide into pools

3.   Pool 1   Pool 2   Pool 3   Pool 4

4.   Extract cDNAs and inject into oocytes

Apply glutamate

5.   Na⁺

Na⁺

Glutamate receptor   No receptor activity

Oocytes producing an inward sodium current on glutamate application identify the pool containing the cDNA coding for the glutamate receptor

Repeat steps 4 and 5 on pool 1 cDNAs, each time using smaller cDNA subpools until a single cDNA giving a positive response in the receptor assay has been identified

Glutamate is one of the most important transmitters in the brain. The gene coding for the ionotropic glutamate receptor, the protein on the surface of neurons that, upon binding glutamate, allows an influx of sodium ions (page 170), remained uncloned for a number of years. Success came with the use of a very clever cloning strategy, based on the function of the receptor. mRNA was isolated from brain cells and used as the template for the production of cDNA molecules. These were inserted into a plasmid expression vector. Following the introduction of these cDNAs into bacteria, a cDNA library representative of all the mRNAs in the brain was produced. The many thousands of cDNA clones in the library were then divided into pools. Each of the many pools was then injected into a frog egg (Xenopus oocyte), which transcribed the cDNAs into RNA and translated the RNA into protein. To see which of the oocytes had been injected with the cDNA for the glutamate receptor, these cells were whole cell-patch clamped (page 153). Glutamate was applied to the oocytes, and the oocyte whose injected pool had included the cDNA for the glutamate receptor responded with an inward current of sodium ions indicating the presence of glutamate receptors in the plasma membrane.

The pool of cDNAs giving this response was further divided into smaller pools. Each of these was rescreened for the presence of glutamate receptor activity. This was followed by several rounds of rescreening. For each round a further subdivision was made of the cDNAs into pools containing fewer and fewer cDNA molecules. Eventually each pool contained only a single cDNA so that the cDNA for the glutamate receptor could be identified. A number of other receptors have now been isolated using the same strategy in which a functional assay is used to identify the cDNA encoding the receptor.

 TABLE 8.1. Vectors Used for Cloning Genomic DNA

Genomic DNA Cloning Vector Size of Insert (kb)	Size of Insert (kb)
Bacteriophage	9–23
Cosmid	30–44
PAC (P1 artificial chromosome)	130–150
BAC (bacterial artificial chromosome)	Up to 300
YAC (yeast artificial chromosome)	200–600

(bacterial artificial chromosome) can hold up to 300 000 bp. PACs and BACs have been used very successfully in the Human Genome Project and in the sequencing of the genomes of other organisms, such as the mouse (page 62). A yeast artificial chromosome (**YAC**) can hold between 200 000 and 600 000 base pairs of foreign DNA. The choice of genomic vector is governed by the size of DNA the scientist needs to clone.

USES OF DNA CLONES

The following techniques need large amounts of identical DNA and therefore can only be performed if one has cloned the gene and can therefore grow up the bacteria containing it in large numbers or, alternatively, if one uses chemically synthesized oligonucleotides that can be produced in large quantity.

Southern Blotting

In 1975 Ed Southern developed an ingenious technique, now known as **Southern blotting,** which can be used to detect specific genes (Figure 8.7). Genomic DNA is isolated and digested with one or more restriction endonucleases. The resultant fragments are separated according to size by agarose gel electrophoresis. The gel is soaked in alkaline solution to break hydrogen bonds between the two DNA strands and then transferred to a nylon membrane. This produces an exact replica of the pattern of DNA fragments in the agarose gel. The nylon membrane is incubated with a cloned DNA fragment tagged with a radioactive label. This **gene probe** is heated before being added to the nylon membrane to make it single-stranded so it will base pair, or hybridize, to its complementary sequences on the nylon membrane. As the gene probe is radiolabeled, the sequences to which it has hybridized can be detected by autoradiography. Modern techniques replace the radioactive label with fluorescent or tagged labels.

Mutations that change the pattern of DNA fragments – for instance, by altering a restriction endonuclease recognition site or deleting a large section of the gene – can easily be detected by Southern blotting. This technique was one of the first to be used in determining whether an individual carries a certain genetic defect. All that is needed is a small DNA sample from white blood cells or, in the case of a fetus, from the amniotic fluid in which it is bathed, or by removing a small amount of tissue from the chorion villus that surrounds the fetus in the early stages of pregnancy.

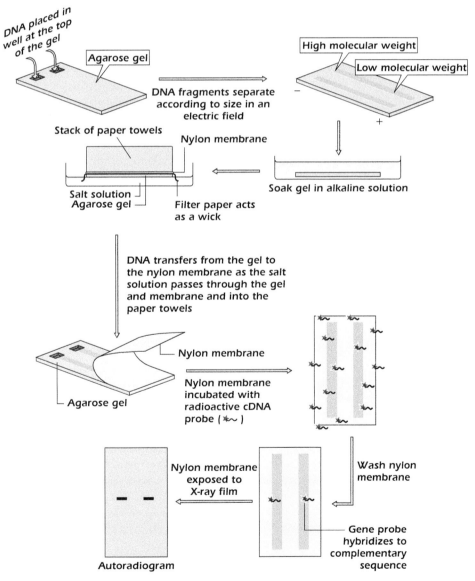

Figure 8.7. The technique of Southern blotting.

Forensic laboratories use Southern blotting to generate DNA fingerprints from samples of blood or semen left at the scene of a crime. A DNA fingerprint is a person-specific Southern blot. The gene probe used in the test is a sequence that is repeated very many times within the human genome – a microsatellite sequence (page 61). Everyone carries a different number of these repeated sequences, and because they lie adjacent to each other on the chromosome they are called **VNTRs (variable number tandem repeats).** When genomic DNA is digested with a restriction endonuclease and then analyzed by Southern blotting, a DNA pattern of its VNTRs is produced. Unless they are identical twins, it is extremely unlikely that two individuals will have the same DNA fingerprint profile. It has been estimated that if eight restriction endonucleases are used, the probability of two people who are not identical twins generating the same pattern is one in 10^{30}.

In-Situ Hybridization

It is possible, using the technique of **in-situ hybridization**, to identify individual cells expressing a particular mRNA. One way to do this is by synthesizing an antisense RNA molecule – an RNA that is complementary in sequence to the mRNA of interest. In a test tube, the appropriate strand of the cloned cDNA is copied into the antisense RNA using RNA polymerase. The RNA is labeled during synthesis by incorporating a modified nucleotide that can subsequently be detected using an antibody and a color reaction. The cDNA must first be cloned into an expression vector that contains a promoter sequence to which RNA polymerase can bind. It is also possible, and less time-consuming, to chemically synthesize an antisense RNA oligonucleotide that is complementary in sequence to a portion of the mRNA of interest. This **antisense oligonucleotide** can be labeled with similarly modified nucleotides and detected in the same way as the antisense RNA produced from the cloned cDNA. A thin tissue section, attached to a glass microscope slide, is incubated with the antisense RNA. The antisense RNA will hybridize, in the cell, to its complementary mRNA partner. Excess antisense RNA is washed off the slide, leaving only the hybridized probe. The color reaction is now carried out so that the cells expressing the mRNA of interest can be identified by bright-field microscopy (page 11). If the modified nucleotide has been fluorescently labeled, it can be detected directly using fluorescence microscopy (page 14). A third method, and the one most commonly used for in-situ hybridization, is to use appropriately labeled cDNA as the probe. For this method to work, the two DNA strands of a cDNA probe must first be separated using heat before being incubated with the tissue section on the glass slide. This is so that the mRNA can hybridize to its complementary single-stranded DNA partner. The cDNA probe can be labeled with detectable nucleotides and the resulting mRNA–DNA hybrids visualized using microscopy.

Northern Blotting

Figure 8.8a shows a blotting technique that can determine the size of an mRNA and tell us about its expression patterns. RNA is denatured by heating to remove any intramolecular double-stranded regions and then electrophoresed on a denaturing agarose gel. The RNA is transferred to a nylon membrane (as described in Figure 8.7 for the transfer of DNA). The nylon membrane is incubated with a radiolabeled, single-stranded cDNA probe, or an antisense RNA probe (page 129). Following hybridization, excess probe is washed off and the nylon membrane exposed to X-ray film. The mRNA is visualized on the autoradiogram because it hybridized to the radioactive probe. By analogy with Southern blotting, this technique is called **northern blotting** (Table 8.2). Figure 8.8b shows a northern blot for a G protein-coupled receptor (page 179) called cP2Y$_1$ in tissues from chick embryo. The blot reveals that the gene is transcribed in brain and muscle but not in heart or liver.

Production of Mammalian Proteins in Bacteria and Eukaryotic Cells

The large-scale production of proteins using cDNA-based expression systems has wide applications for medicine and industry. It is increasingly being used to produce polypeptide-based drugs, vaccines, and antibodies. Such protein products are called **recombinant** because they are produced from a recombinant plasmid. For a mammalian protein to be synthesized in bacteria its cDNA must be cloned into an expression vector (as described on page 132). Insulin was the first human protein to be expressed from a plasmid introduced into bacterial cells and has now largely replaced insulin from pigs and cattle for the treatment of diabetes. Other products of recombinant DNA technology include several vaccines, growth hormone, and factor VIII, a protein used in the treatment of the blood-clotting disorder hemophilia. Factor VIII was previously isolated from donor human blood. However, because of the danger of infection from viruses such as HIV, it is much safer to treat hemophiliacs with recombinant factor VIII. It should, in theory, be possible to express any human protein via its cDNA.

The ability to change the amino acid sequence of a protein by altering the sequence of its cDNA is known as **protein engineering.** This is achieved through the use of a technique known as **site-directed mutagenesis.** A new cDNA is created that is identical to the natural one except for changes designed into it by the scientist. This DNA can then be used to generate protein in bacteria, yeast, or other eukaryotic cell lines.

The first use of protein engineering is to study the protein itself. This technique was used to identify the particular

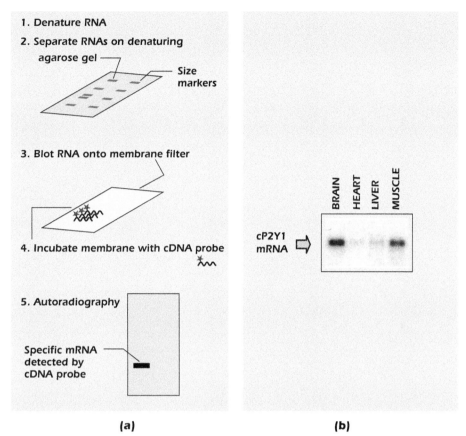

1. Denature RNA

2. Separate RNAs on denaturing agarose gel

Size markers

3. Blot RNA onto membrane filter

4. Incubate membrane with cDNA probe

5. Autoradiography

Specific mRNA detected by cDNA probe

(a)

BRAIN HEART LIVER MUSCLE

cP2Y1 mRNA

(b)

Figure 8.8. (a) The technique of northern blotting. (b) A northern blot reveals that the G protein coupled receptor cP2Y1 is expressed in the brain and muscle of chick embryo, but not in heart or liver. Source: Meyer et al. (1999). Selective expression of purinoceptor cP2Y1 suggests a role for nucleotide signalling in development of the chick embryo. Developmental Dynamics, 214(2), 152–158. doi:10.1002/(sici)1097-0177(199902)214:2<152::aid-aja5>3.0.co;2-I

TABLE 8.2. Blotting Techniques

	What Is Probed	Nature of The Probe	Book Page for Description
Southern blot	DNA	DNA	129
Northern blot	RNA	cDNA or RNA	130
Western blot	Protein	Antibody	88

charged amino acid residues responsible for the selectivity of ion channels (page 152). Now scientists are using protein engineering to generate new proteins as tools, not only for scientific research but for wider medical and industrial purposes.

Subtilisin is a protease and is one of the enzymes used in biological washing powder. The natural source of this enzyme is *Bacillus subtilis*, an organism that grows on pig feces. To produce, from this source, the 6000 tons of subtilisin used per year by the soap-powder industry is a difficult and presumably unpleasant task. The cDNA for subtilisin has been isolated and is now used by industry to synthesize the protein on a large scale in *E. coli*. The wild-type (natural) form of subtilisin is, however, prone to oxidation because of a methionine present at position 222 in the protein. Its susceptibility to oxidation makes it an unsuitable enzyme for a washing powder that must have a long shelf-life and be robust enough to withstand the rigors of a washing machine with all its temperature cycles. Scientists therefore changed the codon for methionine (AUG) to the codon for alanine (GCG). When the modified cDNA was expressed in *E. coli*, the resulting enzyme was found to be active and not susceptible to oxidation. This was excellent news for the makers of soap powder. The mutant enzyme is slightly slower at breaking down proteins, but not by much. By changing a Met to an Ala, a new enzyme has been produced that can do a reasonable job and is stable during storage and in our washing machines.

Green fluorescent protein is found naturally in certain jellyfish. Protein engineering has now created a palette of proteins with different colors. However, the great advantage of these proteins to biologists is that **chimeric proteins** (proteins composed of two parts, each derived from a different protein) incorporating a fluorescent protein are intrinsically fluorescent. This means that our protein of interest can be imaged inside a living cell using a fluorescence microscope (14). The fluorescent part of the chimeric protein tells us exactly where our protein is targeted in the cell and if this location changes in response to signals.

Figure 8.9 illustrates how this approach can be used to determine what concentration of glucocorticoid drug is

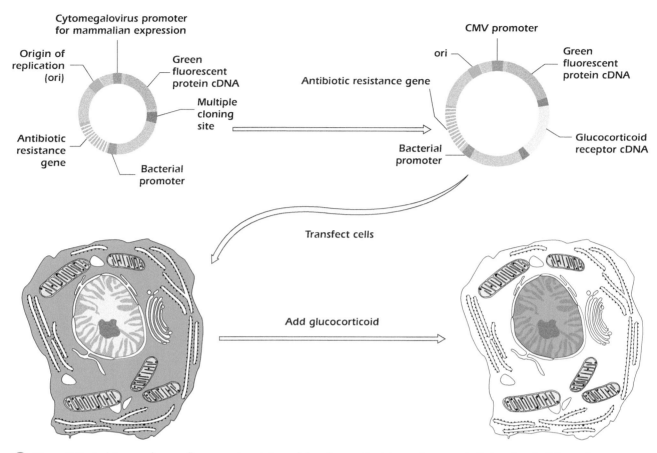

Figure 8.9. A chimera of green fluorescent protein and the glucocorticoid receptor reveals its location in living cells.

required to cause the glucocorticoid receptor to move to the nucleus. The plasmid, like many plasmids designed for convenience of use, contains a **multiple cloning site (MCS),** sometimes called a polylinker, which is a stretch of DNA that contains several restriction endonuclease recognition sites. A convenient restriction endonuclease is used to cut the plasmid (which already contains the sequence that codes for green fluorescent protein) and the cDNA for the glucocorticoid receptor is inserted. The plasmid also contains a promoter sequence, derived from a virus, that will drive the expression of the DNA into mRNA in mammalian cells that have been infected with the plasmid (or **transfected**). This type of plasmid is known as an **expression vector**. The plasmid is grown up in bacteria and then used to transfect mammalian cells. The chimeric protein, part green fluorescent protein and part glucocorticoid receptor, is synthesized in the cells from the mRNA. In the absence of glucocorticoid the protein, and therefore the green fluorescence, is in the cytosol. When enough glucocorticoid is added, it binds to the chimeric protein, which then moves rapidly to the nucleus. Transfection of cDNAs inserted into expression vectors is a common way of expressing a protein of interest inside a mammalian cell to characterize the biological properties of this protein and better understand the role it plays in the function of the cell.

Polymerase Chain Reaction

The PCR is a technique that has revolutionized recombinant DNA technology. It can amplify DNA from as little material as a single cell and from very old tissue, such as that isolated from Egyptian mummies, a frozen mammoth, and insects trapped in ancient amber. A simple mouth swab can yield enough cheek-cell DNA to determine carriers of a particular recessive genetic disorder. PCR is used to amplify DNA from fetal cells or from small amounts of tissue found at the scene of a crime. The tool that makes PCR possible is a thermostable DNA polymerase, an enzyme that can function at extremely high temperatures that would denature (page 119) most enzymes. Thermostable DNA polymerases are isolated from prokaryotes that live in extremely hot deep-sea volcanic environments.

Figure 8.10 shows how PCR uses a thermostable DNA polymerase and two short oligonucleotide DNA sequences called primers. Each primer is complementary in sequence to a short length of one of the two strands of DNA to be amplified. The DNA duplex is heated to 90 °C to separate the two strands (step 1). The mixture is cooled to 60 °C to allow the primers to anneal to their complementary sequences (step 2). At 72 °C the primers direct the thermostable DNA polymerase to copy each of the template strands (step 3). These three steps, which together constitute one cycle of the

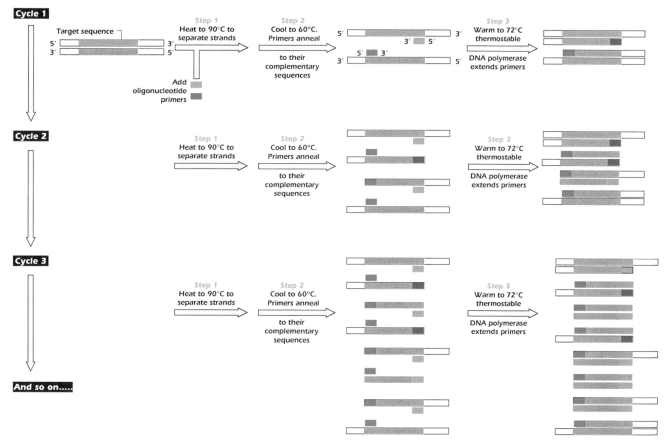

Figure 8.10. Amplification of a DNA sequence using the polymerase chain reaction.

PCR, produce twice the number of original templates. The process of template denaturation, primer annealing, and DNA synthesis is repeated many times in a tube in an automated heater block to yield many thousands of copies of the original target sequence. The expression of specific mRNA transcripts can be quantified by PCR. The mRNA is first converted into cDNA using reverse transcriptase. The PCR is then performed in the presence of a fluorescent dye that binds to double-stranded DNA. This technique is called quantitative PCR (**qPCR**) because the amount of fluorescence detected reflects the amount of starting mRNA.

DNA Sequencing

The ability to determine the order of the bases within a DNA molecule has been one of the greatest technical contributions to molecular biology. DNA is made by the polymerization of the four deoxynucleotides dGTP, dATP, dTTP, and dCTP (collectively, dNTPs). These are joined together when DNA polymerase catalyzes the formation of a phosphodiester link between a free 3'-hydroxyl on the deoxyribose sugar moiety of one nucleotide and a free 5'-phosphate group on the sugar residue of a second nucleotide. However, the artificial dideoxynucleotides ddGTP, ddATP, ddTTP, and ddCTP have no 3'-hydroxyl on their sugar residue (Figure 8.11), and so if

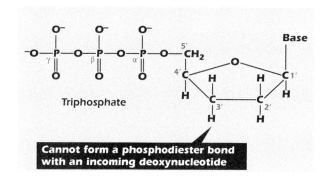

Figure 8.11. General structure of a dideoxynucleotide.

they are incorporated into a growing DNA chain, synthesis will stop. This is the basis of the dideoxy chain termination DNA sequencing technique devised by Frederick Sanger and for which he was awarded a Nobel prize in 1980.

The Human Genome Project required the process of DNA sequencing to be automated. To achieve this, each of the four dideoxynucleotides ddGTP, ddATP, ddTTP, and ddCTP is tagged with a different fluorescent dye. This means that all four of the termination reactions can be carried out in a single reaction tube and loaded onto the same well on the polyacrylamide gel. As the reaction product drips from the bottom of the gel, the fluorescence intensity of each of the four colors corresponding to the four dideoxynucleotides is monitored, and this information is transferred straight to a computer where the data are analyzed. An example of a DNA trace produced using fluorescently tagged dideoxynucleotides is shown in Figure 8.12. Each peak represents a terminated DNA product, so by reading the order of the peaks, the DNA sequence is determined. In this example G is black, A is green, T is red, and C is blue.

Next-Generation Sequencing

Following the automation of Sanger sequencing there has been a rapid development of increasingly advanced sequencing technologies that allow DNA sequences to be obtained more quickly and less expensively (In Depth 4.2, page 62). Next-Generation Sequencing (**NGS**) allows millions of DNA sequences to be obtained from a single run. These sequences are obtained from DNA sequencing libraries that contain the individual DNA fragments of

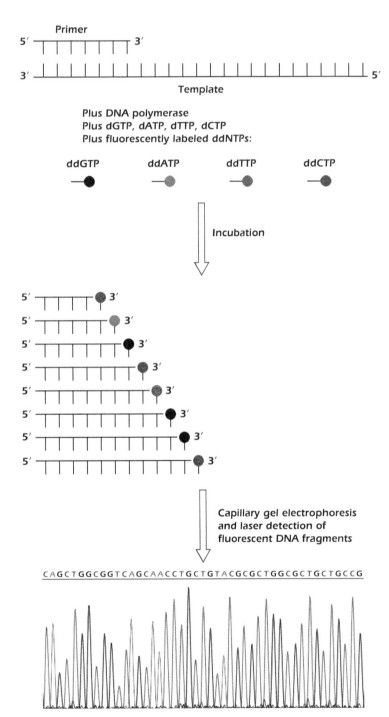

Figure 8.12. DNA sequencing by the dideoxy chain termination method.

Example 8.2 Steps Toward Clinical Gene Therapy

We will see in Chapter 15 (Medical Relevance 15.1) that Luxturna, a gene therapy treatment for Leber congenital amaurosis (**LCA**) has been approved for use in patients. The development of this treatment required validation of the therapeutic in an animal model of the disease. Many of the methods described in this chapter were used for this validation, a necessary step before clinical trials could take place. The virus used for delivery of the therapeutic was engineered by cloning the cDNA that restores normal function into a viral vector, replacing the GFP sequence in the parent vector. The orientation and reading frame of the recombinant plasmid were confirmed by Sanger sequencing and DNA sequence analysis.

A modified Kozak sequence was used to optimize protein translation in target cells and the virus particles used to infect these cells in the animal model of the disease were produced in a mammalian cell line (HEK293). Following the delivery of virus, PCR amplification of cDNA purified from virus-infected target cells was used to confirm expression of the transgene. Western blots were performed to confirm protein production. Mutations in the disease gene can give rise to different clinical phenotypes and NGS has been used to identify genetic variations in patients, making Luxturna a possible treatment for patients who may have an atypical presentation of the disease.

---

interest flanked by specialized adapter sequences. The adapter sequences allow the library fragments to bind to the glass slide where the sequencing reaction takes place. They also contain binding sites for the sequencing primers that are needed to start the DNA sequencing reaction, as well as "barcodes" that are used for the identification of samples within the library.

NGS is based on DNA synthesis, rather than chain termination by dideoxynucleotides. The most popular NGS method is based on optical sequencing. Each base emits a fluorescent signal when it is added to the growing DNA strand. As with Sanger sequencing, these fluorescent emissions are monitored and analyzed computationally. NGS sequence data are filtered for quality and those sequences that pass the quality threshold are retained for further analysis. NGS is a high-throughput technique that tends to be used to analyze >100 genes at a time, to identify novel variants and when only small amounts of DNA are present in a sample. Sanger sequencing is the method of choice for sequencing single genes or a small number of DNA samples and is useful for confirming variants identified by NGS.

 "OMICS"

We encountered genomics earlier in this book (In Depth 4.2 on page 62). It is the study of genomes – the entire set of an organism's genes – and it was the first branch in the scientific discipline that has come to be known informally as "omics." "Omics" is a term that describes the study of the totality of a class of molecule in a cell or in a tissue using high-throughput approaches. Although the rate at which we can characterize individual genomes has accelerated as the throughput of the methodology has increased, proteomics, the characterization of proteomes (In Depth 6.3 on page 98), is also a high-throughput method.

Transcriptomics

Transcriptomics is a high-throughput method that allows us to analyze the transcriptome, the entire complement of RNAs expressed in a cell or tissue. Two main methods have been developed: microarrays and RNA-Seq. The power of both methods is that they can simultaneously analyze multiple RNAs. In comparison, northern blots and PCR analyze a single RNA species. Both microarray and RNA-Seq technology rely on the copying of mRNA into cDNA.

Microarrays

Microarrays, also called gene chips, are tiny glass wafers to which single-stranded cDNAs or short stretches of DNA (oligonucleotides) are attached. These DNA molecules are attached in an ordered array and this produces a microarray of known DNA sequences. The next step is to isolate mRNA from a particular cell type, reverse transcribe the mRNA into cDNA using a fluorescently labeled dNTP and to hybridize this cDNA to the DNAs on the chip. Because the newly synthesized cDNAs are tagged with a fluorescent dye and are single stranded, they will hybridize to their complementary sequence on the chip. If a hybridization event occurs, a fluorescence signal will be detected and this is viewed using a special scanner and microscope. Computer algorithms have been written to analyze the hybridization patterns seen for a particular microarray. The number and type of DNAs used to make a microarray is dependent on the question to be answered. By using different fluorescent dyes, it is possible to compare the gene expression patterns obtained from two different populations on a single microarray. For example, if the mRNA isolated from normal cells is reverse-transcribed in the presence of a red fluorescent dNTP and the mRNA isolated from cancer cells is reverse-transcribed in the presence of a green fluorescent dNTP, then when both populations of cDNA are hybridized to the microarray it is possible

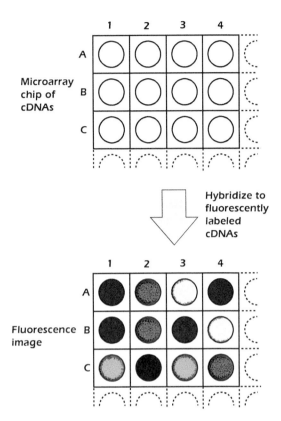

Figure 8.13. Part of a microarray used to compare the transcriptome of two cell populations.

to see the differences in gene expression between normal cells and cancer cells (Figure 8.13). Samples that show red fluorescence correspond to an mRNA that is expressed in normal cells but not in cancer cells. Samples that show green fluorescence correspond to an mRNA that is expressed in cancer cells but not in normal cells. An mRNA that is expressed in both normal and cancer cells gives a yellow fluorescent signal, while no signal is detected if an mRNA is not expressed in either normal or cancer cells.

RNA-Seq

RNA-Seq is the sequencing of cDNA molecules using NGS sequencing. Like microarrays, RNA-Seq can be used for gene expression profiling between samples; however,

RNA-Seq has many advantages over microarrays, which require known target gene sequences to be immobilized on the glass slide. RNA-Seq does not require prior knowledge of target gene sequences and can establish the sequence of all RNAs in the sample without needing any information about the genome of the organism being studied. In RNA-Seq all RNAs (protein coding and noncoding) that are present in a particular cell-type or tissue can be determined. The ENCODE (Encyclopedia of DNA elements) project was set up as an extension of the Human Genome Project to try to understand the role of all the DNA in a genome. The ENCODE project used RNA-Seq to map regions of transcription throughout the genome, which helped lead to the identification of noncoding RNAs (Chapter 5, page 80). A further development of RNA-Seq, Slide-seq, is a method of transferring RNA from tissue sections onto a surface to allow the detection of gene-expression patterns in normal and diseased tissues with a spatial resolution comparable to the size of individual cells.

ChIP-Seq and Epigenomics

Chromatin immunoprecipitation (**ChIP**) is used to detect proteins that bind to DNA in a living cell. This technique can identify, for example, if a known transcription factor is bound to an enhancer region (page 79). When ChiP is combined with NGS (ChIP-Seq) the technique is very powerful, making it possible to identify every DNA fragment to which the transcription factor binds. This builds up a profile of the various sites to which a specific protein binds in a genome.

Epigenetics reflects changes in the way the information in DNA is processed. Changes in transcription can be the result of age, disease, or environmental factors that cause an alteration in the arrangement of nucleosomes and access of proteins to DNA (page 39). DNA methylation and several modifications to histone proteins that include acetylation and phosphorylation contribute to changes in nucleosome organization (page 39). These changes will influence where proteins bind on DNA. ChIP-Seq identifies these epigenetic alterations in protein–DNA-binding patterns, making it possible to obtain a genome-wide understanding of specific protein–DNA interactions, and creating a new "omics": epigenomics.

Medical Relevance 8.2   Microarrays and Cancer Classification

Microarrays are being used to type the mRNAs produced by different cancers in the hope that this will lead to better diagnosis and therefore better treatment. In the case of breast cancer, a number of studies have correlated patterns of gene expression with the probability of disease recurrence after surgery. Patients whose cancer has a particular pattern show no additional benefit from chemotherapy or radiotherapy, meaning that these unpleasant follow-up treatments can be safely omitted. One commercial microarray, Mammaprint, produced by Agendia BV, was approved by the US Food and Drug Administration in 2007.

Other "Omics"

The coupling of existing methodologies for studying cellular functions with NGS sequencing has led to an expansion in the field of "omics." Future developments in the application of NGS are likely to lead to the establishment of new "omics" with which to improve our understanding of the mechanisms that underpin the biological processes taking place inside our cells.

IDENTIFYING THE GENE RESPONSIBLE FOR A DISEASE

Until recently, the starting point for an identification of the gene responsible for a particular inherited disease was a pattern of inheritance in particular families plus a knowledge of the tissues affected. It is very difficult to find the gene responsible for a disease when the identity of the normal protein is unknown. Very often, the first clue is the identification of other genes that are inherited along with the malfunctioning gene and that are therefore likely to lie close on the same chromosome (this is **linkage**, page 245). In the past, **chromosome walking** was then used to identify the disease gene. With the publication of the entire human genome, chromosome walking to generate a series of overlapping clones for testing has become unnecessary, but identifying the gene that is responsible for a particular inherited condition is still a difficult task.

REVERSE GENETICS

Because beginning with an inherited defect in function and working toward identification of a gene is even today a time-consuming task, more and more scientists are now working the other way: they take a gene with a known sequence but unknown function and deduce its role. Since we know the complete genome of many different species, we can sit at a computer and identify genes that look interesting – for example, because their sequence is similar to a gene of known function. The gene of interest can be mutated and reinserted into cells or organisms, and the cells or organisms tested for any altered function. Protein-encoding genes can be manipulated to remove or reduce the amount of protein produced. The consequences of the absence or reduction in the levels of the protein are then established. This approach is called **reverse genetics.**

TRANSGENIC AND KNOCKOUT MICE

A transgenic animal is produced by introducing a foreign gene into the nucleus of a fertilized egg (Figure 8.14a). The egg is then implanted into a foster mother and the offspring are tested to determine whether they carry the foreign gene. If they do, a transgenic animal has been produced. The first transgenic mice ever made were used to identify an enhancer sequence that activates the metallothionein gene when an animal is exposed to metal ions in its diet. The 5′ flanking sequence of the metallothionein gene was fused to

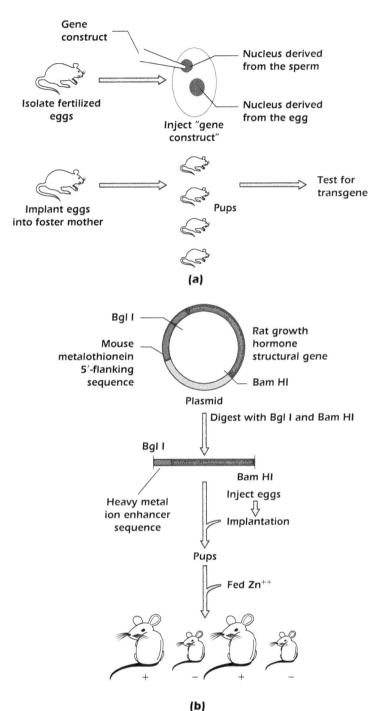

Gene construct

Isolate fertilized eggs

Nucleus derived from the sperm

Nucleus derived from the egg

Inject "gene construct"

Implant eggs into foster mother

Pups

Test for transgene

(a)

Bgl I

Mouse metalothionein 5′-flanking sequence

Rat growth hormone structural gene

Bam HI

Plasmid

Digest with Bgl I and Bam HI

Bgl I

Bam HI

Heavy metal ion enhancer sequence

Inject eggs

Implantation

Pups

Fed Zn^{++}

+   −   +   −

(b)

Figure 8.14. (a) Transgenic mouse carrying a foreign gene. (b) The metallothionein gene contains a heavy-metal ion-enhancer sequence. The + mice carry the transgene while littermates without the transgene are indicated by −.

the rat growth hormone gene (Figure 8.14b). This DNA construct, the transgene, was injected into fertilized eggs. When the mice were a few weeks old, they were given drinking water containing zinc. Mice carrying the transgene grew to twice the size of their littermates because the metallothionein enhancer sequence, stimulated by zinc, had increased growth hormone production.

Transgenic farm animals – such as sheep synthesizing human factor VIII in their milk – have been created. This is

an alternative to producing human proteins in bacteria or eukaryotic cells.

Genetically modified mice are increasingly being used to prove a protein's function. To do this the gene sequence is modified so that protein function is knocked out. In this case a **knockout mouse** is produced. This is done by either inserting a piece of foreign DNA into the gene, or by deleting the gene from the mouse genome. The effect of knocking-out the gene can then be analyzed.

It is possible to generate tissue-specific knockout mice in order to study the effect of the gene in a specific tissue or organ.

 # RNA INTERFERENCE (RNAi)

RNAi is a normal biological process. We have seen in Chapter 5, for example, that miRNAs are able to regulate the stability or translation of an mRNA, which in turn influences the amount of protein produced (page 80). However, RNAi can also be engineered for use in the laboratory to regulate the expression of specific genes both in cultured cells and in organisms, making it an effective means of disrupting gene expression without the need for producing transgenic animals. RNAi is controlled by the RISC complex (page 80) and can be initiated either by endogenous miRNAs or by a small interfering RNA (**siRNA**) or small hairpin RNA (**shRNA**) that is produced from an exogenous double-stranded RNA that is specifically engineered to reduce, or knock down, the levels of the mRNA that encodes the target protein. The RNAi pathway is being used to treat human disease. In 2018, the first siRNA drug, patisiran (Onpattro®), was approved for use in patients suffering from the rare and often fatal disease hereditary transthyretin-mediated amyloidosis. In 2019, a second siRNA drug, givosiran (Givlaari®), was approved for the clinical treatment of acute hepatic porphyria. A number of siRNA drug candidates are currently in Phase 3 clinical trials.

 ### BrainBox 8.1   Andrew Fire and Craig Mello

Andrew Fire and Craig Mello. Source: Nobel Prize website.

In 2006 Andrew Fire and Craig Mello were awarded the Nobel Prize in Physiology and Medicine for their discovery of RNA interference: post-transcriptional gene silencing. In 1998 while studying the nematode worm *Caenorhabditis elegans* they made the discovery that double-stranded RNA blocks mRNA, preventing the information in the mRNA from being transferred to protein. This "silencing" of genes is an important way of regulating gene expression within a genome and can, using recombinant DNA technology, be used to regulate the expression of specific genes in cells and organisms.

 # CRISPR/CAS9

CRISPR-Cas9 is a naturally occurring part of the prokaryotic adaptive immune system that, using recombinant DNA technology, has been used as a gene-editing tool to understand biological processes in eukaryotic cells, in plants and in animals. (**CRISPR** stands for Clustered Regularly Interspaced Short Palindromic Repeats.) Cas9, a CRISPR-associated endonuclease, makes double-stranded breaks in

DNA at specific target sequences. These double-stranded breaks are then repaired using a guide RNA as a template. The design of this guide RNA allows the introduction of changes to the genome including gene knockout (by introducing frameshift or nonsense mutations), gene deletion, the introduction of missense mutations or mutations to correct a defective gene sequence, and the addition of cassettes containing exogenous gene sequences. This application of the technique is a potential major breakthrough in gene therapeutics to correct genetic disorders. CRISPR-Cas9 has been used in a proof of principle trial in patients to correct sickle cell disease and transfusion-dependent β-thalassemia, conditions caused by an E6V mutation in the hemoglobin β subunit (page 46). In addition to its clinical applications, CRISPR can be used to generate genome-wide knockout screens, providing an opportunity to conduct genetic perturbation screens to identify genes involved in fundamental cellular processes. Perturb-seq is a method that combines these

CRISPR-mediate gene inactivations with single-cell RNA sequencing (scRNA-Seq), an NGS technology, to obtain the gene expression patterns for each perturbation. This provides information about the functional consequences of each gene knockout at the level of the transcriptome.

ETHICS OF DNA TESTING FOR INHERITED DISEASE

The applications of recombinant DNA technology are exciting and far-reaching. However, the ability to examine the base sequence of an individual raises important ethical questions. Would you want to know that you had inherited a gene that will cause you to die prematurely? Some of you might feel fine about this and decide to live life to the full. We suspect most people would not want to know their fate. But what if you have no choice and DNA testing becomes obligatory should you wish to take out life insurance? There is

---

BrainBox 8.2   Emmanuelle Charpentier and Jennifer Doudna

Emmanuelle Charpentier and Jennifer Doudna. Source: Nobel Prize website.

In 2020, the Nobel Prize in Chemistry was awarded to Emmanuelle Charpentier and Jennifer Doudna "for the development of a method for genome editing." The CRISPR/Cas9 system is a sophisticated but simple technology that has become very popular since it was first discovered in 2012. Like restriction enzymes, the first genetic scissors to be described and deployed as a biotechnology tool, the CRISPR/Cas9 genetic scissors are part of an ancient immune system in bacteria that cleaves the DNA

of foreign invaders. Unlike previous methods for genome editing, which were time-consuming and expensive, CRISPR/Cas9 modifications can be made in a matter of weeks, bringing extraordinary opportunities and exciting advances to genome editing in the life and medical sciences. This award marks the first time in its 119-year history that the Nobel Prize in Chemistry has been awarded to two women in the same year, bringing the total number of women to receive a Nobel Prize in Chemistry from 5 to 7.

much ongoing debate on this issue. In the United Kingdom, a voluntary practice has been agreed between the Association of British Insurers (**ABI**) and the UK Government on the use of genetic tests in underwriting insurance policies. Insurers can ask if a genetic condition runs in a family, but they are not allowed to ask for family members' predictive genetic test results (a test to show whether you are likely to get a certain health condition in the future).

SUMMARY

1. DNA sequences can be cloned using reverse transcriptase, which copies mRNA into DNA to make a hybrid mRNA–DNA double-stranded molecule. The mRNA strand is then converted into DNA using the enzymes ribonuclease H and DNA polymerase. The new double-stranded DNA molecule is called complementary DNA (cDNA).

2. Restriction endonucleases cut DNA at specific sequences. DNA molecules cut with the same enzyme can be joined together. To clone a cDNA, it is joined to a cloning vector – a plasmid or a bacteriophage. Genomic DNA clones are made by joining fragments of chromosomal DNA to a cloning vector. When the foreign DNA fragment has been inserted into the cloning vector, a recombinant molecule is formed.

3. Recombinant DNA molecules are introduced into bacterial cells by the process of transformation. This produces a collection of bacteria (a library) each of which contains a different recombinant DNA molecule. The DNA molecule of interest is then selected from the library using an antibody or a nucleic acid probe, or by functional screening.

4. There are many important medical, forensic, and industrial uses for recombinant DNA clones including. These include:
   - Determination of the base sequence of the cloned DNA fragment.
   - In-situ hybridization to detect specific cells making RNA complementary to the clone.
   - Synthesis of mammalian proteins in bacteria or eukaryotic cells.

   - Changing the DNA sequence to engineer a new protein.
   - Generation of fluorescent protein chimeras for subsequent microscopy on live cells.

5. The polymerase chain reaction (PCR) allows multiple copies of a section of DNA to be synthesized if short sequences at each end of the section are known. The expression of specific mRNA transcripts can be quantified by PCR.

6. "Omics" are the study of all the molecules of a particular class in a cell or tissue. For example, transcriptomics is the study of all the molecules of RNA that are present in a cell or tissue. "Omics" involves the use of high-throughput data acquisition and sophisticated data analysis and management. Transcriptomics data are collected using microarray or RNA-Seq techniques.

7. Transgenic organisms have been engineered to contain foreign genes. Transgenic organisms may be used to study the function of a gene, or to produce useful proteins.

8. Knockout cells or organisms have a gene deleted or otherwise rendered ineffective. Alternatively, RNA interference can be used to knock down mRNA concentrations.

9. CRISPR/Cas9 cuts DNA at sites defined by a guide RNA designed by the scientist. It has been used as a therapeutic tool in patients.

Answer to thought question: Transform *E. coli* and plate the bacteria on agar containing the appropriate selectable antibiotic. Cells transformed with any of the recombinant plasmid will grow colonies. If the vector containing the glucocorticoid cDNA has a different selectable marker from all of the other recombinant plasmids, then the only plasmids able to grow on agar containing that antibiotic will be those that contain the glucocorticoid cDNA. If all cDNAs have been inserted into the same vector and, therefore, all have the same antibiotic resistance marker, then this strategy will not work. Instead, select individual bacterial colonies and use these to isolate plasmid DNA. At this point you have several different options. You can digest each sample with restriction enzymes to identify a colony with the pattern you expect from your knowledge of the glucocorticoid cDNA sequence. You can perform a Southern blot using an oligonucleotide probe that corresponds to a portion of the glucocorticoid sequence, or you could perform PCR with glucocorticoid receptor-specific primers. The simplest and least expensive solution, however, would be to use chain-termination sequencing to obtain a short stretch of DNA sequence from several different clones and analyze the results to identify which of these sequences corresponds to the cDNA sequence of the glucocorticoid receptor.

FURTHER READING

Aldridge, S. and Teichmann, S.A. (2020). Single cell transcriptomics comes of age. *Nature Communications* 11 (1): 4307. https://doi.org/10.1038/s41467-020-18158-5. PMID: 32855414; PMCID: PMC7453005.

Bobbin, M.L. and Rossi, J.J. (2016). RNA interference (RNAi)-based therapeutics: delivering on the promise? *Annual Review of Pharmacology and Toxicology* 56: 103–122. https://doi.org/10.1146/annurev-pharmtox-010715-103633. PMID: 26738473.

Doudna, J.A. (2020 Feb). The promise and challenge of therapeutic genome editing. Nature. 578 (7794): 229–236. https://doi.org/10.1038/s41586-020-1978-5. Epub 2020 Feb 12. PMID: 32051598.

Doudna, J.A. and Charpentier, E. (2014). Genome editing. The new frontier of genome engineering with CRISPR-Cas9. *Science* 346 (6213): 1258096. https://doi.org/10.1126/science.1258096. PMID: 25430774.

Gibson, D., Young, L., Chuang, R.Y. et al. (2009). Enzymatic assembly of DNA molecules up to several hundred kilobases.

Nat Methods 6: 343–345. https://doi.org/10.1038/nmeth.1318.

International Human Genome Sequencing Consortium (2004). Finishing the euchromatic sequence of the human genome. *Nature* 431: 931–945.

Mullis, K.B. (1990). The unusual origin of the polymerase chain reaction. *Scientific American* 262: 56–65.

van Dijk, E.L., Jaszczyszyn, Y., Naquin, D., and Thermes, C. (2018 Sep). The third revolution in sequencing technology. Trends Genet. 34 (9): 666–681. https://doi.org/10.1016/j.tig.2018.05.008. Epub 2018 Jun 22. PMID: 29941292.

Wang, K.C. and Chang, H.Y. (2018 Apr 27). Epigenomics: technologies and applications. Circ Res. 122 (9): 1191–1199. https://doi.org/10.1161/CIRCRESAHA.118.310998. PMID: 29700067; PMCID: PMC5929475.

Watson, J.D., Caudy, A.A., Myers, R.M., and Witkowski, J.A. (2007). Recombinant DNA: Genes and Genomes – A Short Course, 3e. W.H. Freeman.

 ## REVIEW QUESTIONS

8.1   Theme: A mammalian expression plasmid

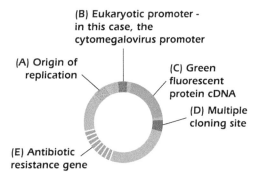

Figure 14.14 on page 248 shows fluorescent images of human cells containing a chimera of green fluorescent protein and cytochrome c. We tabulate below the steps that the experimenters went through to generate these cells. Identify the elements in the plasmid shown above that allow each step to be performed.

1. DNA encoding cytochrome *c* is inserted into the plasmid. Which element in the plasmid makes this possible?

2. To generate large amounts of the recombinant plasmid, the plasmid is grown up in bacteria. The plasmid is used to transform competent *E. coli* which are then cultured in such a way that only bacteria containing the plasmid survive. Which element of the plasmid allows the survival of host bacteria when sister bacteria are dying?

3. The transformed bacteria divide repeatedly, producing colonies each derived from a single transformed progenitor. What element in the plasmid allows it to be copied in parallel with the host bacterium's DNA?

4. Some clones will contain nonrecombinant plasmid, without the cytochrome *c* insert. However, the recombinant plasmid can be recognized by its higher relative molecular mass. Clones containing the recombinant plasmid are grown up further and then lysed, allowing purification of large amounts of recombinant plasmid. The purified plasmid is then used to transfect human cells, which synthesize the green fluorescent protein: cytochrome *c* chimera. Which element in the plasmid allows the chimaeric protein to be expressed in the HeLa cells, even though it was not expressed in the bacteria?

5′TACGGATCCCTTTGCAGGATCCAG – – – – – – – – –TTCTGCAGACGCTGCAGTAGGCA 3′

3′ATGCCTAGGGAAACGTCCTAGGTC – – – – – – – – –AAGACGTCTGCGACGTCATCCGT 5′

8.2 Theme: Choosing an oligonucleotide for a specific task

The first two questions refer to the DNA sequence shown at the top of the page. Note that we show only the sequence at the ends of the double-stranded DNA molecule.

A 5′ TTTTTTTTTTTTTTTT 3′
B 5′ TGCCTACTGCAGCGTCTGCA 3′
C 5′ TACGGATCCCTTTGCAGGATGAATTC 3′
D 5′ TTCTGCAGACGCTGCAGTAG 3′
E 5′ GAATTCTACGGATCCCTTTGCAGGAT 3′
F 5′ GTGCATCTGACTCCTGTGGAGAAGTCT 3′
G 5′ GACTGCCATCGTAAGCTGAC 3′

From the above table of DNA sequences, select the one described in each of the descriptions below.

1. In the PCR: indicate the oligonucleotide that should be used together with the oligonucleotide 5′ TACGGATCCCTTTGCAGGAT 3′ to amplify the double-stranded DNA molecule shown at the top of the page.

2. You wish to use the PCR to create, using the double-stranded DNA molecule shown at the top of the page, a DNA product that can then be cloned into the EcoR1 site of a plasmid. Indicate the oligonucleotide that you would use in place of 5′ TACGGATCCCTTTGCAGGAT 3′ in the PCR reaction mix.

3. An oligonucleotide that could be used to prime the synthesis of DNA from most of the mRNAs present in tissue from a patient, in order to generate a cDNA library.

4. An oligonucleotide that could be used in a Southern blot to identify carriers of sickle cell anemia. Note that this disease is caused because an A in the sequence 5′ GTGCATCTGACTCCTG$\underline{A}$GGAGAAGTCT 3′ in the normal β globin gene is mutated to a T to generate the sequence 5′ GTGCATCTGACTCCTG$\underline{T}$GGAGAAGTCT 3′.

5. An oligonucleotide that could be used for northern blotting to detect mRNA containing the sequence 5′GUCAGCUUACGAUGGCAGUC3′.

8.3 Theme: Uses of cDNA clones

A a pair of oligonucleotides, one complementary to a short length of one strand of a DNA molecule, the other complementary to a short length, up to about 4000 base pairs distant, of the *other* strand

B a pair of oligonucleotides, one complementary to a short length of one strand of a DNA molecule, the other complementary to a short length, up to about 4000 base pairs distant, of the *same* strand

C a radiolabeled oligonucleotide complementary to a known sequence within the molecule of interest

D an oligonucleotide complementary to the bases at the 3′ end of a partially known, but largely unknown, DNA sequence

E an oligonucleotide complementary to the bases at the 5′ end of a partially known, but largely unknown, DNA sequence

From the above list of oligonucleotides select the option appropriate for each of the techniques described below.

1. amplifying a known or partially known DNA sequence using the PCR
2. automated DNA sequencing by the dideoxy chain termination method or NGS methods
3. detection of specific DNA sequences by Southern blotting, for example to differentiate DNA from two human subjects
4. investigation, by northern blotting, of the degree to which a gene of interest is transcribed in a particular tissue

THOUGHT QUESTION

You have been provided with a plasmid mixture. The mixture contains four different recombinant DNA plasmids, each containing a different receptor cDNA, including a recombinant plasmid containing the cDNA for the glucocorticoid receptor. What can you do to separate from your mix the plasmid that contains the glucocorticoid receptor cDNA?

SECTION 3

CELL COMMUNICATION

Cells respond to their environment. Single-celled organisms swim or crawl toward food or away from danger. The cells in animals and plants not only sense the outside world but are in constant communication with each other to mold their development and behavior. Cell:cell communication allows the whole organism to organize itself, respond to stimuli, move and function in an integrated way.

In this section, we will introduce one of the fundamental properties of cells whose change allows rapid cell responses: the voltage across the plasma membrane. We will then describe how movements of calcium ions control a wide range of cell behaviors and finally describe slower, biochemically based signaling pathways.

9

CARRIERS, CHANNELS, AND VOLTAGES

We described in Chapter 2 how membranes are composed of phospholipids arranged so that their hydrophobic tails are directed toward the center of the membrane, while the hydrophilic head groups face out. Membranes are a barrier to the movement of many solutes. In particular, important solutes such as small ions and sugars are hydrophilic, meaning that they will not leave an aqueous environment for a hydrophobic one. Simple diffusion across a membrane involves passing through the hydrophobic interior and is therefore impossible for these solutes (Figure 2.2 on page 22). Two consequences follow from the fact that membranes are barriers. First, the composition of the liquid on one side of a membrane can be different from the composition of the liquid on the other side. Indeed, by allowing cells to retain proteins, sugars, ATP, and many other solutes, the barrier property of the plasma membrane makes life possible. Table 9.1 shows typical concentrations of some important ions in extracellular medium and cytosol. Second, because the plasma membrane is a barrier to the movement of small hydrophilic solutes such as sugars and ions, cells have to put in place proteins to carry out the transport of any such solutes that do need to be moved from the extracellular medium to the cytosol or vice versa. Almost all such transport proteins have the same basic structure: a protein spans the membrane and forms a tube linking the two sides. However, transport proteins fall into two clear

groups. **Channels** can adopt a shape in which the tube is open all the way through the membrane. Solute can pass from one side to the other without the protein changing shape. The potassium channel (Figure 9.1a) is one example. In contrast **carriers** are only ever open to one side or the other. The glucose carrier (Figure 9.1b) is a simple example. Solute enters the tube from one side of the membrane. The carrier then changes shape so that it is open to the other side, and the solute can leave.

We have already described one example of a channel: the gap-junction channel (page 26). This has the property that it is **gated** shut unless it can find a partner on a neighboring cell and link up to form a tube connecting the cytosol of one cell with the cytosol of a neighbor. The channels and carriers we will describe in this chapter, in contrast, mediate the movement of solute between the extracellular medium and the cytosol.

CARRIERS

We will introduce the wide range of complexity of carriers by describing three: the glucose carrier, the sodium/calcium exchanger, and the sodium/potassium ATPase. All play critical roles in our cells.

Cell Biology: A Short Course, Fourth Edition. Stephen Bolsover, Andrea Townsend-Nicholson,
Greg FitzHarris, Elizabeth Shephard, Jeremy Hyams and Sandip Patel.
© 2022 John Wiley & Sons Ltd. Published 2022 by John Wiley & Sons Ltd.
Companion website: www.wiley.com/go/bolsover/cellbiology4

Medical Relevance 9.1   Cytochrome *c* – Vital but Deadly

All the channels described elsewhere in this chapter lie in the plasma membrane and allow solutes to flow between the cytosol and the extracellular medium. However, intracellular membranes also contain channels. One critical one is a large channel called VDAC (for voltage-dependent anion channel) in the outer mitochondrial membrane. VDAC allows solutes up to relative molecular mass 7000 through. VDAC allows ATP, Mr. ~500, made inside the mitochondrion to pass out into the cytosol.

Cytochrome *c*, relative molecular mass 12 000, is a soluble protein that resides in the intermembrane space between the outer and inner mitochondrial membranes and helps the electron transport chain create the hydrogen ion electrochemical gradient across the mitochondrial inner membrane. It is too big to escape into the cytosol through VDAC.

Although cytochrome *c* is essential for mitochondrial function, it has another, deadly role. If cytochrome *c* comes into contact with a class of cytosolic enzymes called caspases, it activates them, turning on the process of cell suicide called apoptosis (page 246). Under certain conditions, VDAC can associate with other proteins to form a channel of larger diameter; when this happens, cytochrome *c* can leak out and the cell dies by apoptosis. This process seems to occur in the heart during a heart attack, and in the brain in Alzheimer's disease and during a stroke. Screening has identified lead compounds called VBITs that prevent this process but to date we are not aware of any drug trials.

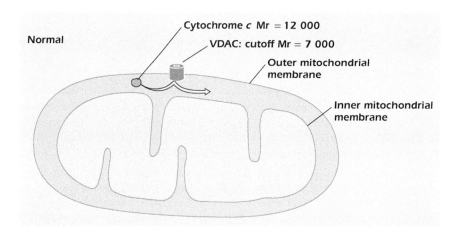

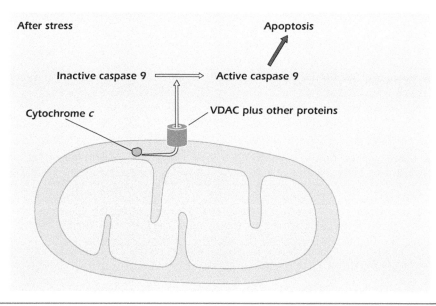

TABLE 9.1. Typical concentrations for five important ions in mammalian cytosol and extracellular medium.

Ion	Cytosol	Extracellular medium
Sodium Na^+	10 mmole liter^{-1}	140 mmole liter^{-1}
Potassium K^+	140 mmole liter^{-1}	5 mmole liter^{-1}
Calcium Ca^{2+}	100 nmole liter^{-1}	1 mmole liter^{-1}
Chloride Cl^-	10 mmole liter^{-1}	100 mmole liter^{-1}
Hydrogen ions H^+ (really H_3O^+)	60 nmole liter^{-1} or pH 7.2	40 nmole liter^{-1} or pH 7.4

Note that the unit n for nano (10^{-9}) is one million times smaller than m for milli (10^{-3}).

(a) The potassium channel

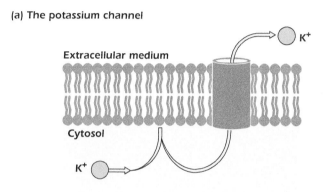

(b) The glucose carrier

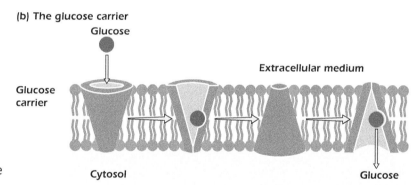

Figure 9.1. Channels form a tube open all the way across a membrane while carriers change shape to allow solute across.

The Glucose Carrier

One of the simplest carriers is the glucose carrier (Figure 9.1b). It switches freely between a form that is open to the cytosol and a form that is open to the extracellular medium. Inside the tube is a site to which a glucose molecule can bind. On the left, a glucose molecule is entering the tube and binding to the site. Sometimes glucose leaves the binding site before the carrier switches shape. In Figure 9.1b we see the other possibility: the carrier switches shape before the glucose leaves. The binding site is now open to the cytosol, and the glucose can enter the cell. It has been carried across the plasma membrane. In most of

the cells of our body, the net movement of glucose through the glucose carrier is inward as glucose moves in down its concentration gradient to be metabolized inside the cell. However, the glucose carrier is just as capable of mediating a net outward movement. For example liver cells can convert stored glycogen to glucose which then leaves the cells, via the glucose carrier, down a concentration gradient into the blood.

The glucose carrier is very simple, whereas other carriers are more complex. We will next consider two carriers, the sodium/calcium exchanger and the sodium/potassium ATPase, which do similar jobs: they push ions up concentration gradients.

Example 9.1   The Glucose Carrier Is Essential

The cells of our bodies are bathed in a glucose-rich solution. However, glucose cannot cross the plasma membrane by simple diffusion because it is strongly hydrophilic: it can only get in via the glucose carrier. Some cells have glucose carriers in their membranes all the time, but others, such as muscle and fat cells, translocate the glucose carrier to the plasma membrane only in the presence of insulin. Insulin-dependent diabetics cannot produce their own insulin, and unless they inject synthetic insulin, their muscles and fat cells cannot take up glucose and therefore run out of energy, even though the concentration of glucose in the blood becomes very high. This is why muscular weakness is one symptom of insulin-dependent diabetes.

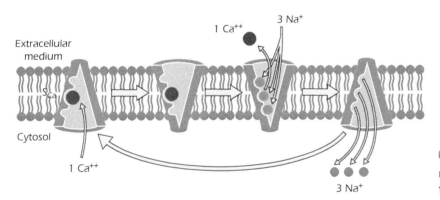

Figure 9.2. The sodium/calcium exchanger undergoes more complicated shape changes to drive calcium ions out of the cell.

The Sodium/Calcium Exchanger

Figure 9.2 shows that the sodium/calcium exchanger, like the glucose carrier, can exist in two shapes, one open to the extracellular medium and one open to the cytosol. Inside the tube are three sites. All three are able to bind sodium ions but one, S_{Ca}, has a much higher affinity for calcium. However when S_{Ca} binds calcium the protein is distorted so that the other two sites are no longer able to bind sodium. The sodium/calcium exchanger is not free to switch between its two shapes at any time. Instead, it switches only if either all three binding sites are occupied by sodium or if calcium is bound to S_{Ca}.

On the left of the figure the carrier is open to the cytosol. Sodium ions are at about $10 \, \text{mmole liter}^{-1}$ in the cytosol, while calcium is about $100 \, \text{nmole liter}^{-1}$, five orders of magnitude less concentrated. However S_{Ca} is a much better fit for calcium than for sodium so calcium binds, and this distorts the protein so that sodium is unable to bind to the other two sites. The channel is now able to switch shape so that it is open to the extracellular medium where sodium is $140 \, \text{mmole liter}^{-1}$ rather than $10 \, \text{mmole liter}^{-1}$. At this higher concentration sodium can drive calcium off S_{Ca} and, once the calcium has left, all three sites become occupied by sodium ions. Once all three sites are occupied by sodium ions the carrier becomes once again able to switch shape so that it is open to the cytosol. In the low-sodium environment of the cytosol the sodium ions tend to leave, freeing S_{Ca} to bind calcium so

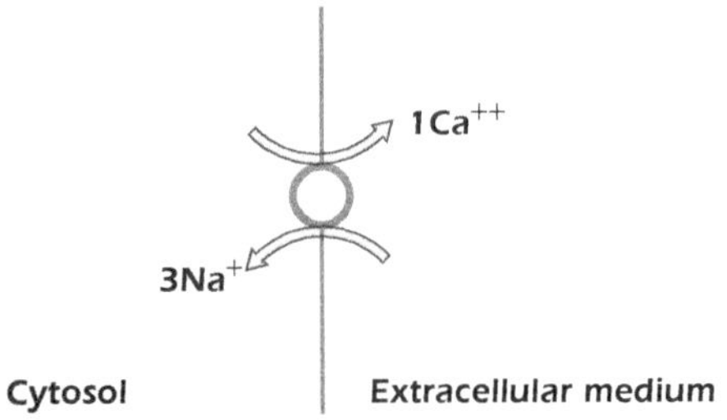

Figure 9.3. Action of the sodium/calcium exchanger.

that the cycle can begin again. Figure 9.3 shows a simple way of representing what is happening. The circle represents one cycle of operation, from open to cytosol back to open to cytosol again. A simple rule about when the carrier can switch shape has produced a machine that can push calcium ions up a concentration gradient out of the cell, at the expense of letting sodium in down its concentration gradient. So now we need to explain how the cell maintains low sodium concentrations in the cytosol.

The Sodium/Potassium ATPase

The sodium/potassium ATPase is a carrier with an additional level of complexity in that it carries out an enzymatic action as well as a carrier function. Hydrolysis of ATP provides energy that drives two solutes up their concentration gradients. Sodium ions are pumped out of the

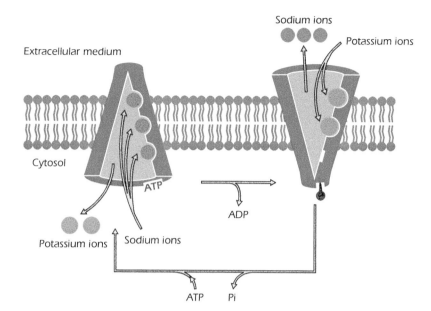

Figure 9.4. The sodium/potassium ATPase undergoes a cycle of phosphorylation and dephosphorylation. These drive shape changes that in turn push sodium ions out of the cell and potassium ions in.

cell, from about $10\,mmole\ liter^{-1}$ in the cytosol to about $140\,mmole\ liter^{-1}$ in the extracellular medium, and potassium ions are pumped in from the extracellular medium at about $5\,mmole\ liter^{-1}$ to cytosol where the concentration is about $140\,mmole\ liter^{-1}$.

In the state shown on the left of Figure 9.4 an ATP molecule has bound to a site on the cytosolic aspect of the carrier. The strains and forces set up by this binding cause the carrier to adopt a shape in which the tube is open to the cytosol and three ion-binding sites with an affinity for sodium are revealed inside the tube. Once three sodium ions have bound, an intrinsic enzymatic action of the protein attaches the γ phosphate of ATP to an aspartate residue on the carrier, and the rest of the ATP leaves as ADP. The forces and strains on the protein are therefore altered and it switches to the very different shape shown on the right. The tube is now open to the extracellular medium but, even more strikingly, the ion-binding sites have changed. One is eliminated completely and the other two change shape so that they now have an affinity not for sodium but for the larger potassium ion. Sodium ions therefore leave the tube into the extracellular medium, and potassium ions enter and bind to the two ion sites. Once the two potassium ions have bound a second catalytic action of the protein hydrolyzes the bond between aspartate and phosphate and the phosphate leaves. The whole ATP-binding site is now available for ATP to bind; when it does, the ion-binding sites switch to a shape that is unsuitable for potassium, which leaves into the cytosol, but which is suitable for sodium, and the cycle begins again.

Figure 9.5 summarizes what is happening. The circle represents one cycle of operation. One molecule of ATP is hydrolyzed to ADP and inorganic phosphate, three sodium ions move out of the cell, and two potassium ions move in.

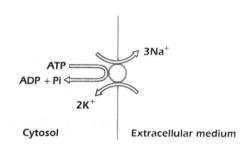

Figure 9.5. Action of the sodium/potassium ATPase.

The Calcium ATPase

The sodium/calcium exchanger is the major mechanism by which calcium is kept low in the cytosol of many cells. However other cells use a pump called the calcium ATPase. This is related to the sodium/potassium ATPase and has a very similar structure and mode of operation. It moves one calcium ion out of the cell for every ATP hydrolyzed (Figure 9.6). Every cell in our bodies has one of these two proteins, the sodium/calcium exchanger or the calcium ATPase, and many have both.

The actions of the sodium/potassium ATPase, the sodium/calcium exchanger, and the calcium ATPase explain the first three rows of Table 9.1. They create the large gradients of sodium, potassium, and calcium across the plasma membrane. The reason why chloride, on the fourth row, is much less concentrated in the cytosol than outside the cell is because the cytosol of all cells has a negative voltage with respect to the outside. Chloride ions, being negatively charged, are repelled out. We will next explain why the inside of cells is negative, which will then lead on to a description of how one class of cells – electrically excitable cells – use the voltage across the plasma membrane to transmit signals.

Medical Relevance 9.2   Poisoned Hearts Are Stronger

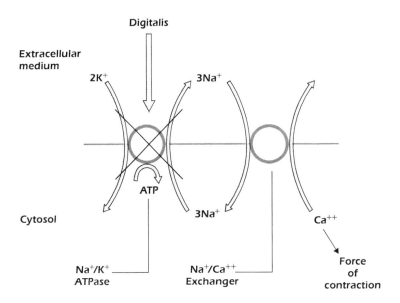

Digitalis is used to treat heart failure. Digitalis inhibits the sodium/potassium ATPase and is extremely toxic. Nevertheless, a small dose, which inhibits the sodium/ potassium pump just a little, causes the heart muscle to beat more strongly. The reason is that partially inhibiting the sodium/potassium pump causes a small increase of cytosolic sodium concentration. Because the sodium/calcium exchanger has three binding sites for sodium, its activity is extremely sensitive to sodium concentration, and even a small increase of cytosolic sodium reduces its activity significantly. The calcium concentration in the cytosol therefore rises. The mechanical motor that drives heart contraction (Chapter 13) is controlled by calcium, so that a small increase of cytosolic calcium makes the heart beat more strongly.

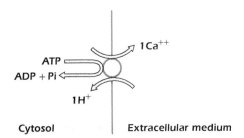

Figure 9.6. Action of the calcium ATPase.

THE POTASSIUM GRADIENT AND THE RESTING VOLTAGE

Ions are electrically charged. This fact has two consequences for membranes. First, the movement of ions across a membrane will tend to change the voltage across that membrane. If positive ions leave the cytosol, the cytosol will be left with a negative voltage, and vice versa. Second, a voltage across a membrane will exert a force on all the ions present. If the cytosol has a negative voltage, then positive ions such as sodium and potassium will be attracted in from the extracellular medium. We will begin to address the question of how ions and voltages interact by considering the effect of potassium movements on the voltage across the plasma membrane, hereafter called the **membrane voltage.**

Potassium Channels Make the Plasma Membrane Permeable to Potassium Ions

We have already introduced potassium channels (Figure 9.1) as tubes that link the cytosol with the extracellular medium. Potassium ions, which cannot pass through the lipid bilayer of the plasma membrane, pass through potassium channels easily. Other ions cannot go through. The precise shape of the tube, and the position of charged amino acid side chains within the tube, blocks their movement. The channels are selective for potassium. Potassium channels are found in the plasma membrane of almost all cells. At first glance this seems perverse. The sodium/potassium ATPase expends ATP to pull potassium into the cell – why does it not rush back out again through the potassium channels? To explain why, we must think about the effects of ion movement on membrane voltage.

IN DEPTH 9.1   MEASURING THE MEMBRANE VOLTAGE

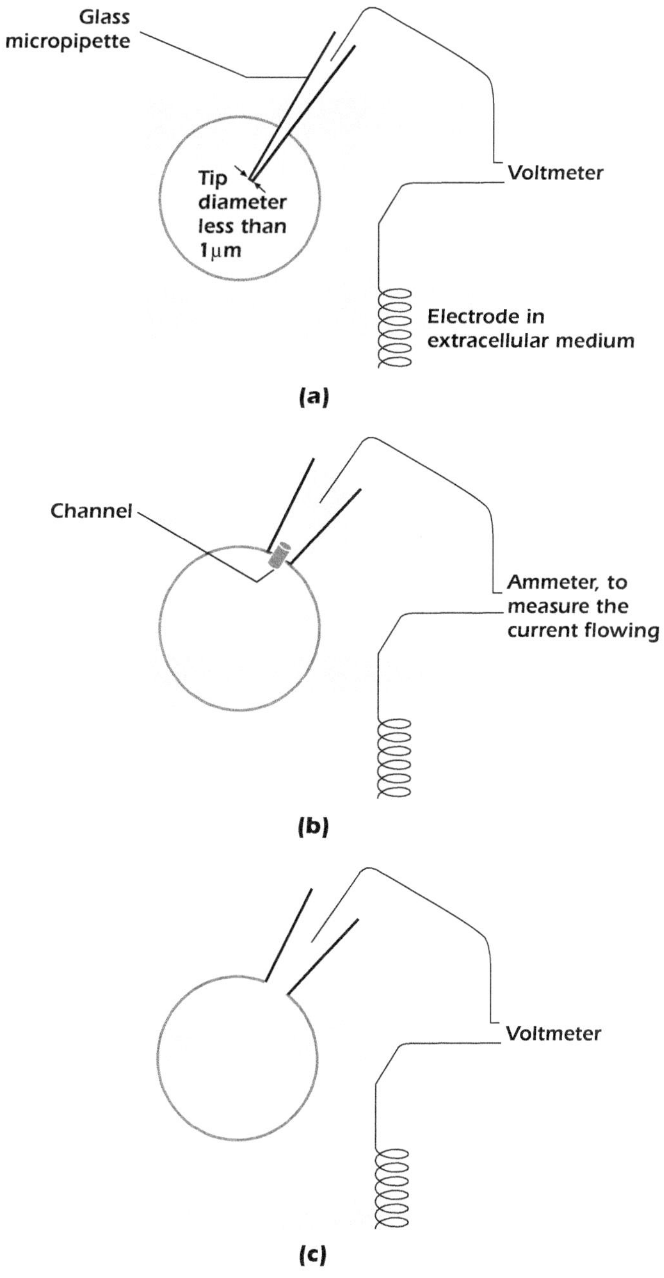

Glass
micropipette

Tip
diameter
less than
1 μm

Voltmeter

Electrode in
extracellular medium

(a)

Channel

Ammeter, to
measure the
current flowing

(b)

Voltmeter

(c)

In 1949 Gilbert Ling and Ralph Gerard discovered that when a fine glass micropipette filled with an electrically conducting solution impaled a cell (panel a), the plasma membrane sealed to the glass, so that the membrane voltage was not discharged. The voltage difference between a wire inserted into the micropipette and an electrode in the extracellular medium could then be measured. By passing current through the micropipette, the membrane voltage could be altered.

Twenty-five years later Erwin Neher and Bert Sakmann showed that the micropipette did not have to impale the cell. If it just touched the cell, a slight suction caused the plasma membrane to seal to the glass (panel b). The technique, called cell-attached patch clamping, can measure currents through the few channels present in the tiny patch of membrane within the pipette.

Stronger suction bursts the membrane within the pipette (panel c). The membrane voltage can now be measured. Alternatively, current can be passed through the micropipette to change the membrane voltage – this is the whole cell patch-clamp technique. In 1991, Neher and Sakmann received the Nobel Prize in Medicine.

Concentration Gradients and Electrical Voltage Can Balance

Figure 9.7a shows a glial cell, one of the two main cell types in the nervous system. For many glial cells most of the time, the only plasma membrane channels open and allowing ion movement are potassium channels. For these cells, the cytosol is about −90 mV relative to the extracellular medium. Potassium ions are acted upon by two forces. They would leave the cell under the influence of the concentration gradient, but are pulled in by the negative voltage of the cytosol. For every ion present on both sides of a membrane, it is possible to calculate the membrane voltage that would exactly balance the concentration gradient (In Depth 9.2). This voltage is the **equilibrium voltage** for that ion at that membrane. When cytosolic and extracellular potassium concentrations are 140 and 5 mmole liter⁻¹ respectively the potassium equilibrium voltage at human body temperature is −90 mV. Thus for the glial cell with a membrane voltage of −90 mV, the forces on potassium ions exactly balance and the cell neither gains nor loses potassium. This in turn means that the cytosol is neither gaining nor losing charge, so the membrane voltage does not change. The condition shown, in which the membrane voltage is equal to the equilibrium voltage of potassium, is a stable one. We can see that the condition is stable by thinking about what would happen if the voltage was for some reason perturbed to, for example, −80 mV. At this new voltage the electrical force pulling the potassium ions in is not strong enough to oppose the concentration force favoring potassium loss. Potassium ions would leave and as they did so they would carry their positive charge out,

so the cytosol voltage would move in a negative direction. This would continue until the concentration and voltage forces again balanced, which would be when the voltage had returned to the potassium equilibrium voltage. In a similar but opposite way, if the membrane voltage were artificially perturbed to a value more negative than the potassium equilibrium voltage, potassium ions would move in until the resulting movement of charge had returned the membrane voltage to the potassium equilibrium voltage. In general, a cell whose plasma membrane is permeable to one ion only will settle down to have a membrane voltage equal to the equilibrium voltage of that ion.

We use the term **electrochemical gradient** to describe the sum total of all the forces on an ion due to both voltage and its concentration gradient. For potassium at the plasma membrane of a glial cell, the electrochemical gradient is zero, because the electrical and concentration forces exactly balance. In contrast, sodium and calcium both have large inward electrochemical gradients in glial cells and indeed in all other eukaryotic cells. As Table 9.1 shows, for both ions the concentration inside is much lower than the concentration outside and, in addition, the ions are positively charged and so experience an inward force due to the negative voltage inside the cell.

In many cells the situation is more complicated than in glial cells. In neurons, for instance (Figure 9.8), the voltage of an unstimulated cell is about −70 mV (Figure 9.7b). This is because in addition to potassium channels these cells have a second type of channel called the **voltage-gated sodium channel** that allows sodium ions (and only sodium ions) to pass. As its name suggests, opening of the voltage-gated

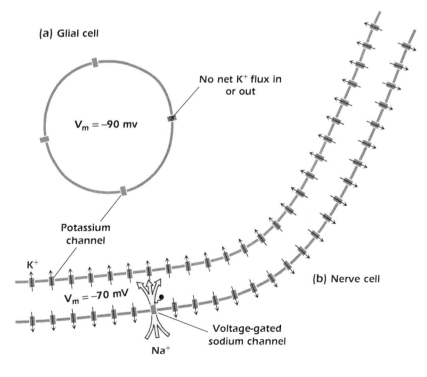

Figure 9.7. The resting voltage of (a) glial and (b) nerve cells.

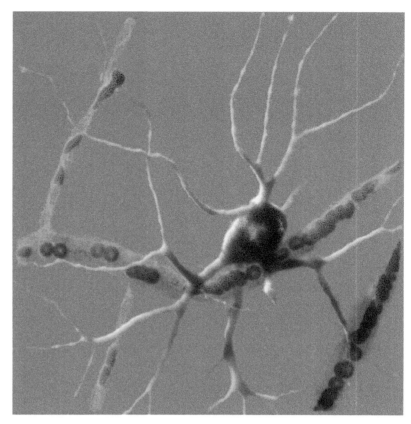

Figure 9.8. A nerve cell from the retina viewed with its nutritive capillaries. Red blood cells are an example of cells whose membrane voltage varies little, while nerve cells are specialized to transmit electrical signals over long distances. Source: Image by Professor David Becker, University College London; used with permission.

sodium channel is controlled by the membrane voltage. At −70 mV, the vast majority of voltage-gated sodium channels are shut: only one in every 4000 are open at any one time. Nevertheless the few channels that are open carry a relatively large sodium flux since the electrochemical gradient for sodium is large and inward. The result is that the membrane voltage of an unstimulated neuron settles down to a **steady-state** voltage in which the inward current through voltage-gated sodium channels is balanced by an outward current of potassium. Potassium flows out because the cytosol is no longer negative enough to hold these ions in against their outward concentration gradient, that is, there is now a small outward electrochemical gradient for potassium. Even in neurons, though, the membrane is more permeable to potassium than to any other ion, so the voltage of the unstimulated cell does not deviate very far from the potassium equilibrium voltage. However, in neurons and in all other cells in which the plasma membrane has a significant permeability to sodium, the sodium/potassium ATPase has to work constantly to maintain the concentration gradients across the plasma membrane. As we will see, the voltage across the neuron membrane changes dramatically when the cell is stimulated and transmits the electrical signals for which it is specialized. The term **resting voltage** is used for the voltage across the plasma membrane of the unstimulated cell. We also talk about the resting voltage of cells where the membrane voltage never changes, so we say that the resting voltage of a glial cell is −90 mV.

IN DEPTH 9.2   THE NERNST EQUATION

Ion I at concentration $[I]_{outside}$

Ion I at concentration $[I]_{inside}$

Cytosol at voltage V

An ion that can pass across a membrane is acted on by two forces. The first derives from the concentration gradient. The ion tends to diffuse from a region where it is at high concentration to one where it is at low concentration. The second force derives from the membrane voltage. In the case of positively charged ions such as Na^+ and K^+, the ions tend to move toward a negative voltage. Negatively charged ions such as Cl^- tend to move toward a positive voltage. For each ion there is a value of the membrane voltage for which these forces balance and the ion will not move. The ion is said to be at equilibrium, and this value of the membrane voltage is called the equilibrium voltage for that ion at that membrane.

When the forces balance, then ions that move in will neither gain nor lose energy. This way of describing equilibrium is useful because it allows us to set equivalent the effects of the two very different gradients, concentration and voltage. For concentration, the free energy possessed by a mole of ions I by virtue of its concentration is

$$G = G^{o'} + RT \log_e [I] \text{ joules}$$

where $G^{o'}$ is the standard free energy, R is the gas constant $(8.3 \text{ J mol}^{-1} \text{ degree}^{-1})$, and T is the absolute temperature.

A mole of I passing in therefore moves from a region where it had a free energy of

$$G_{outside} = G^{o'} + RT \log_e [I_{outside}] \text{ joules}$$

to one where its free energy is

$$G_{inside} = G^{o'} + RT \log_e [I_{inside}] \text{ joules}$$

One mole of ions I moving inward therefore gains by virtue of the concentration gradient free energy equal to

$$RT \log_e [I_{inside}] - RT \log_e [I_{outside}] \text{ joules}$$

Now consider the electrical force. The definition of a volt means that one coulomb of charge moving across a membrane with a membrane voltage of V volts gains V joules of free energy. However, we are working in moles, not coulombs. One mole of ions has a charge of zF coulombs, where z is the charge on the ion. For Na^+ and K^+ z is 1; for Ca^{++} z is 2; and for Cl^- z is -1. The term F is a number that relates the coulomb to the mole. It has the value 96 500. One mole of ions I moving inward gains by virtue of the membrane voltage free energy equal to

$$zFV \text{ joules}$$

This does not mean that an ion always gains energy from the membrane voltage when it moves inward: the term zFV can just as easily be negative as positive.

When the effects of concentration and voltage just balance, a mole of ions moving inward neither gains nor loses free energy. Hence, at equilibrium

$$RT \log_e [I_{inside}] - RT \log_e [I_{outside}] + zFV_{eq} = 0$$

This can be simplified to

$$V_{eq} = \left(\frac{RT}{F}\right)\left(\frac{1}{z}\right) \log_e \left(\frac{[I_{outside}]}{[I_{inside}]}\right) \text{ volts}$$

This is the Nernst equation. The value of $\left(\frac{RT}{F}\right)$ is 0.025 at a room temperature of 22 °C. At human body temperature $\left(\frac{RT}{F}\right)$ is 0.027.

"In" and "out" can refer to any two solutions separated by a membrane. At the plasma membrane *in* is the cytosol and *out* is the extracellular medium.

⬤ THE ACTION POTENTIAL

For many types of cell the voltage across the plasma membrane never changes. However, neurons and muscle cells use changes of membrane voltage as a method of rapid signaling over long distances.

The Pain Receptor Neuron

We will illustrate the operation of neurons using pain receptors. Figure 9.9 shows the cell responsible for the sensation of pain in a finger. The cell body is close to the spinal cord and extends an **axon** that branches to the spinal cord and out to the body. The particular pain receptor illustrated in Figure 9.9 sends its axon for almost a meter to the tip of a finger, an extraordinary distance for a single cell. The axon terminal in the skin is the **distal,** or far away, terminal and that in the spinal cord is the **proximal** one. Potentially

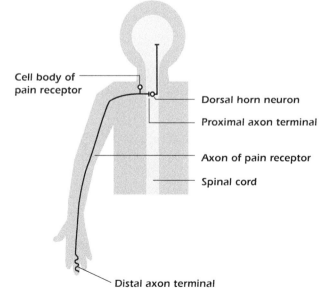

Cell body of pain receptor

Dorsal horn neuron

Proximal axon terminal

Axon of pain receptor

Spinal cord

Distal axon terminal

⬤ Figure 9.9. The connections of a pain-receptor nerve cell.

damaging events, such as high temperatures, are detected at the finger, and the message is passed on to another neuron (a dorsal horn neuron) in the spinal cord. We will now explain how this function is performed (Figure 9.10).

The plasma membrane of the distal terminal contains an ion channel called TRPV that is closed at a normal body temperature of 37 °C but which spends a greater and greater fraction of time open at higher temperatures. The channel is a nonselective cation channel, that is, it allows sodium, potassium, and calcium ions to pass. Since, even in neurons, potassium ions are close to equilibrium, the dominant flow through open TRPV channels is an inward movement of sodium and calcium, both of which have a large inward electrochemical gradient.

Figure 9.11a shows what happens when the finger is passed through the warm air from a hand dryer. The warmth

causes some of the TRPV channels to open, and sodium and calcium ions move in. This inward current causes the membrane voltage in the distal axon terminal to **depolarize.** We use the word depolarization to mean any positive shift in the membrane voltage, whatever its size or cause. Because the cytosol is less negative than it was, and hence pulls potassium ions inward less strongly, the outward flux through potassium channels increases. The membrane reaches a new steady state in which the inward current through TRPV channels is balanced by the increased outward current through potassium channels.

Figure 9.11b shows what happens when the finger is passed through the hot air from a hair dryer. More TRPV channels open, so the distal axon terminal depolarizes further. However, at a voltage of −40 mV, a new phenomenon appears: a massive but temporary depolarization to +30 mV.

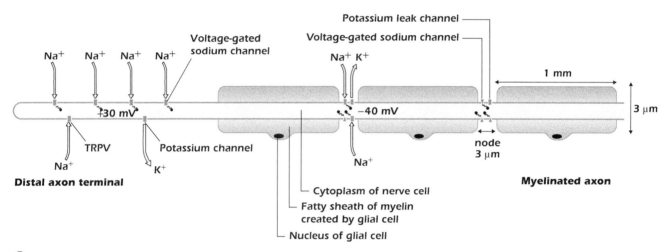

Figure 9.10. One type of pain receptor has a distal terminal in the skin and is connected to the central nervous system by a myelinated nerve fiber.

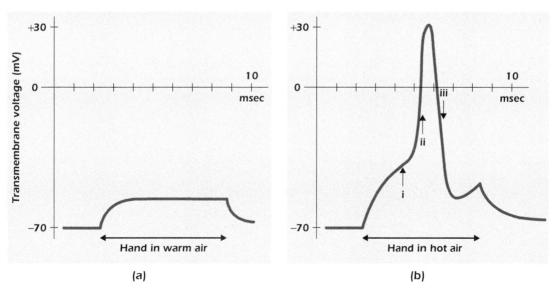

Figure 9.11. Electrical events in a pain-receptor nerve cell.

Example 9.2   Peppers and Pain

Capsaicin, the active ingredient in chili peppers, activates one isoform of TRPV channels and so causes the hot, burning sensation in the mouth. In contrast sanshool, the active ingredient in Szechuan peppers, activates pain receptors and other neurons by closing potassium channels. This depolarizes the cells to a new steady-state voltage. At this new voltage, the outward current of potassium once again has the same amplitude as the inward current through voltage-gated sodium channels. However, since there are fewer open potassium channels, the new steady-state voltage must lie further away from the potassium equilibrium voltage such that each potassium channel carries a larger outward current. The depolarization caused is great enough to reach the threshold for action potential generation.

This is the **action potential.** Cells capable of generating action potentials are said to be **electrically excitable** and include neurons and muscle cells. To explain the action potential we must return to the voltage-gated sodium channel and describe it in more detail.

The Voltage-Gated Sodium Channel

Figure 9.12 shows how the voltage-gated sodium channel works. On the left is the shape of the protein when the membrane voltage is −70 mV. The protein forms a tube, but it is not open all the way through the membrane. Positively charged arginine and lysine side chains are attracted by the negative cytosol, favoring a state in which the tube is gated closed. If the membrane is depolarized the positive side chains are pulled inwards more weakly and the channel can switch to the open state. Sodium ions can now pass through the channel. A protein domain in the cytosol called the inactivation plug is constantly jiggling about at the end of a flexible link and can bind to the inside of the open channel. The resulting blockage is **inactivation,** and it occurs after about one millisecond. As long as the plasma membrane remains depolarized, the voltage-gated sodium channel will remain inactivated. When the plasma membrane returns to the resting voltage (**repolarizes**), the positive charges are attracted back toward the inside of the cell, squeezing the plug out. In summary:

1. When the membrane voltage is −70 mV, the voltage-gated sodium channel is gated shut.
2. When the plasma membrane is depolarized, the channel opens rapidly and after about 1 ms inactivates.
3. After the channel has gone through this cycle, it must spend at least 1 ms with the membrane voltage at the resting voltage before it can be opened by a second depolarization.

The voltage-gated sodium channel goes through this process whether or not sodium ions are present. To understand the action potential, we must now think about the effect that movements of sodium ions through this channel have on the membrane voltage.

The Sodium Action Potential

Even the slight depolarization in the warm air opened some voltage-gated sodium channels in the distal terminal of the pain receptor, but the additional inward current that these channels passed was neutralized by the additional outward potassium current, so the membrane voltage stabilized at a new depolarized value. In contrast, when the membrane is depolarized past the **threshold voltage,** which for this neuron is about −40 mV, sufficient voltage-gated sodium channels open such that the inward current through them (plus the current through the open TRPV channels) is greater than the outward current through the potassium channels. This is

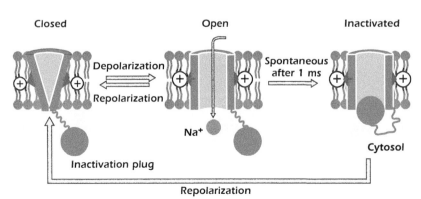

Figure 9.12. The voltage-gated sodium channel.

Example 9.3 Chewing off the Inactivation Plug

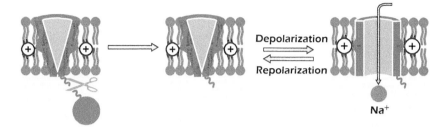

Proteases are enzymes that hydrolyze the peptide bonds within proteins. In 1976 Emilio Rojas and Bernardo Rudy investigated the effect of introducing a protease into squid axons. The membrane of squid axons contains voltage-gated sodium channels that, like ours, normally inactivate about 1 ms after they are opened by a depolarization. However, after introduction of the protease, depolarization caused the voltage-gated sodium channels to open and remain open indefinitely, although the channels would close if the membrane was repolarized. Back in 1976 Rojas and Rudy had no idea why this should be so. Now we can understand the result – the protease cuts the linker between the main part of the voltage-gated sodium channel and the inactivation plug. The plug then floats off and is not available to block the open channel.

the situation at time (i) in Figure 9.11b. This state is not a stable one because the cytosol is gaining charge and is therefore becoming more positive. More voltage-gated sodium channels therefore pop into the open state, letting more sodium ions pour inward, depolarizing the membrane further. Very rapidly the cell reaches state (ii), where a large fraction of the voltage-gated sodium channels are open and the membrane is rapidly depolarizing.

There are so many voltage-gated sodium channels in the membrane that the voltage would approach the sodium equilibrium voltage (about +70 mV) but for the fact that the channels inactivate, stopping the sodium influx. By time (iii) more and more of the voltage-gated sodium channels have inactivated, so that the only channels that can carry current are the potassium channels. Each of these is now carrying a large outward current since the membrane voltage is much more positive than the potassium equilibrium voltage. The membrane is rapidly **repolarizing.** Once the membrane voltage has returned to negative values, the voltage-gated sodium channels will begin to recover to the closed (but ready to open) state, and the cell would soon be ready to generate a second action potential if the TRPV channels were still open.

Figure 9.11 shows a critical feature of an action potential: it is all or nothing. The warm air did not depolarize the pain receptor sufficiently to evoke an action potential. The hot air depolarized the cell sufficiently to start the process, and the action potential then took off in an explosive, self-amplifying way until a large fraction of all the voltage-gated sodium channels were open and the plasma membrane had depolarized greatly. In all excitable cells there is a threshold for initiating an action potential; in the pain receptor it is about −40 mV. Depolarizations to voltages below the threshold elicit nothing; depolarizations to voltages more positive than the threshold elicit the complete action potential. At the heart of the action potential is a positive feedback loop. Depolarization causes voltage-gated sodium channels to open, and open sodium channels cause depolarization.

The Strength of a Signal Is Coded by Action Potential Frequency

At the distal axon terminal of the pain receptor, the intensity of the stimulus is represented by the amount of depolarization produced. The hotter the air blowing on the skin, the more positive the membrane voltage of the terminal becomes. Other sensory cells behave in the same way. For instance, in Chapter 11 we will meet the scent-sensitive neurons in the nose that detect smell chemicals. The higher the concentration of the smell chemical, the more these cells depolarize. However, the signal that passes toward the brain is composed of action potentials. The strength of the stimulus cannot be encoded in the amplitude of the action potentials, because they are discrete all-or-nothing events. Rather, the strength of the stimulus is encoded by the frequency of the action potentials.

Figure 9.13 shows how this happens. In Figure 9.13a the stimulus is too weak to elicit an action potential. The graph on the right shows that, away from the terminal, the membrane voltage of the pain receptor does not deviate from the resting voltage. Figure 9.13b represents what happens when the stimulus is over threshold, but not by much. After each action potential, the membrane voltage at the distal terminal

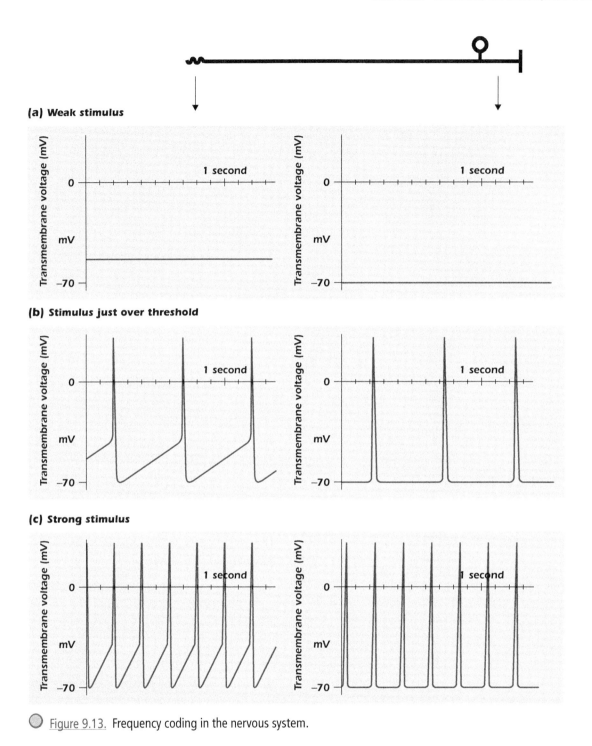

Figure 9.13. Frequency coding in the nervous system.

depolarizes slowly to threshold; the cell therefore fires action potentials at low frequency. The graph on the right shows that the action potentials, but not the TRPV-mediated depolarization, propagate up the axon. Figure 9.13c shows what happens when the stimulus is strong. Because more TRPV channels are open, carrying in positive charge, the depolarization after each action potential is more rapid. Therefore the time between successive action potentials is short and the frequency of action potentials is greater.

Amplitude modulation (AM) and frequency modulation (FM) are familiar terms from radio. Here we are seeing exactly the same two coding strategies in the nervous system. AM is used in the distal terminals of the pain receptor. The strength of the stimulus is coded for by the amplitude of the depolarization. FM is used in almost all axons, including those of the pain receptor. The strength of the stimulus is coded for by the frequency of action potentials, each of which has the same amplitude.

Example 9.4   Local Anesthetic, Overall Wellbeing

Nerve conduction is exploited when a dentist uses local anesthetic. A patient feels pain from the drill because pain receptors in the tooth are depolarized and transmit action potentials toward the brain. The site at which they are being depolarized is inside the tooth and therefore inaccessible to drugs until the drill has made a hole. However, the axons of the pain receptors run through the gum. Local anesthetics injected into the gum close to the axon bind to the neuron membrane at the node and prevent the opening of voltage-gated sodium channels. Drilling into the tooth still depolarizes the pain-receptor membrane, and action potentials begin their journey toward the brain, but they cannot pass the injection site because the nodes there cannot generate an action potential. The patient therefore feels no pain.

Myelination and Rapid Action Potential Transmission

The axon of the pain receptor transmits its signal to the central nervous system rapidly because for most of its length it is insulated by a fatty sheath called **myelin** made by glial cells (the glial cells outside the brain and spinal cord are called Schwann cells). Figure 9.10 shows part of the axon of the pain receptor and its associated glial cells, each of which wraps around the axon to form an electrically insulating sheath. Only at short gaps called nodes is the membrane of the neuron exposed to the extracellular medium. Note that the vertical and horizontal scales of this diagram are completely different. The axon together with its myelin sheath is only 3 μm across, but the section of axon between nodes is 1 mm long, a large distance by normal cell standards. Between nodes the axon is essentially an insulated electrical cable: just as voltage changes at one end of an insulated metal wire have an almost instantaneous effect on the voltage at the other end, so a change in membrane voltage at one node has an almost instantaneous effect at the next node along. The plasma membrane at the nodes has voltage-gated sodium channels and can therefore generate action potentials.

Figure 9.10 shows the moment at which the action potential is jumping from the axon terminal to the first node. Current has flowed along the axon cytosol from the depolarized terminal to the node, in the same way as current flows down a wire from a point at a positive voltage to a negative one. Current flowing into the node has depolarized it, and the membrane at the node has now reached its threshold for action potential generation. Now, at the node itself, more current is flowing in through voltage-gated sodium channels than is leaving through potassium channels, and the node membrane will rapidly depolarize to about +30 mV. The same thing will now happen along the next millimeter of axon. Current flowing along the axon cytosol will depolarize the next node, and soon it too will generate an action

Answer to thought question: The resting voltage will depolarize, that is, become more positive. There will be two effects on voltage-gated sodium channels: they will be more likely to open but there will also be more channels in a permanently inactivated state.

Under normal conditions, with extracellular potassium at 5 mmole liter⁻¹ and cytosolic potassium at 140 mmole liter⁻¹, the resting voltage of cardiac muscle cells is about −80 mV. This is the voltage at which the outward current through the many open potassium channels is equal and opposite to the inward current through the few open voltage-gated sodium channels. In turn what allows an outward potassium current is the fact that the voltage is significantly (about 10 mV) more positive than the potassium equilibrium voltage of −90 mV.

If the extracellular potassium increases then the potassium equilibrium voltage will shift to be more positive. For example, when the extracellular potassium is 10 mmole liter⁻¹ then the potassium equilibrium voltage will be −71 mV. In order for there still to be an outward potassium current to balance the inward sodium current, the resting voltage must be about 10 mV more positive than the potassium equilibrium voltage, so that now means about −61 mV. Thus the increase of extracellular potassium depolarizes the muscle cell.

Initially, the effect of the depolarization will be to increase the probability that voltage-gated sodium channels open, and therefore to increase the frequency of action potentials in the heart: this is the condition of a rapid heartbeat or tachycardia. However, recovery of voltage-gated sodium channels from the inactivated state back to the closed state requires that the membrane voltage be negative. If the resting voltage is −60 mV rather than −80 mV, many voltage-gated sodium channels will not recover but will remain in the inactivated state. If too few recover to the closed state, the cardiac muscle cells become electrically inexcitable: they do not generate action potentials at all and do not contract, a situation that is rapidly lethal.

IN DEPTH 9.3   VOLTAGE-GATED POTASSIUM CHANNELS

In this chapter we use the simple term "potassium channels." We are in fact referring to potassium leak channels, which are found in the membranes of almost all animal cells and are responsible for the resting voltage. It is the potassium leak channels that are responsible for repolarizing the membrane voltage of mammalian nodes of Ranvier.

A different sort of potassium channel is found in many locations in the nervous system and on muscle cells. Voltage-gated potassium channels are gated in a very similar way to voltage-gated sodium channels; they open when the plasma membrane is depolarized and inactivate after a delay. They respond more slowly than voltage-gated sodium channels, so that the cell gets a chance to depolarize because of the sodium flooding in before the voltage-gated potassium channels open allowing potassium to flood out and repolarize the membrane.

---

potential. The action potential is jumping from node to node. This process is named after the Latin for "to jump," *saltare*, so it is **saltatory conduction.** For a pain fiber conducting at $15\,m\,s^{-1}$ the action potential takes $67\,\mu s$ to jump each one millimeter distance. The fastest axons in our bodies conduct at $60\,m\,s^{-1}$, with each node-to-node jump taking only $17\,\mu s$.

TRPV channels are found only in pain-receptor neurons. Other sensory neurons are depolarized by other stimuli such as cold or touch, while the neurons that send commands out from the brain, or process information within it, are triggered to fire action potentials by other mechanisms that we will describe in the next chapter. However, all these cells express voltage-gated sodium channels and so all share the

mechanism of action potential propagation that we have described.

The combination of axon plus myelin sheath is a nerve fiber. The nerve fiber shown in Figure 9.10 has an overall diameter of $3\,\mu m$. The human body contains myelinated nerve fibers with diameters ranging from $1\,\mu m$, conducting at $6\,m\,s^{-1}$, to $10\,\mu m$, conducting at $60\,m\,s^{-1}$. Some neurons have axons that are not myelinated and generally conduct more slowly than myelinated fibers. The axons of scent-sensitive neurons (page 181) are an example. Getting scent information from the nose to the brain as fast as possible is apparently not a significant advantage, given that a scent can take many seconds to waft from its source to our nose.

SUMMARY

1. Carriers form a tube across the membrane that is always closed at one end or the other. Solutes can move into the tube through the open end. When the carrier changes shape, so that the end that was closed is open, the solute can leave to the solution on the other side of the membrane.

2. The glucose carrier is present in the plasma membrane of all human cells.

3. The sodium/calcium exchanger is a carrier that uses the energy released by sodium entering the cell to push calcium ions out of the cell. Some cells use a calcium ATPase instead, some cells have both carriers.

4. The sodium/potassium ATPase is both a carrier and an enzyme. Energy released by the hydrolysis of ATP is used to drive sodium ions out of the cell and to pull potassium ions in.

5. Channels are membrane proteins with a central water-filled hole open at both ends. Hydrophilic sol-

utes, including ions, can pass from one side of the membrane to the other. Changes in protein structure may act to gate the channel but are not required for movement of the solute.

6. The equilibrium voltage for an ion at a membrane is the voltage at which the tendency for the ion to move one way because of a concentration gradient would be exactly balanced by the tendency to move the other way because of voltage.

7. In many cells almost the only open ion-permeable channel present in the plasma membrane is a potassium channel. In such cells the membrane voltage will settle down to be equal to the potassium equilibrium voltage.

8. Some cells have plasma membranes that are permeable to sodium as well as potassium, although potassium is still the most permeable. In such cells the membrane voltage will settle down to a steady-state value more positive than the potassium equilibrium

voltage and the sodium/potassium ATPase must run constantly to expel sodium from the cell and pull potassium back in.

9. Cells that can generate action potentials are said to be electrically excitable. Many such cells use voltage-gated sodium channels to drive the action potential.

10. The voltage-gated sodium channel is shut at the resting voltage but opens upon depolarization. After about 1 ms the channel inactivates.

11. While the voltage-gated sodium channel is open, sodium ions pour into the cell down their electrochemical gradient. The mutual effect of current through the channel on membrane voltage, and of membrane voltage on current through the channel, constitutes a positive feedback system. If the membrane is initially depolarized to threshold, the positive feedback of the sodium current system ensures that depolarization

continues in an all-or-nothing fashion. Repolarization occurs when the voltage-gated sodium channel inactivates. The entire cycle of depolarization and repolarization is called an action potential.

12. Action potentials are all-or-nothing events. The strength of the signal transmitted by a neuron is encoded not by the amplitude of the action potentials but by their frequency.

13. Long neuronal processes called axons transmit action potentials at speeds up to 60 m per second. They can transmit the signal at such high rates because myelin, a fatty sheath, insulates the 1 mm distances between nodes.

14. A number of axons are not myelinated; these generally conduct action potentials more slowly than do myelinated axons and are found where high conduction speed is not required.

FURTHER READING

Ashcroft, F.M. (2012). *The Spark of Life: Electricity in the Human Body*. London: Penguin Books.

Levitan, I.B. and Kaczmarek, L.K. (2015). *The Neuron*, 4e. New York: Oxford University Press.

REVIEW QUESTIONS

9.1 Theme: Cytosolic and extracellular concentrations of important ions

A $\geq 10^{-1}$ mol liter^{-1}
B 5×10^{-2} mol liter^{-1}
C 10^{-2} mol liter^{-1}
D 10^{-3} mol liter^{-1}
E 10^{-4} mol liter^{-1}
F 10^{-5} mol liter^{-1}
G 10^{-6} mol liter^{-1}
H $\leq 10^{-7}$ mol liter^{-1}

From the above list of concentrations, choose the one that most closely approximates the concentration described in the statements below. Answer for a general human cell.

1. cytosolic calcium
2. extracellular calcium
3. cytosolic chloride
4. extracellular chloride
5. cytosolic H$^+$ (H$_3$O$^+$)
6. extracellular H$^+$ (H$_3$O$^+$)
7. cytosolic potassium
8. cytosolic sodium
9. extracellular sodium

9.2 Theme: Pathways for solute movement across the plasma membrane

A sodium/potassium ATPase
B connexon
C glucose carrier
D potassium channel
E sodium/calcium exchanger
F TRPV
G voltage-gated sodium channel

From the above list of channels and carriers, select the protein corresponding to each of the descriptions below.

1. an enzyme that carries out a hydrolytic reaction
2. a carrier with no enzymic activity that carries one solute in one direction and a second solute in the opposite direction

3. a channel that can pass glucose
4. a protein whose expression in almost all human cells is responsible for the fact that their cytosol is at a negative voltage with respect to the extracellular medium
5. a channel expressed in pain receptor neurons which opens at damagingly high temperatures
6. a channel expressed in almost all neurons, even those that are not sensitive to painful stimuli, but which is not expressed in most non-neuronal human cells (it is not found, for example, in liver or blood cells)

9.3   Theme: Ion fluxes in a neuron

A Cl^-
B H^+ (H_3O^+)
C K^+
D Na^+

From the above list of ions, select the ion corresponding to each of the descriptions below.

1. Although gradients of this ion play a crucial role at other cell membranes, the concentration gradient of this ion across the plasma membrane is small: the concentration in the cytosol is the same, within a factor of two, as the concentration in the extracellular medium.
2. In a resting neuron, this ion is constantly leaking into the cell, and must be removed by the action of an ATP-consuming carrier.
3. In a resting neuron, this ion is constantly leaking out of the cell, and must be pumped back in by the action of an ATP-consuming carrier.
4. The more a neuron is depolarized, the greater the electro-chemical gradient favoring entry of this ion into the cell.
5. Although this ion is about 10 times more concentrated inside than out, it is close to equilibrium across the plasma membrane of a resting neuron.

⬤ THOUGHT QUESTION

In kidney failure, the concentration of potassium in the extracellular medium can rise from its normal value of 5 mmole liter^{-1} to 10 mmole liter^{-1} or even higher. What will be the effect of this on the resting voltage of heart muscle cells? What will be the consequent effect on voltage-gated sodium channels in the heart?

10

SIGNALING THROUGH IONS

The behavior of cells is not constant. Cells need to be able to alter their behavior in response to internal changes or to external events. The signaling mechanisms that allow cells to do so are varied and complex. We will begin by continuing the story begun in Chapter 9 and explain how, in neurons, calcium ions carry a signal from the plasma membrane to vesicles within the cytosol resulting in the release of transmitters. We go on to describe how calcium ions can also cause contraction of muscle cells and how nerves and muscle work together to bring about meaningful physiological change.

CALCIUM AS A SIGNALING ION

Calcium ions are present at a very low concentration (about $100\,\mathrm{nmol\,l^{-1}}$) in the cytosol of a resting cell. An enormous number of processes in many types of cells are activated when the concentration of calcium rises. Calcium can move into the cytosol from two sources: the extracellular medium or intracellular organelles.

Calcium Can Enter Cells from the Extracellular Medium

In Chapter 9, we saw how heating the hand causes action potentials to travel from the hand to the spinal cord along the axons of pain receptors. When the action potential reaches the proximal axon terminal, the amino acid glutamate is released onto the surface of dorsal horn neurons (Figure 10.1). Glutamate is a transmitter that stimulates the dorsal horn neuron, so that the message that the finger is being damaged is passed toward the brain. Whereas movement of sodium and potassium ions underpins the actional potential, it is the movement of calcium ions that drives the release of glutamate.

The proximal axon terminal is unmyelinated, so that the plasma membrane is exposed to the extracellular medium. The membrane contains not only voltage-gated sodium channels, but also **voltage-gated calcium channels.** Voltage-gated calcium channels are close relatives of voltage-gated sodium channels but are selective for calcium. Like voltage-gated sodium channels, they open and then inactivate in response to depolarization, but the inactivation process is considerably slower. They are closed in a resting cell, but they open when an action potential travels in from the skin and depolarizes the plasma membrane of the proximal axon terminal. Calcium ions pour in, increasing their concentration in the cytosol by at least 10 times, from the normal concentration of $100\,\mathrm{nmol\,l^{-1}}$ to $1\,\mathrm{\mu mol\,l^{-1}}$. At the proximal axon terminal, the cytosol of the neuron contains vesicles filled with glutamate. In response to the increased concentration of cytosolic calcium the vesicles move to the plasma membrane and fuse with it, releasing their contents into the extracellular medium. The glutamate then diffuses across the gap to

Cell Biology: A Short Course, Fourth Edition. Stephen Bolsover, Andrea Townsend-Nicholson, Greg FitzHarris, Elizabeth Shephard, Jeremy Hyams and Sandip Patel.
© 2022 John Wiley & Sons Ltd. Published 2022 by John Wiley & Sons Ltd.
Companion website: www.wiley.com/go/bolsover/cellbiology4

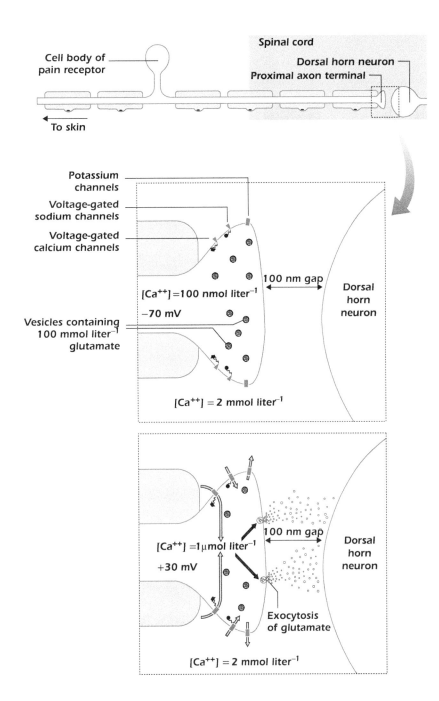

Figure 10.1. Calcium ions entering the cytosol from the extracellular medium activate regulated exocytosis in the proximal axon terminal of a pain receptor nerve cell.

the dorsal horn neuron, stimulating it. As the action potential in the axon terminal of the pain receptor cell is over in 1 ms, the voltage-gated calcium channels do not have time to inactivate. They simply return to the ready-to-open state and can be reopened immediately by the next action potential that arrives.

In the process that we have described, calcium ions act as a link between the depolarization of the plasma membrane and the vesicles within the cytosol. Calcium ions are thus **intracellular messengers.** The controlled release of

vesicle contents into the extracellular medium is called **regulated exocytosis.** In most cells, the trigger that activates regulated exocytosis is an increase of cytosolic calcium concentration. Regulated exocytosis will be covered in Chapter 12.

Calcium Can Be Released from Organelles

As discussed, calcium is present at high concentrations in the extracellular fluid such that when calcium channels in the plasma membrane open, calcium floods into the cell. But calcium is also present at high concentrations inside organelles. Opening of calcium channels present on these organelles is a second way of increasing calcium levels in the cytosol.

In muscle cells, the **sarcoplasmic reticulum** or SR is a major intracellular store of calcium. The sarcoplasmic reticulum is similar to the endoplasmic reticulum but is specialized to facilitate contraction. It contains a second type of calcium channel known as the **ryanodine receptor.** This is so named because it binds the plant toxin ryanodine. Ryanodine receptors were first identified in skeletal muscle. In these cells they are physically linked at membrane contact sites (page 26) to voltage-gated calcium channels in the plasma membrane (left-hand side of Figure 10.2). When the voltage-gated calcium channels switch to their open configuration upon depolarization, they induce the ryanodine receptors below them to also switch to an open configuration. This allows calcium to flow out of the sarcoplasmic reticulum into the cytosol.

Ryanodine receptors are also found in cardiac muscle. Here, there is no such physical linkage between ryanodine receptors and plasma membrane channels. Instead, ryanodine receptors are opened when the concentration of calcium ions in the cytosol rises above a critical level. This happens when voltage-gated calcium channels in the plasma membrane open upon depolarization (right-hand side of Figure 10.2). The calcium ions that enter through this channel bind to the cytosolic aspect of the ryanodine receptor, causing it to open too, allowing calcium to flow out of the SR into the cytosol.

Thus, in both skeletal and cardiac muscle, depolarization of the cells results in the sequential activation of voltage-gated calcium channels in the plasma membrane and ryanodine receptors in the SR. The resulting increase in calcium causes the muscle cell to contract. This chain of events is known as **excitation-contraction coupling.**

Processes Activated by Cytosolic Calcium Are Extremely Diverse

The targets activated by an increase in the concentration of cytosolic calcium differ between different cells. An increase of calcium is a crude signal that says "do it" but contains no information about what the cell should do. This depends on what the cell was designed to do. Cells designed for regulated exocytosis (e.g. axon terminals) exocytose when cytosolic calcium increases. Cells designed to contract (e.g. muscle cells) contract when

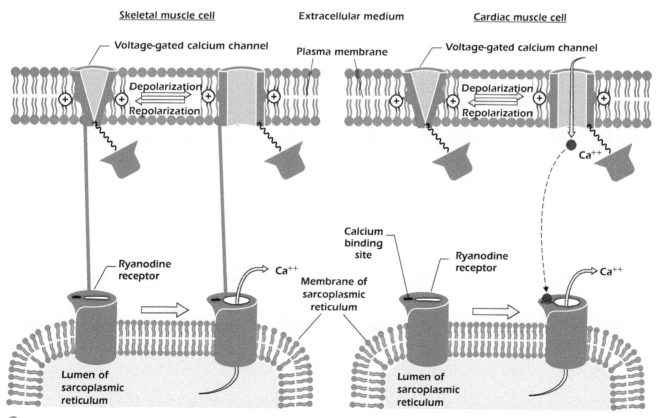

Example 10.1   Visualizing Calcium Signals

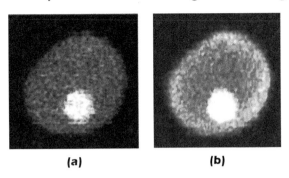

(a)   (b)

Calcium signals in cells can be visualized by using dyes that fluoresce brightly when they bind calcium. The images in the figure show the cell body of a pain receptor cell that has been filled with a calcium indicator dye. In the first image, at rest, the cell is dim. The second image was acquired after 100 ms of depolarization. In this the edges of the cell are bright because calcium ions have entered through voltage-gated calcium channels. The bright object in the lower part of the cell is the nucleus.

calcium increases, and so on. In each case, the calcium ions bind to a calcium-binding protein and it is the calcium ion–calcium-binding protein complex that activates the target process.

In neurons, calcium ions entering the presynaptic terminal bind to a protein called synaptotagmin. Synaptotagmin in turn binds to the SNARE proteins (Chapter 12) that mediate exocytosis of synaptic vesicles containing transmitters.

In skeletal muscle cells the escape of calcium from the sarcoplasmic reticulum activates several processes (Figure 10.3). First, calcium ions bind to a protein called troponin that is attached to the cytoskeleton. This causes

the cytoskeleton to contract, using the energy released by ATP hydrolysis to do mechanical work (Chapter 13). Second, calcium passes down its electrochemical gradient from the cytosol into the mitochondrial matrix through a calcium channel. Once there, calcium stimulates the mitochondria to increase the production of ATP. Lastly, the calcium binds to the protein calmodulin, which in turn activates enzymes that break down the energy store **glycogen** to fuel metabolism.

The same simple intracellular messenger, calcium, has many different actions inside the skeletal muscle cell. Under its influence, the cytoskeleton begins to use ATP, glycogen is broken down, and the mitochondria produce more ATP. The

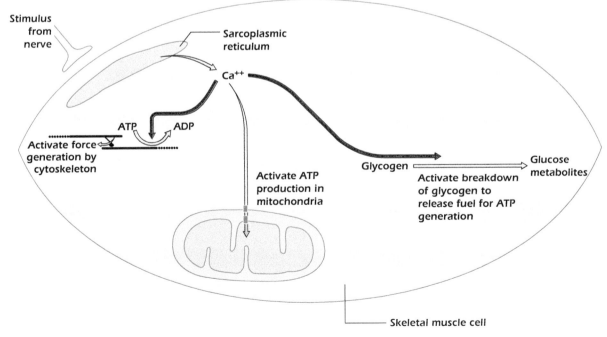

Figure 10.3. Calcium activates multiple processes in a skeletal muscle cell.

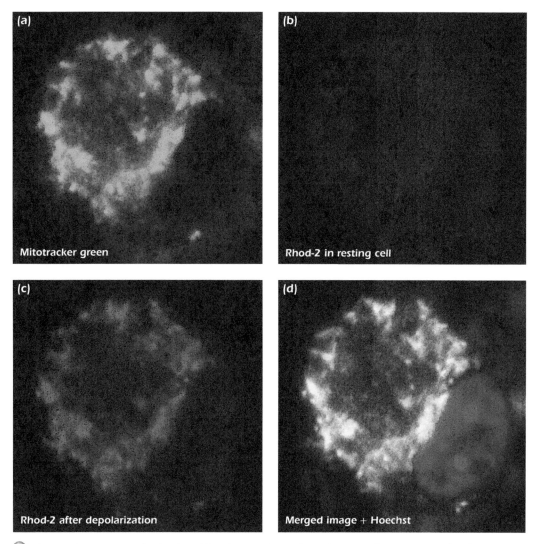

Figure 10.4. Nerve cell mitochondria take up calcium from the cytosol. Experiment of William Coatesworth and Stephen Bolsover. Source: Coatesworth, W. & Bolsover, S. (2006). Spatially organised mitochondrial calcium uptake through a novel pathway in chick neurones. *Cell Calcium* 39 (3): 217–225.

skeletal muscle cell is an example of how diverse mechanisms inside a cell can be integrated by the action of a single intracellular messenger.

Figure 10.4 shows an experiment that used fluorescent microscopy to follow the uptake of calcium by mitochondria. The cell body of a living pain receptor neuron was stained with a dye to reveal the mitochondria (Figure 10.4a) and with Hoechst 33342 to reveal the nucleus (Figure 10.4d). It was also stained with the calcium indicator dye Rhod-2, which concentrates inside mitochondria. In the unstimulated cell the Rhod-2 emits little light, indicating that mitochondrial calcium is low (Figure 10.4b). However, when the plasma membrane was depolarized for one second, opening voltage-gated calcium channels, calcium flowed in from the extracellular medium to the cytosol, and from there into the mitochondria. This caused the Rhod-2 to fluoresce brightly (Figure 10.4c).

Return of Calcium to Resting Levels

As soon as a stimulus disappears, cytosolic calcium concentration falls again. Calcium ions are transported up their electrochemical gradients from the cytosol into the extracellular medium or into the organelles by carriers. The Na^+/Ca^{2+} exchanger (page 150) uses the energy released by sodium ion movement into the cytosol to move calcium out of the cytosol into the extracellular medium. The Ca^{2+} ATPase (page 151) uses the energy released by ATP hydrolysis to do the same job. Another type of Ca²⁺ ATPase is present on the sarcoplasmic reticulum and endoplasmic reticulum. This protein is similar to the plasma membrane Ca²⁺ ATPase and uses ATP to pump Ca²⁺ out of the cytosol into the lumen of intracellular organelles.

PROPAGATING THE SIGNAL

The millions of cells that make up a multicellular organism can work together only because they continually exchange the chemical messages called **transmitters.** We have already come across the transmitter glutamate in relaying the pain signal. Glutamate is in the exocytotic vesicles of many neurons, but other neurons release different transmitters. Transmitters are sensed by neighboring cells that continue the broadcast.

Transmitters Are Released at Synapses

Many neurons have their axon terminals close to a second cell and release their transmitter onto it. In such cases, the complete unit of axon terminal, gap, and the part of the second cell that receives the transmitter is called a **synapse.** The part of the axon terminal that releases the transmitter is the **presynaptic terminal,** and the cell upon which the transmitter is released is the **postsynaptic cell.** The released transmitter acts on the postsynaptic cell and thereby relaying the signal.

Neurons often form synapses with other neurons. In the brain, there exists a dense network of such connections forming the very fabric of our being. But neurons can also form synapses with muscle cells. Here the muscle is considered postsynaptic. Neuron-muscle synapses form the **neuromuscular junction.** Just as a transmitter released by a neuron instructs another neuron, it can also instruct muscle cells to contract on demand. This is what happens during excitation-contraction coupling in skeletal muscle.

Ligand-Gated Ion Channels Respond to Transmitters

Ligand-gated ion channels (also called ionotropic receptors) open when a specific chemical binds to the extracellular face of the channel protein. The **ionotropic glutamate receptor** found on postsynaptic neurons (Figure 10.5) is one example. In the absence of the glutamate in the extracellular medium, the channel is closed. When glutamate is released into the synapse, for example during heat sensation, it binds to the protein, the channel opens and allows sodium and potassium ions to pass through. The electrochemical gradient pushing sodium ions into the cell is much greater than that pushing potassium ions out of the cell, so when the channel opens, sodium ions pour in carrying positive charge and depolarizing the plasma membrane. Glutamate is therefore an **excitatory transmitter.**

The amino acid γ-aminobutyric acid or GABA (Figure 3.7a on page 43) is another transmitter. It activates a ligand-gated ion channel known as the **GABA_A receptor.** When GABA_A receptors open, they allow the flow of negatively charged chloride ions into the cell. GABA is therefore an **inhibitory transmitter.**

The **nicotinic acetylcholine receptor** is our final example of a ligand-gated ion channel. As its name suggests, this ion channel binds the transmitter acetylcholine. Like glutamate, acetylcholine is an excitatory transmitter. This is because when it opens nicotinic acetylcholine receptors, there is a flow of sodium into the cell causing depolarization. Nicotinic acetylcholine receptors are found at the neuromuscular junction.

Example 10.2   A Toxic Glutamate Analogue

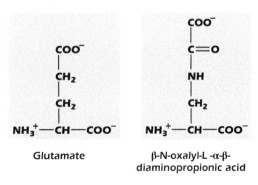

Glutamate

β-N-oxalyl-L -α-β-diaminopropionic acid

The grass pea is a protein-rich crop that has been cultivated since ancient times and is an important source of calories and protein in India, Africa, and China. The peas contain the amino acid β-N-oxalyl-L-α-β-diaminopropionic acid. If untreated peas are eaten, the toxin binds to and opens the glutamate receptor on neurons. The resulting long-lasting depolarization damages and finally kills the neurons. Boiling the peas during cooking destroys the toxin; but in times of famine, when fuel is scarce, many people are poisoned, and the resulting brain damage is irreversible.

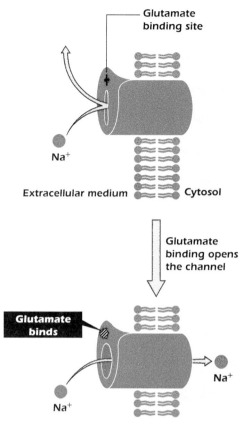

Figure 10.5. The ionotropic glutamate receptor is an ion channel that opens when glutamate in the extracellular medium binds.

RAPID COMMUNICATION: FROM NEURONS TO THEIR TARGETS

We have already introduced the synapse between the pain receptor and dorsal horn neurons (page 165). An action potential in the axon terminal of the pain receptor raises cytosolic calcium from $100 \, \text{nmol} \, l^{-1}$ to $1 \, \mu\text{mol} \, l^{-1}$ and evokes release of the transmitter glutamate. The pain receptor neuron is said to be **glutamatergic.** Free glutamate survives only a few milliseconds in the extracellular medium and is quickly taken back up into cells by a carrier protein. However, its target – the dorsal horn neuron – is only $100 \, \text{nm}$ away, so it has time enough to bind to the ionotropic glutamate receptors present in the plasma membrane of the postsynaptic cell. These open, evoking a small depolarization (Figure 10.6). However, the depolarization is not enough to take the membrane voltage to threshold, which in these cells is about $-40 \, \text{mV}$. As soon as glutamate is removed from the extracellular medium, the membrane voltage returns to the resting level. In the axon of the dorsal horn neuron, some distance from the synapse, the membrane voltage does not change, and no message passes on to the brain. The subject does not feel pain.

A subject does feel pain when they move their finger into the hot air from a hair dryer (Figure 10.7). This is because the hot air heats a large area and causes many pain receptors to fire action potentials. Many of the pain receptors synapse onto one postsynaptic cell, which therefore receives many doses of glutamate. Because of this **spatial summation,** enough glutamate receptor channels open to depolarize

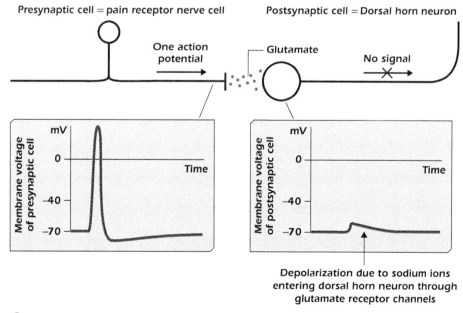

Figure 10.6. Opening of ionotropic glutamate receptors depolarizes the postsynaptic cell.

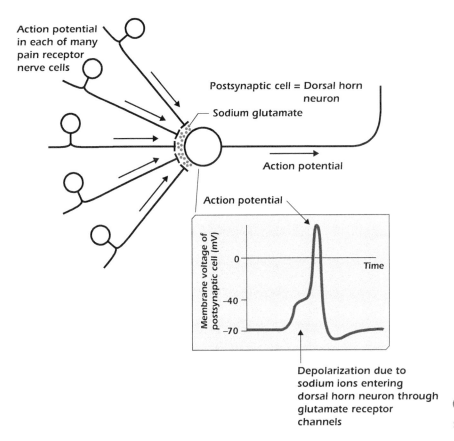

Figure 10.7. Spatial summation at a synapse.

the dorsal horn neuron to threshold. Once threshold is reached the mechanism we described in Chapter 9, driven by sodium influx through voltage-gated sodium channels, takes over and an action potential travels rapidly up the axon of the postsynaptic cell to the brain so that the subject becomes aware of the painful event.

An intense stimulus to a small area is also painful. When the subject is jabbed with a needle, only one pain receptor is activated, but that receptor is intensely stimulated and fires a rapid barrage of action potentials, each of which causes the release of glutamate and an extra depolarization of the dorsal horn neuron (Figure 10.8). Soon the membrane voltage of the postsynaptic cell reaches threshold, and an action potential travels along its axon toward the brain and the subject feels pain. Such **temporal summation** only occurs if the presynaptic action potentials are frequent enough to ensure that the depolarizations produced in the postsynaptic cell do not have time to recover.

Inhibitory Transmission

The pain relay system is powerfully modulated by the brain. One component of this is mediated through the effects of GABA. Figure 10.9 illustrates a single neuron bearing both glutamate and GABA receptors.

At A on the graph, an action potential in the GABA-secreting (**GABAergic**) axon releases GABA onto the surface of the postsynaptic cell, causing the GABA receptor channels to open. Although chloride ions could now move into or out of the cell, they do not, because their tendency to travel into the cell down their concentration gradient is balanced by the repulsive effect of the negative voltage of the cytosol: chloride ions are at equilibrium (page 154). The opening of GABA channels therefore causes no ion movements and therefore does not alter the membrane voltage.

At B, action potentials occur simultaneously in six glutamate-secreting axons. In this example, this activity provides enough glutamate to depolarize the postsynaptic cell to threshold, and the postsynaptic neuron fires an action potential.

At C, action potentials occur simultaneously in six glutamate-secreting axons and also in the GABA-secreting axon. The same number of glutamate receptor channels open as before, and about the same number of sodium ions flow into the postsynaptic neuron, depolarizing it. However, as soon as the membrane voltage of the postsynaptic cell deviates from the resting value, chloride ions start to enter through the GABA receptor channels because the cytosol is no longer negative enough to prevent them from entering the cell down their concentration gradient. The inward movement of negatively charged chloride ions neutralizes

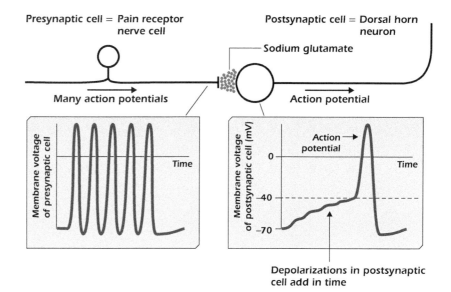

Figure 10.8. Temporal summation at a synapse.

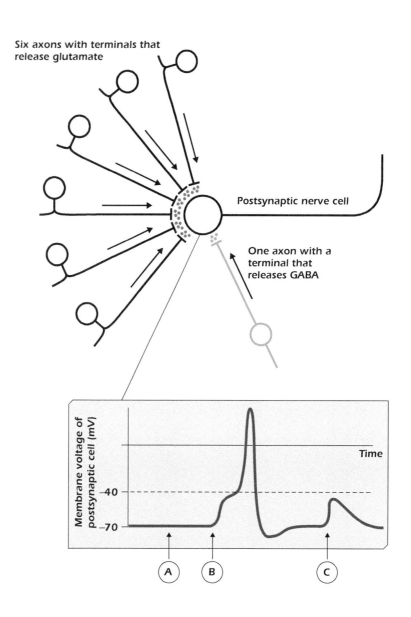

Figure 10.9. Inhibition by a GABAergic synapse.

some of the positive charge carried in by sodium ions moving in through the glutamate receptor channels. The postsynaptic neuron therefore does not depolarize as much and does not reach threshold. No action potential is generated in the postsynaptic neuron.

Glutamate and GABA are respectively the most important excitatory and inhibitory transmitters throughout the nervous system, and drugs that affect their operation have dramatic effects on neural processing. For example, antianxiety drugs such as Valium act on the GABA receptor and increase the chance of its channel opening. Neurons exposed to the drug are less likely to depolarize to threshold. Valium therefore reduces action potential activity in the brain, calming the patient.

IN DEPTH 10.1   ION CHANNEL STRUCTURE: VARIATIONS ON A THEME

Many ion channels belong to one of two large families, members of which share a similar modular structure. One family is the P-loop channels, so called because of a loop of polypeptide chain that forms part of the lining of the pore. All members of this family show a fourfold structure. In their simplest form, illustrated by the glutamate receptor (and both ROMK and voltage-gated potassium channels, see Example 7.1 on page 107 and In Depth 9.3 on page 162), the complete channel is formed when four subunits, each a separate polypeptide, come together. The channel is therefore an example of quaternary protein structure (page 118). In contrast, voltage-gated sodium (Chapter 9) and calcium channels are single polypeptide chains containing four repeats of the voltage-sensing and pore domains arranged one after the other. They presumably arose through two rounds of gene duplication from a simpler ancestor. Potassium channels (specifically, potassium leak channels) show an intermediate step in this evolutionary process: they are formed of two subunits, each of which contains two repeats of the voltage-sensing and pore domains arranged in tandem.

The other large family of channels is the cys-loop receptor family. They are named so because of a conserved disulphide bond in the extracellular part of the protein. In contrast to the P-loop channels, they form pentamers. The nicotinic acetylcholine receptor and the GABA$_A$ receptor belong to this family.

By no means all channels belong to one of these two families. For example, the CFTR, the channel that malfunctions in cystic fibrosis (Chapter 15) is a maverick member of what is otherwise a family of ATP-powered carriers.

Signaling at the Neuromuscular Junction

We can use the gastrocnemius muscle to illustrate more of the concepts in this chapter (Figure 10.10). This is the calf muscle at the back of the lower leg. When it contracts, it pulls on the Achilles tendon so that the toes push down on the ground. Most of the bulk of the muscle is made up of one type of cell, skeletal muscle cells. The plasma membrane of the skeletal muscle cell contains voltage-gated sodium channels so, like neurons, they can generate action potentials.

All skeletal muscle cells are relaxed until they receive a command to contract from neurons called **motor neurons.** Motor neuron cell bodies are in the spinal cord, while the myelinated axons run to the muscle. To press the foot down, action potentials travel from the spinal cord down the moto-neurone axon to its terminal, opening voltage-gated calcium channels in the plasma membrane, and transmitter is released

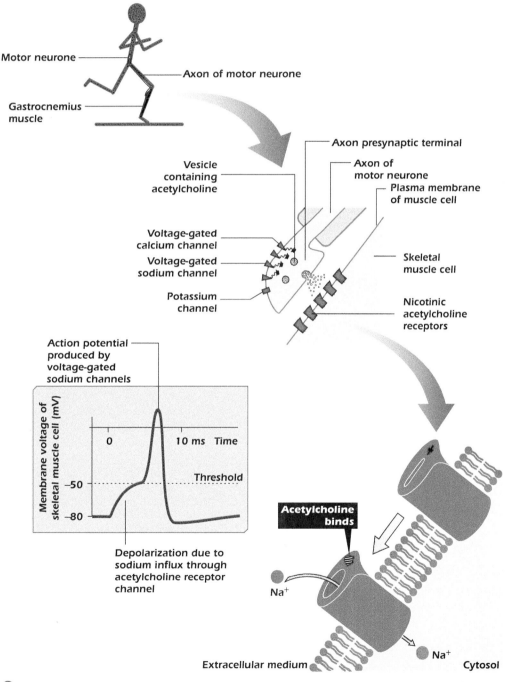

Figure 10.10. Motor neurons release the transmitter acetylcholine that binds to nicotinic receptors on skeletal muscle cells. The plasma membrane of the muscle cell is depolarized to threshold and fires an action potential.

into the synaptic cleft. Motor neurons use acetylcholine as their transmitter. The plasma membrane of the skeletal muscle cell thus contains nicotinic acetylcholine receptors. Like ionotropic glutamate receptors, nicotinic acetylcholine receptors cause a depolarization in the skeletal muscle cell. Indeed, just one action potential in a motor neuron elicits a

depolarization so large that it depolarizes the skeletal muscle cell to threshold, which is about −50 mV in these cells. The resulting action potential in the muscle cell in turn causes the release of calcium from the sarcoplasmic reticulum, and the increase of calcium concentration in the cytosol of the muscle cell causes it to contract (Chapter 13).

Answer to thought question: the outward voltage force is opposed by a large inward concentration force. Even when the inside of the cell is at +50 mV, the concentration gradient is so large as to push the calcium ions in. We can calculate when the forces would balance by using the Nernst equation (page 149). Extracellular calcium is about 1 mmol l⁻¹. In a resting cell, cytosolic calcium is about 100 nmol l⁻¹ (page 149) so that there is a 10000-fold concentration difference favoring calcium entry. However, we have stated that during an action potential calcium in axon terminals can increase to 1 μmol l⁻¹, that is, the concentration gradient is now only 1000-fold. It would be more appropriate to find the equilibrium voltage under those conditions. The general statement of the Nernst equation is

$$V_{eq} = \left(\frac{RT}{F}\right)\left(\frac{1}{z}\right)\log_e\left(\frac{[I_{outside}]}{[I_{inside}]}\right) \text{ volts}$$

For this calculation, $z = 2$, $[Ca_{outside}] = 10^{-3}$ mol l⁻¹, $[Ca_{inside}] = 10^{-6}$ mol l⁻¹, and if we work at human body temperature RT/F has the value 0.027. Thus under these conditions the equilibrium voltage for calcium is +0.093 V or +93 mV. This means that even when the cytosolic calcium concentration has increased to 1 μmol l⁻¹, the net electrochemical gradient for calcium will be inward unless the membrane depolarizes all the way to +93 mV, which will never happen under physiological conditions.

SUMMARY

1. Calcium ions form the basis of an important signaling system.

2. Increases in cytosolic calcium may be derived from two sources: the extracellular medium and the sarcoplasmic or endoplasmic reticulum.

3. Cytosolic calcium levels are reset by pumps and exchangers.

4. In neurons, influx of calcium from the extracellular medium through voltage-gated calcium channels causes exocytosis of transmitters by binding to synaptotagmin.

5. Transmitters are released at a synapse from a presynaptic neuron onto a postsynaptic cell.

6. Ligand-gated ion channels are ion channels on postsynaptic cells that open in response to released transmitters. The effect upon the target cell is electrical.

7. Synapses between neurons do not generally mediate a 1-to-1 transmission of the action potential. The presynaptic signal must show summation in time or space to elicit a postsynaptic action potential.

8. Stimulation of ionotropic cell surface receptors that pass chloride ions makes it more difficult for a neuron to be depolarized to threshold.

9. In muscle, release of calcium from the sarcoplasmic reticulum through ryanodine receptors causes contraction through activation of troponin. It also activates energy production.

10. The gastrocnemius muscle provides examples of how nerves and muscle operate in concert to fit the operation of the tissue to the requirements of the organism.

REVIEW QUESTIONS

10.1 Theme: The role of calcium at the neuro-muscular junction

A calcium ATPase
B sodium/calcium exchanger
C ryanodine receptor
D troponin
E voltage-gated calcium channel

From the above list of proteins, choose one or more that applies to the following statements:

1. a channel that allows extracellular calcium to enter the cytosol of presynaptic neurons during an action potential
2. a sensor that detects voltage changes across the plasma membrane of skeletal muscle cells and is physically linked to calcium channels on the sarcoplasmic reticulum
3. a carrier that pumps calcium ions into the sarcoplasmic reticulum
4. a calcium channel in the membrane of the sarcoplasmic reticulum
5. a protein attached to the cytoskeleton of skeletal muscle cells that triggers contraction when it binds calcium

10.2 Theme: Pathways for calcium movement

A calcium ATPase
B ryanodine receptor
C sodium/calcium exchanger
D sodium/potassium ATPase
E voltage-gated calcium channel

From the above list, choose the channel or carrier that provides a route for calcium to cross a membrane and fits each of the descriptions below:

1. a channel in the plasma membrane of a number of cell types
2. a channel in the membrane of the sarcoplasmic reticulum of skeletal muscle cells

3. a carrier that moves calcium ions up their concentration gradient but which does not use ATP
4. a carrier in the membrane of the sarcoplasmic and endoplasmic reticulum
5. a channel that is caused to open by an increase in the cytosolic calcium concentration

10.3 Theme: Synapses

A a depolarization that does not reach the threshold for action potential generation
B a depolarization that exceeds the threshold for action potential generation
C a long-lasting (more than one second in duration) switch to a membrane voltage more positive than 0 mV
D a reduction in action potential frequency, or the complete cessation of action potentials
E no voltage change

From the above list of electrical changes in postsynaptic cells, choose the change that would be evoked by each of the stimuli below.

1. a burst of activity in a presynaptic GABAergic neuron for a postsynaptic cell that is receiving steady excitatory input from a number of presynaptic glutaminergic neurons
2. a rapid burst of action potentials in a presynaptic pain receptor neuron caused by a painful stimulus such as a pin prick
3. a single action potential in a motor neuron
4. a single action potential in a presynaptic GABAergic neuron, at a time when no other synapses onto the postsynaptic cell are active
5. a single action potential in a presynaptic glutaminergic neuron, at a time when no other synapses onto the postsynaptic cell are active

THOUGHT QUESTION

When an action potential invades an axon terminal, the membrane voltage depolarizes to values in the range +30 mV to +50 mV and voltage-gated calcium channels open. We have stated that calcium ions flow into the cell through the calcium channels. How can this happen: why are they not pushed out of the cytosol by the positive voltage of the cell interior? If you can, use a calculation to support your answer.

11

SIGNALING THROUGH ENZYMES

In the previous chapter, we described how ions can be used for signaling. The physiological events we covered in neurons and muscle cells are rapid, occurring in milliseconds. But in many tissues and cells, signaling occurs over longer time frames and involves slow-acting enzymes. In this chapter, we consider some of the signaling pathways that act on this timescale. These include examples where enzymes are activated to make intracellular messengers or deactivated to control transcription, and where the primary proteins that respond to a specific stimulus have intrinsic enzyme activity. Stimuli that initiate these pathways take the form of peptides, proteins, and even gases.

 ## G PROTEIN-COUPLED RECEPTORS AND SECOND MESSENGERS

The presence of a chemical in the extracellular medium is often detected by integral membrane proteins known as **receptors.** Receptors recognize a particular chemical with high affinity. In Chapter 10, we came across several receptors in the form of ion channels that bind transmitters. But there are many other types of receptors that are not ion channels but rather proteins that participate in a more general mechanism, the end result of which is a change in the behavior of the cell.

G Protein-Coupled Receptors Are an Abundant Class of Cell Surface Receptors

The most abundant and diverse type of receptor found in the body is the **G protein-coupled receptors** or GPCRs. There are more than 800 in the human genome. GPCRs share a common structure comprising seven transmembrane regions. GPCRs exert their effects by being guanine nucleotide exchange factors (GEFs) for a particular glass of GTPases (page 90), **heterotrimeric G proteins.** When a particular GPCR binds a specific chemical, a specific type of heterotrimeric G protein is activated. This in turn often results in a change in the levels of a **second messenger.** Second messengers are small molecules that continue to broadcast the signal by regulating the activity of target proteins. Like calcium (Chapter 10) they are a type of intracellular messenger. As such, they relay information from the outside world. In the following text, we will illustrate how different physiological stimuli bring about cellular change through G protein-coupled receptors and second messengers.

Inositol Trisphosphate Controls Secretion in the Exocrine Pancreas

Acinar cells in the pancreas secrete digestive enzymes into the gut. Together with the pancreatic duct, they form the

Cell Biology: A Short Course, Fourth Edition. Stephen Bolsover, Andrea Townsend-Nicholson, Greg FitzHarris, Elizabeth Shephard, Jeremy Hyams and Sandip Patel.
© 2022 John Wiley & Sons Ltd. Published 2022 by John Wiley & Sons Ltd.
Companion website: www.wiley.com/go/bolsover/cellbiology4

exocrine pancreas. Pancreatic acinar cells use the release of calcium from the endoplasmic reticulum as one step in the process of enzyme secretion (Figures 11.1 and 11.2). Two new mechanisms are involved in calcium release from the endoplasmic reticulum.

The first is the production of the second messenger **inositol trisphosphate** (**IP$_3$** for short) (Figure 11.1). When we eat, digestive enzymes are secreted by the pancreas into the duodenum. This happens in response to the hormone cholecystokinin (CCK) produced when partially digested food enters the duodenum. The plasma membrane of the acinar cell contains a receptor that binds CCK, that is, CCK is its ligand. When the CCK has bound, the receptor becomes a guanine nucleotide exchange factor (page 91) for a heterotrimeric G protein called G$_q$. Like the GTPases IF2, EF-Tu, and EF-G that we met earlier (page 90), heterotrimeric G proteins are GTPases that activate target proteins when they have GTP bound, but turn

themselves off by hydrolyzing the GTP to GDP. Heterotrimeric G proteins have a slight additional complication in that, as the name indicates, they are composed of three subunits. The α subunit is homologous to the GTPases we have met before, while the β and γ subunits dissociate from the α subunit when it has GTP bound and only reassociate when the GTP has been hydrolyzed. GTP-loaded G$_q$ activates the β isoform of an enzyme called phosphoinositide phospholipase C, which we will call **phospholipase C** or **PLC** for short. PLC specifically hydrolyzes **phosphatidylinositol bisphosphate** (**PIP$_2$** for short), a phospholipid in the plasma membrane. Hydrolysis releases IP$_3$ to diffuse freely in the cytosol, leaving behind a residual lipid called diacylglycerol, or DAG. DAG, like IP$_3$, is also a second messenger.

The second new mechanism we must describe is the **inositol trisphosphate-gated calcium channel** (Figure 11.2). As IP$_3$ diffuses through the cytosol, it reaches the

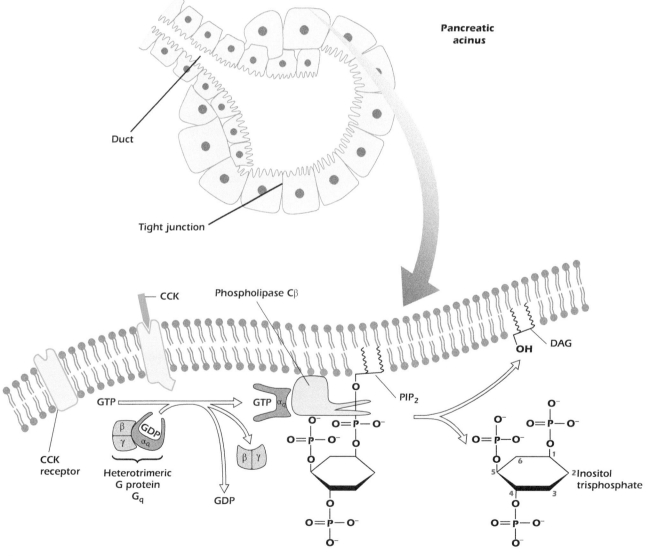

Figure 11.1. CCK activates G$_q$ and hence phospholipase Cβ in pancreatic acinar cells.

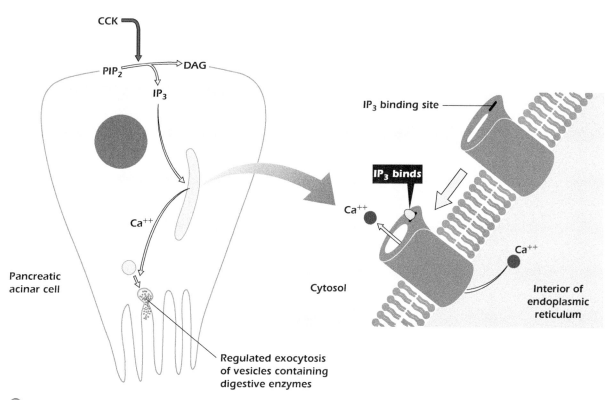

Figure 11.2. The inositol trisphosphate-gated calcium channel releases calcium from the endoplasmic reticulum.

endoplasmic reticulum. Here it binds to the cytosolic face of the inositol trisphosphate-gated calcium channels. Like the ryanodine receptor (Chapter 10), this ion channel allows calcium ions to pass. When inositol trisphosphate-gated calcium channels open, and calcium ions pour out of the endoplasmic reticulum into the cytosol, the increase of calcium concentration triggers the regulated exocytosis of granules containing the digestive enzymes to fuse with the membrane. The enzymes are secreted into the lumen of the duct and eventually make their way to the duodenum to aid digestion.

Phosphatidylinositol bisphosphate, G_q, PLCβ, and the inositol trisphosphate-gated calcium channel are found in almost all eukaryotic cells, but the distribution of the CCK receptor is much more restricted. Only cells that express the CCK receptor will show an increase of cytosolic calcium concentration in response to CCK. Other cells respond to other chemicals. Each produces a receptor specific for that chemical, which then activates G_q. Over 100 such receptors are known.

Cyclic Adenosine Monophosphate Helps Us Smell

We have already met the nucleotide cyclic adenosine monophosphate, or cAMP (Figure 5.9 on page 73) in the context of the regulation of the *lac* operon in bacteria (page 71). In eukaryotes, cAMP is also important and acts as an intracellular messenger in a great many cells, including the scent-sensitive neurons in our nose (Figure 11.3). These

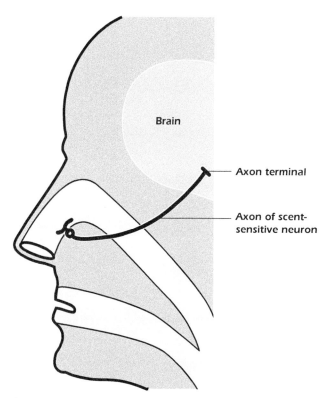

Figure 11.3. Scent-sensitive nerve cells send axons to the brain.

cells have their cell bodies in the skin of the air passages in the nose. Each cell sends an axon into the brain, and shorter processes, called dendrites, into the mucus lining the air passages. Scent-sensitive neurons are stimulated by scents in the air. Particular chemicals stimulate these cells because the cells have protein receptors that specifically bind the scent (Figure 11.4). When the scent binds, the receptor becomes a guanine nucleotide exchange factor for a heterotrimeric G protein called G_s. The GTP-loaded α subunit of G_s activates the enzyme **adenylate cyclase** that converts ATP to cAMP. The next stage in the detection of scents is a channel in the plasma membrane called the cAMP-gated channel. Like the inositol trisphosphate-gated calcium channel, the cAMP-gated channel is opened by a cytosolic solute, in this case cAMP. When the channel is open, it allows sodium and potassium ions to pass through. Potassium ions are close to equilibrium while there is a large inward electrochemical gradient for sodium. Thus, when the cAMP-gated channel opens, sodium ions pour in, carrying their positive charge and depolarizing the plasma membrane. The plasma membrane also contains voltage-gated sodium channels. Thus, when enough cAMP-gated channels open, the membrane voltage reaches threshold for the generation of sodium action potentials. These then propagate along the axon to the brain, and the person becomes aware of the particular scent for which that scent-sensitive neuron was specific.

Like G_q and phospholipase Cβ, G_s and adenylate cyclase are found in many cells, but only scent-sensitive neurons are

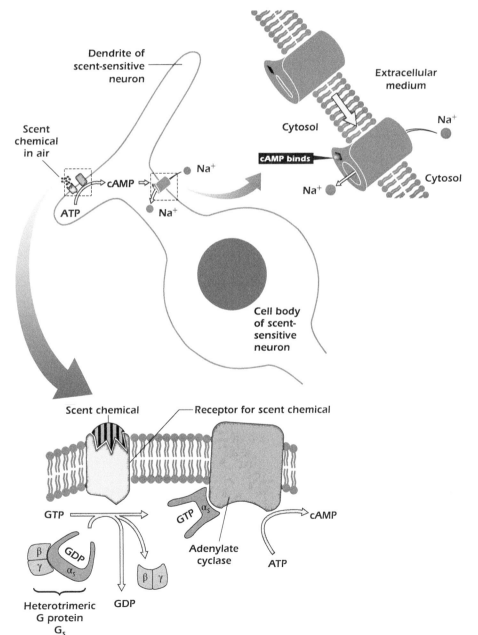

Figure 11.4. Scent chemicals activate G_s and hence adenylate cyclase in scent-sensitive nerve cells. cAMP then opens a nonselective cation channel in the plasma membrane.

sensitive to scents, because only they have specific scent receptors in their plasma membranes. Other cells that use cAMP as an intracellular messenger are sensitive to other specific chemicals because each makes a receptor that binds the chemical, whatever it may be, and then activates G_s. Once the stimulating chemical has gone, the cAMP concentration returns to resting levels. The enzyme **cAMP phosphodiesterase** hydrolyzes cAMP to AMP, which is inactive at cAMP-gated channels and other cAMP-binding proteins.

Most of the symptoms of the deadly disease cholera are caused by a toxin released by the gut bacterium *Vibrio cholera*. The toxin is an enzyme that enters the cytosol of the cells lining the gut and attaches an ADP ribosyl group (compare Example 6.2 on page 97) to the catalytic domain of G_s, preventing it from hydrolyzing GTP. G_s is therefore locked in the active state and activates adenylate cyclase nonstop.

The cAMP concentration in the cytosol then shoots up. CFTR chloride channels in the plasma membrane are opened by the increase of cAMP (page 257), allowing ions to leak from the cells into the gut contents. If untreated this loss of ions and of the water that accompanies them (by osmosis: In Depth 2.1 on page 23) leads to death from dehydration.

The interactions between G protein-coupled receptors and the α subunit of heterotrimeric G proteins are an example of how the specificity of protein–protein interactions defines cell responses. The scent receptors in the nose bind and are GEFs for the α subunit of G_s; they do not bind G_q. CCK receptors bind and are GEFs for the α subunit of G_q; they do not bind G_s. The precision of the cell's response to external signals is made possible by the precision of these protein–protein interactions.

IN DEPTH 11.1   HETEROTRIMERIC G PROTEINS: A FORTUITOUS DISCOVERY

"Enter the letter G, lucky seven in the alphabet the starting point for everything surely profound. . ." This quote is from the 1994 Nobel Banquet Speech of Martin Rodbell, a pioneer of signal transduction who determined that GTP was a driving force for the transfer of signals across the cell membrane. Rodbell's laboratory was studying how the hormone glucagon caused an activation of adenylate cyclase and the production of cyclic AMP from ATP. Noting that assays using certain preparations of ATP were more effective than others in eliciting cAMP production led to the intriguing discovery that it was not the

ATP that was the biologically active component in the assay but, rather, that cyclase activity was due to the presence of a GTP contaminant in the preparation. No hormone activation of adenylate cyclase was observed with purified ATP unless GTP was also added to the reaction. The fortuitous discovery led to the identification of heterotrimeric G proteins as an essential component of the signal transduction pathway and the missing link between extracellular hormones (for which Earl Sutherland had received the Nobel Prize in Physiology or Medicine in 1971) and second messengers inside the cell.

RECEPTOR TYROSINE KINASES AND THE MAP KINASE CASCADE

Platelets are common in the blood. They are small fragments of cells and contain no nucleus, but they do have a plasma membrane and some endoplasmic reticulum. Blood platelets use the release of calcium from the endoplasmic reticulum as one step in the mechanism of blood clotting. They also trigger the damaged blood vessel to repair itself. They do this by regulated exocytosis of a protein called **platelet-derived growth factor (PDGF).** The plasma membranes of the cells of the blood vessel contain receptors for this protein (Figure 11.5). PDGF receptors (Figure 7.6 on page 110) comprise an extracellular domain, which binds PDGF, a single polypeptide chain that crosses the plasma membrane, and a cytosolic domain that can phosphorylate tyrosine residues (page 108). The catalytic activity of the monomeric protein is low. PDGF can bind to two receptor molecules, drawing them close to each other,

causing a shape change that dramatically increases the tyrosine kinase activity. One of the first proteins to be phosphorylated is the receptor itself.

A number of cytoplasmic proteins have a domain called **SH2** that is just the right shape to bind to phosphorylated tyrosine. At the bottom of a deep pocket in the protein surface is a positively charged arginine. Although proteins can be phosphorylated on amino acids other than tyrosine, only tyrosine is long enough to insert down the pocket so that the negative phosphate can interact with the positive arginine. Proteins with SH2 domains therefore bind to dimerized PDGF receptors. In contrast, SH2 domains do not bind to solitary PDGF receptors whose tyrosines do not carry the negatively charged phosphate group.

One protein that has an SH2 domain is **growth factor receptor binding protein number 2 (Grb2).** Grb2 has no catalytic function but recruits a second protein called **SOS.** SOS is a guanine nucleotide exchange factor for a GTPase called **Ras,** allowing it to discard GDP and bind GTP. Active

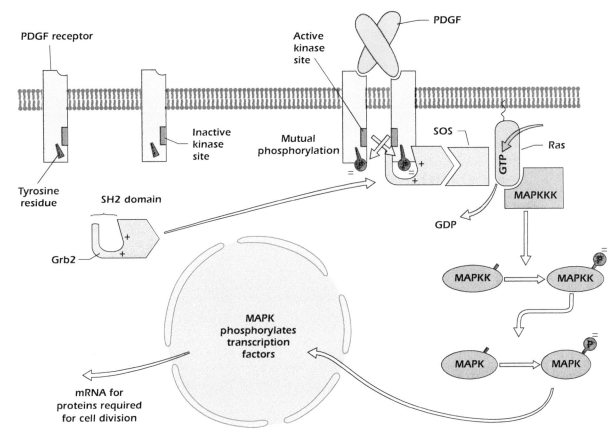

Figure 11.5. The PDGF receptor, like other growth factor receptors, activates the GTPase Ras and therefore the MAP kinase cascade.

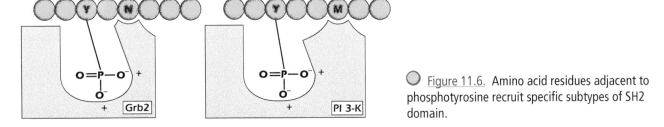

Figure 11.6. Amino acid residues adjacent to phosphotyrosine recruit specific subtypes of SH2 domain.

Ras turns on the first of a series of protein kinases, each of which phosphorylates and hence activates the next, culminating in a kinase called **mitogen-activated protein kinase (MAP kinase** or **MAPK).** The kinase that phosphorylates MAP kinase is **MAP kinase kinase (MAPKK),** while the one that phosphorylates MAPKK is **MAP kinase kinase kinase (MAPKKK)!** MAPKKK is at the top of the cascade and thus activated directly by Ras.

The term "mitogen" means a chemical that tends to cause mitosis, reflecting the fact that MAP kinase is activated by transmitters such as PDGF that turn on cell division (page 239). It phosphorylates many targets on serine and threonine residues. When MAP kinase is phosphorylated, it moves to the nucleus and phosphorylates transcription factors that in turn stimulate the transcription of the genes for **cyclin D** (page 238) and other proteins that are required for DNA synthesis and cell division.

Platelet-derived growth factor is one of many growth factors. All work in much the same way: their receptors are tyrosine kinases that are triggered to dimerize and phosphorylate their partners on tyrosine when the growth factor binds. The general term for this type of receptor is **receptor tyrosine kinase.** Phosphorylated tyrosine then recruits proteins with SH2 domains including Grb2, which in turn allows activation of Ras and the MAP kinase pathway, leading to DNA synthesis and cell division. In fact, the situation is more complicated than a simple "SH2 domains bind phosphotyrosine" rule would suggest. Different SH2 domain proteins have specific requirements for the amino acids flanking the phosphotyrosine (Figure 11.6). The PDGF receptor can recruit Grb2 because one of its phosphorylated tyrosines has an asparagine two amino acids away in the C terminus direction. Grb2 will only bind to phosphotyrosines that meet this criterion.

Medical Relevance 11.1   Blocking Growth Factor Receptors

Active growth factor receptors cause cell division by activating the MAP kinase pathway, and keep cells alive by activating Akt (page 248). Turning off growth factor receptors therefore tends both to stop cells dividing and to kill them. The drugs cetuximab and panitumumab are effective in slowing down the progression of bowel cancer because they prevent a growth factor, called epidermal growth factor (EGF), from binding to its receptor tyrosine kinase. This slows cell division and promotes cell death in the cancer cells.

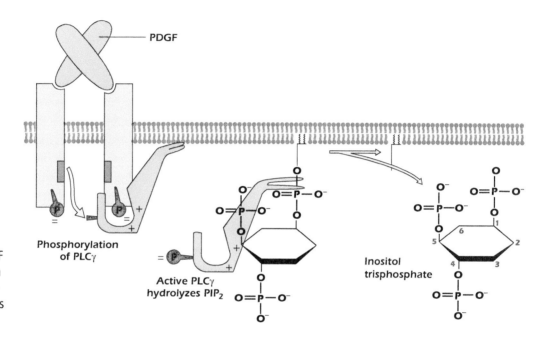

Figure 11.7. The PDGF receptor, like other growth factor receptors, phosphorylates and hence activates phospholipase Cγ.

Growth Factors Can Trigger a Calcium Signal

A second protein that contains an SH2 domain and that is therefore recruited to phosphorylated receptor tyrosine kinases is another isoform of phospholipase C, **PLCγ**. Binding to the phosphorylated tyrosine holds PLCγ at the growth factor receptor long enough for it itself to be phosphorylated (Figure 11.7), and this activates its enzymatic action. Active PLCγ hydrolyzes PIP_2 into diacylglycerol and inositol trisphosphate, and in turn inositol trisphosphate triggers the release of calcium from the smooth endoplasmic reticulum.

Akt and the Glucose Carrier: How Insulin Works

Figure 11.8 shows the **insulin receptor.** Like other receptor tyrosine kinases, the insulin receptor has an extracellular domain that can bind the transmitter, in this case the protein insulin, a single polypeptide chain that crosses the plasma membrane, and a cytosolic domain with tyrosine kinase activity. Unlike growth factor receptors, the insulin receptor exists as a dimer even in the absence of its ligand. When insulin binds, the shape and orientation of the individual insulin receptor subunits changes a little, and this allows each subunit to phosphorylate its partner upon tyrosine. An associated protein called the **insulin receptor substrate number 1 (IRS-1)** is also phosphorylated on tyrosine.

In particular, nine of the tyrosines on IRS-1 that become phosphorylated have a methionine three amino acids away in the C terminus direction. This is the preferred sequence for another SH2 domain protein, **phosphoinositide 3-kinase (PI 3-kinase)** (Figure 11.6). Binding to the phosphorylated tyrosines holds PI 3-kinase close to the insulin receptor long enough for it to become phosphorylated, and this activates its enzymatic action. PI 3-kinase operates on the same substrate, PIP_2, that PLC does, but instead of hydrolyzing it, PI 3-kinase adds another phosphate group to the inositol head group, generating the intensely charged lipid **phosphatidylinositol trisphosphate** (PIP_3) (Figure 11.8). In contrast, none of the phosphorylated tyrosines in the insulin receptor itself, or on IRS-1, are good at recruiting PLCγ, so insulin does not generally evoke a calcium signal in cells.

Just as phosphotyrosine is bound by a wide variety of proteins that contain SH2 domains, so highly phosphorylated inositols, such as those in PIP_3, are bound by a domain called the **PH domain** found on many proteins. Most

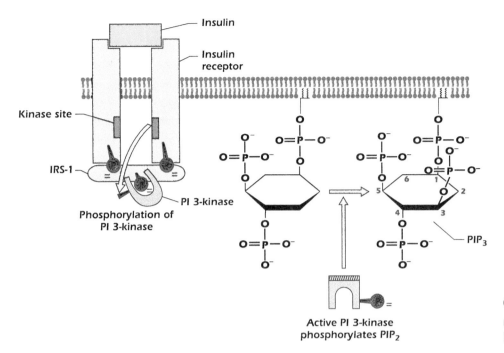

Figure 11.8. The insulin receptor phosphorylates and hence activates PI 3-kinase.

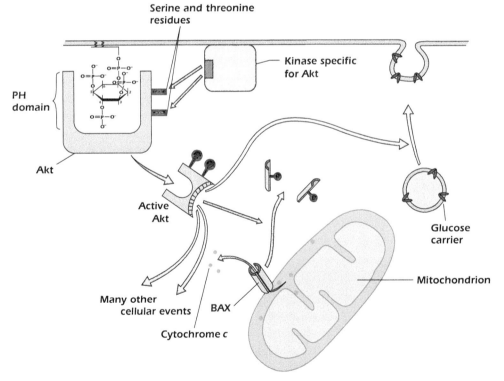

Figure 11.9. PIP$_3$ recruits Akt (protein kinase B) to the plasma membrane where it is activated. Active Akt has many effects, including exocytosis of vesicles containing glucose carriers. The action of Akt on the protein BAX will be described in Chapter 14.

important among these is **Akt** or protein kinase B (Figure 11.9). Akt, which phosphorylates targets on serine and threonine residues, is itself activated by phosphorylation, but the kinase that does this is located at the plasma membrane and therefore only gets a chance to phosphorylate Akt when Akt is held at the plasma membrane through its binding to PIP$_3$.

In many cell types, particularly fat cells and muscle cells, the final stage in the trafficking of the glucose carrier (page 149) from the Golgi apparatus to the plasma membrane requires active Akt. At the same time, glucose carriers are being internalized and are only returned to the plasma membrane if Akt is active. When we eat a large meal, insulin concentrations in the blood increase. Activation of the insulin receptor therefore causes an increase in the activation of Akt and hence a translocation of glucose carriers to the plasma membrane. This allows muscle and fat cells to take up large amounts of glucose from the extracellular medium. The action of Akt on the protein BAX, shown in Figure 11.9, will be described in Chapter 14.

 CYTOKINE RECEPTORS

Some of the most complicated interactions between cells occur in the immune system of vertebrates. The complex cell–cell interaction is mediated in part by a large family of protein transmitters called **cytokines.** A large and disparate collection of cell-surface receptors bind cytokines and activate downstream processes. Some are G protein-coupled receptors related to the CCK receptor. Others are receptor tyrosine kinases related to the PDGF and insulin receptors.

However, many cytokines act through a distinct family of receptors known as the type 1 cytokine receptor family. A type 1 cytokine receptor, such as the receptor for interleukin II (Medical Relevance 12.2 on page 214), comprises a number of integral membrane proteins, each of which crosses the plasma membrane only once (Figure 11.10). The extracellular aspects of these proteins form a binding site for the cytokine. Associated with the cytosolic part of the receptor are proteins called **JAK.** JAKs are tyrosine kinases, but unlike the receptor tyrosine kinases that we have met before JAKs are not integral membrane proteins but are peripheral membrane proteins (page 21) that remain at the membrane only because they associate with the cytosolic aspect of the cytokine receptor.

When the receptor binds its cytokine ligand the cytosolic aspect changes shape. This in turn alters the orientation of the two JAKs, allowing them to phosphorylate each other on their own tyrosine residues. Phosphorylation greatly increases

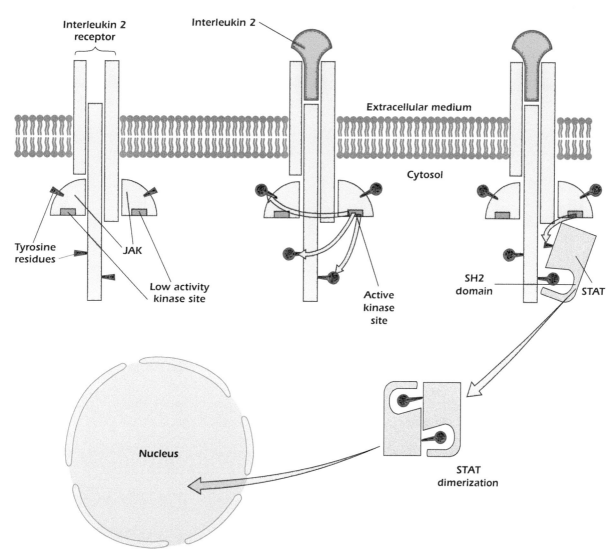

Figure 11.10. Signaling from type 1 cytokine receptors.

the catalytic activity of the JAKs, which can now phosphorylate tyrosines on the other components of the complex. These phosphotyrosines can activate many of the signaling pathways that we have already discussed in the context of receptor tyrosine kinases. For example, active interleukin II receptors can signal via Grb2 to Ras, and via phosphoinositide 3-kinase to Akt. However, type 1 cytokine receptors have a uniquely direct signaling route to the nucleus. This operates through a family of transcription factors called **STATs** (for Signal Transducers and Activators of Transcription) which in unstimulated cells are found mainly in the cytosol. STATs have SH2 domains and are recruited to activated type 1 cytokine receptors, where they are themselves phosphorylated by the active JAKs. This induces STATs to dimerize, with the SH2 domain of one STAT binding the phosphotyrosine on its partner. Dimerized STATs translocate to the nucleus and turn on the transcription of genes that promote cell division such as cyclin D (Chapter 14).

● SIGNALING THROUGH PROTEOLYSIS

The various signaling pathways we have covered thus far involve active switching on of various proteins by the stimulus. For example, the activity of enzymes that produce second messengers is normally very low and only increases when the stimulus is present. But in some cases, the signaling cascade is actively switched off in the absence of the stimulus. In these cases, the stimulus turns on the system by disabling the off switch. Here, we consider two such examples where **proteolysis** is used as a way to signal to the nucleus.

Wnt Proteins Signal Through Receptors that Prevent Proteolysis of Beta Catenin

During early development, a single fertilized egg transforms into a multicellular embryo with a clearly defined body plan. Much signaling is at play here to ensure this. The **Wnt proteins** are a family of secreted proteins that play key roles during patterning. Many Wnt proteins signal through the protein β **catenin.**

In the absence of Wnt proteins, β catenin is part of a destruction complex that promotes its degradation (Figure 11.11). The destruction complex is a complex of proteins that includes axin, a scaffold protein, and glycogen synthase kinase 3 or GSK3. In this resting state, β catenin is phosphorylated and this targets it for ubiquitination and degradation by the proteasome. β catenin is thus actively switched off and no transcription occurs.

When Wnt proteins are secreted, they bind to **Frizzled receptors.** Frizzled receptors form a complex with a co-receptor. In the presence of Wnt, another protein called

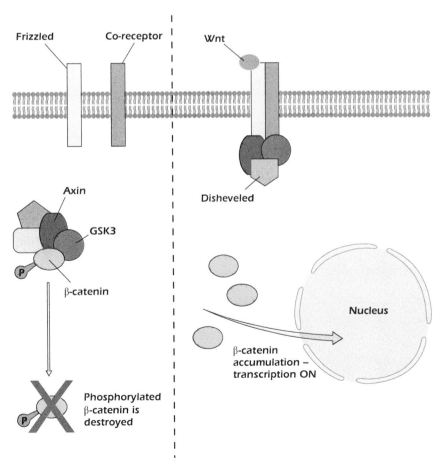

● <u>Figure 11.11.</u> The Wnt signaling pathway.

Dishevelled is recruited. And so too are axin and GSK3 that had been part of the β catenin destruction complex. Under these conditions, β catenin is no longer degraded and its levels increase. β catenin translocates to the nucleus and initiates transcription of target genes including many that promote cell growth and division.

Wnt signaling regulates axis formation during development, helping to define both the dorsal–ventral (back to front) and anterior–posterior (head to tail) axes. The activity of Wnt proteins is graded throughout the embryo, and it is this gradient that underpins differential gene expression, which in turn defines different parts of the embryo.

Low Oxygen Levels Are Sensed by Preventing Proteolysis of Hypoxia-Inducing Factor

Oxygen is central to many forms of life. Consequently, a reduction in oxygen supply or **hypoxia** is life-threatening. Animals detect decreased oxygen levels through the

hypoxia-inducible factor or HIF. HIF is a transcription factor. Like beta-catenin, it is actively degraded until it is needed (Figure 11.12).

HIF is composed of an α and β subunit. At normal oxygen concentrations the α subunit undergoes **hydroxylation.** Hydroxylation is a post-translational modification catalyzed by oxygen-dependent enzymes that add hydroxyl groups on specific amino acid residues.

Hydroxylation of HIF-α on proline residues results in the binding of HIF-α to **von Hippel–Lindau tumor suppressor** or pVHL. pVHL is a ubiquitin ligase (page 98). Ubiquitination of HIF-α by pVHL therefore promotes degradation of HIF-α by the proteasome.

When oxygen levels drop, oxygen-dependent hydroxylases are inhibited. Consequently, hydroxylation of HIF-α is prevented and it is no longer degraded. HIF-α accumulates in the cytoplasm and then translocates to the nucleus together with HIF-β. Here, the dimer binds to **hypoxia-response elements** (HRE) on DNA resulting in transcription of target

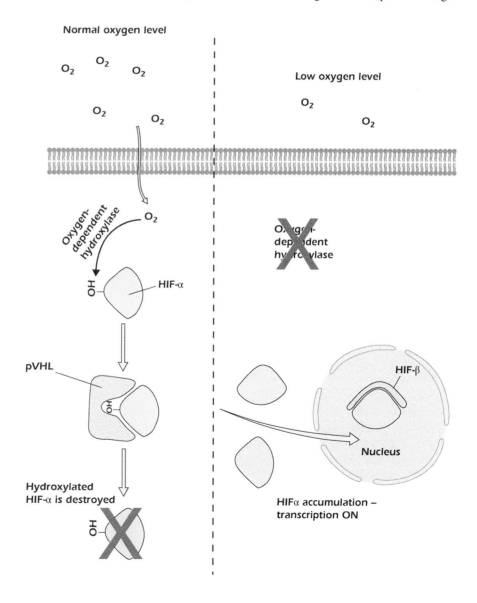

Figure 11.12. The HIF signaling pathway.

genes. These target genes regulate processes that increase oxygen supply and decrease oxygen consumption. For example, the production of erythropoietin, a hormone that increases production of red blood cells (Example 12.3 on page 207), is upregulated by HIF.

The developing mammalian embryo is a naturally occurring hypoxic environment. Local oxygen levels decrease further as the embryo grows and its metabolic demands increase. This induces formation of the embryonic blood vessels. The HIF pathway plays a key role in this process. An important target of HIF is the gene encoding vascular endothelial growth factor or VEGF. Like PDGF, VEGF is a growth factor that binds to a specific receptor tyrosine kinase. VEGF acts on the primitive vascular network to promote its development, thus increasing oxygen delivery. In summary, HIF is an important mechanism to match the oxygen supply to the needs of the body.

 INTRACELLULAR RECEPTORS

The receptors we have discussed thus far are all found in the plasma membrane and bind to transmitters in the extracellular fluid. Intracellular receptors in contrast lie within the cell (in the cytosol or in the nucleus) and bind transmitters that diffuse through the plasma membrane. They can exert their effects by activating enzymes. The receptors for **nitric oxide** and **steroid hormones** are two examples.

Guanylate Cyclase Is a Receptor for Nitric Oxide

Nitric oxide, or NO, is a transmitter in many tissues. It is not stored ready to be released like glutamate (Chapter 10) but is made at the time it is needed. Because it is a small uncharged molecule, NO diffuses easily through the plasma membrane and binds to various cytosolic proteins that are NO receptors. One particularly important NO receptor is the enzyme **guanylate cyclase,** which in the presence of NO converts the nucleotide GTP to **cyclic guanosine monophosphate,** or cGMP.

cGMP is a second messenger similar to cAMP. For example, like cAMP, cGMP can directly activate ion channels in sensory neurons. This happens in the light-sensitive neurons known as photoreceptors. However, most of the effects of cGMP are mediated because it binds to and activates the enzyme **cGMP-dependent protein kinase** or **protein kinase G.** Protein kinase G then phosphorylates target proteins to either activate them or turn them off. Like cAMP, cGMP is degraded by a phosphodiesterase, in this case **cGMP phosphodiesterase.**

Many Steroid Hormone Receptors Are Transcription Factors

Steroid hormones play key roles throughout the body. They include **corticosteroids,** such as cortisol and aldosterone which are involved in stress and ion homeostasis, and **sex steroids,** such as testosterone and estrogen. They all have intracellular receptors, such as the glucocorticoid receptor (page 79). In the absence of hormone, this receptor remains in the cytosol and is inactive because it is bound to an inhibitor protein. However, when the glucocorticoid hormone binds to its receptor, the inhibitor protein is displaced. The complex of the glucocorticoid receptor with its attached hormone now moves into the nucleus. Here, two molecules of the complex bind to a 15-bp sequence known as the HRE, which lies upstream of the TATA box (page 79). The HRE is a transcriptional enhancer sequence. The binding of the glucocorticoid hormone receptor to the HRE stimulates transcription.

 CROSSTALK – SIGNALING PATHWAYS OR SIGNALING WEBS?

Cell biologists often talk of signaling pathways in which a transmitter activates a chain of events culminating in an effect. The vertical arrows in Figure 11.13 show four signaling pathways that we have discussed in this chapter. We have already seen one example of how these pathways can interact: growth factor receptors can trigger a calcium signal by activating phospholipase Cγ. In fact, the pathways can interact in many ways and at many levels. The red arrows in Figure 11.13 show some of the most important. SOS is recruited to phosphorylated IRS-1 and can in turn activate Ras (arrow 1 in Figure 11.13). PI 3-kinase can bind to phosphotyrosine on growth factor receptors via its SH2 domain and be activated by phosphorylation so that growth factors, as well as insulin, will activate Akt (arrow 2 in Figure 11.13). In Chapter 14 we will see that this process is quite literally vital for our cells – if it stops, they die (page 248). A second route by which growth factors can activate PI 3-kinase is shown as arrow 3 in Figure 11.13: active Ras can activate PI 3-kinase.

Many of the actions of cAMP are mediated through it activating a protein kinase known as **cAMP-dependent protein kinase** or **protein kinase A** for short. This is similar to the activation of protein kinase G by cGMP. In particular, cAMP can activate transcription of many eukaryotic genes (arrow 4 in Figure 11.13), but it does so by causing the phosphorylation of transcription factors, and hence activating them. This is in sharp contrast to the cAMP–CAP system in *Escherichia coli* (page 72).

Calcium ions can activate a number of protein kinases, notably **protein kinase C** and **calcium–calmodulin-dependent protein kinase.** These phosphorylate target proteins on serine and threonine residues and recognize many of the same targets as protein kinase A, hence many of the downstream events are the same (arrows 5 in Figure 11.13). Paradoxically, calcium–calmodulin also activates the phosphatase **calcineurin,** which dephosphorylates and hence activates the transcription factor NFAT (Medical Relevance 12.2 on page 214).

The more we know about intracellular signaling systems, the more they come to resemble a web of interactions rather than a set of discrete pathways.

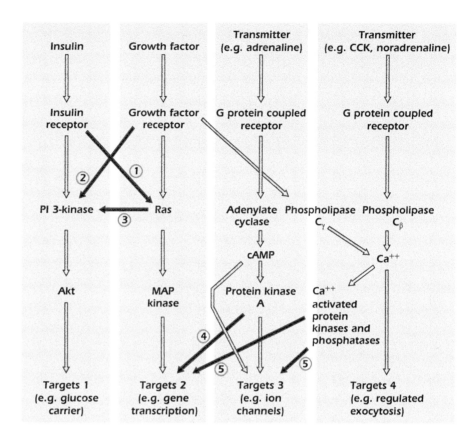

Figure 11.13. Interactions of signaling pathways.

BrainBox 11.1   Stephen Mayo

Stephen Mayo. Source: Caltech website at https://www.bbe.caltech.edu/people/stephen-l-steve-mayo.

Stephen Mayo has spent much of his research career at Caltech in the USA, focusing on protein design using sophisticated computer systems. One aspect of his work has been to look at how calcium, calmodulin, and its targets interact under the conditions that operate in real cells. He mutated calmodulin (page 115) so that only two of the EF hands were still able to bind calcium, in order to model the state of the protein in the calcium conditions found in active neurons, and showed that the mutant protein can still activate calcium–calmodulin-dependent protein kinase (page 190). He has been active in the design of novel fluorescent proteins with colors in the red and even far-red, to add to the palette provided by green fluorescent protein and its variants (page 15). A long-term aim of his work is the exciting and ambitious one of designing stable proteins and even functioning enzymes from scratch, rather than by modifying existing, functioning proteins. Mayo was elected to the National Academy of Sciences (USA) in 2004. He is currently the Bren Professor of Biology and Chemistry at Caltech.

SIGNALING IN THE CONTROL OF MUSCLE BLOOD SUPPLY

While transmission between neurons, and between neurons and skeletal muscle cells (Chapter 10), is extremely rapid, communication between the various cells within a tissue takes seconds and minutes rather than milliseconds. We can illustrate this and many of the signaling pathways discussed above by considering the control of blood flow to the muscle (Figure 11.14). The inner surface of the blood vessel is a thin

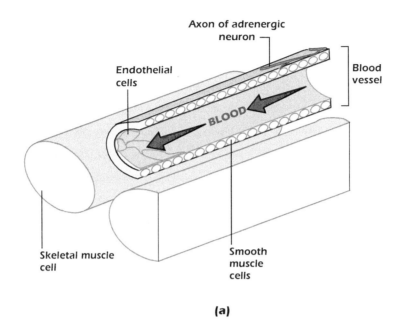

(a)

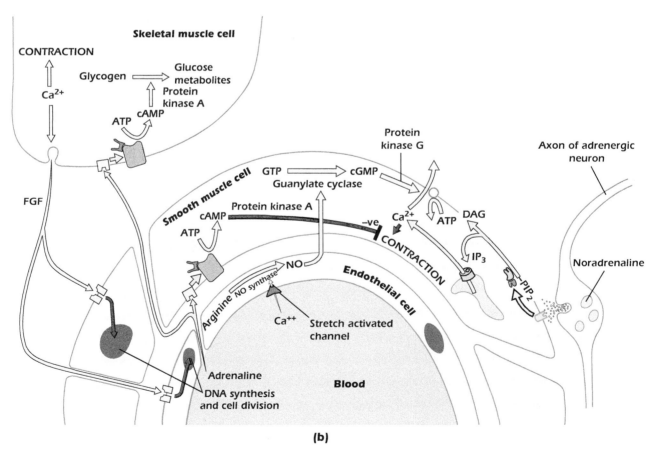

(b)

Figure 11.14. Transmitters regulate the blood supply to muscles.

layer of epithelium called **endothelium.** Wrapped around the endothelium are muscle cells of a different type, smooth muscle cells. Both endothelial cells and smooth muscle cells are much smaller than skeletal muscle cells. A small blood vessel may be only as large as a single skeletal muscle cell. The flow of blood is controlled by the relaxation and contraction of the smooth muscle cells, which in turn is regulated both by local factors and by signals originating from outside the tissue.

The Blood Supply Is Under Local Control

Exercise increases the blood flow to the muscles. A number of local mechanisms operate to cause this. Nitric oxide plays an important role. Physical stress of the plasma membranes of endothelial cells opens **stretch-activated channels** that allow calcium ions to flow into the cells. Inside the endothelial cells is the enzyme **NO synthase** that makes nitric oxide, and this enzyme is activated by calcium. Once made, the nitric oxide easily passes through the plasma membranes of both endothelial and smooth muscle cells and reaches its receptor within the smooth muscle cells. Here it activates guanylate cyclase, causing an increase of cGMP concentration and the activation of protein kinase G.

One of the targets of protein kinase G is the calcium ATPase (page 151). When this is phosphorylated by protein kinase G it works harder, reducing the concentration of calcium ions in the cytosol. This has the effect of relaxing the smooth muscle cells. The blood vessel therefore dilates, delivering more oxygen to the active tissue. Nitric oxide lasts for only about four seconds before being broken down. It is therefore a transmitter, able to diffuse through and relax all the smooth muscle cells of the blood vessel, but without lasting long enough to pass into more remote tissues.

Example 11.1 Nitroglycerine Relieves Angina

The discovery in 1987 that nitric oxide was a transmitter explained why nitroglycerine (more familiar as an explosive than as a medicine) relieved angina pectoris. Angina is a pain felt in an overworked heart. Nitroglycerine spreads throughout the body via the bloodstream and slowly breaks down, releasing nitric oxide that then dilates blood vessels. The heart no longer has to work so hard to drive the blood around the body.

Example 11.2 Viagra

cGMP is destroyed by cGMP phosphodiesterase. There are a number of isoforms of this enzyme in different human tissues. The drug sildenafil, sold as Viagra, inhibits the form of the enzyme found in the penis. If cGMP is not being made, this has little effect on blood flow to the region. However, when cGMP is made in response to a local production of NO, its concentration in blood vessel smooth muscle increases much more than would otherwise occur because its hydrolysis to GMP is blocked. This in turn causes a greater activation of protein kinase G, a greater activation of the calcium ATPase, a lower cytosolic calcium concentration, a greater relaxation of blood vessel smooth muscle, and therefore greater blood flow.

The Blood Supply Is Under Nervous System Control

Superimposed on this local mechanism is control from outside the tissue itself. The nervous system can act to divert blood from organs that are not in heavy use to areas that need it, such as active muscles. This is achieved by the transmitter **noradrenaline.** Action potentials in so-called adrenergic neurons cause exocytosis of the noradrenaline onto the surface of the smooth muscle cells. The smooth muscle cells have α-adrenergic receptors in their plasma membranes. The α-adrenergic receptor (Figure 11.15) is another receptor that causes cytosolic calcium concentration to increase. Binding of noradrenaline to α-adrenergic receptors activates PLCβ, which generates IP_3, which in turn releases calcium from the endoplasmic reticulum into the cytosol. The increase of cytosolic calcium concentration causes the smooth muscle cells to contract, constricting the blood vessel and reducing the flow to select organs.

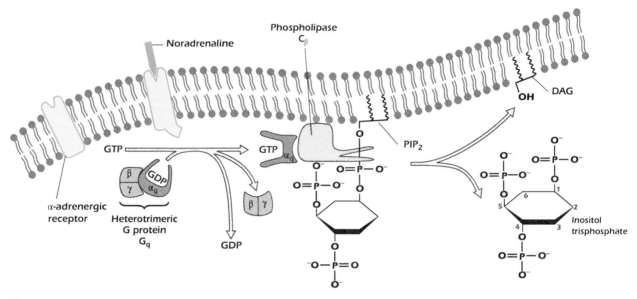

⊙ Figure 11.15. Noradrenaline activates G_q and hence phospholipase C_β in many cells including smooth muscle.

The Blood Supply Is Under Hormonal Control

The hormone adrenaline is chemically related to noradrenaline but is more stable, lasting a minute or so in the extracellular medium before being broken down. It is released from an endocrine gland (the adrenal gland) during times of stress and spreads around the body in the blood. The smooth muscle cells of blood vessels have β-adrenergic receptors connected to adenylate cyclase. When cAMP rises, it activates cAMP-dependent protein kinase. This in turn phosphorylates proteins that relax the smooth muscle cell. The action of adrenaline is therefore to increase the blood supply to all the muscles of the body in preparation for **fight or flight.**

The α- and β-adrenergic receptors are distinct proteins that bind similar transmitters, noradrenaline and adrenaline, but which signal to different heterotrimeric G proteins and therefore different downstream targets. To simplify the issue somewhat, we can say that noradrenaline acts mainly on α receptors and adrenaline acts mainly on β receptors. Because the α and β receptors are not the same, it is possible to design drugs (α and β blockers) that interfere with one or the other.

New Blood Vessels in Growing Muscle

All the phenomena we have discussed so far occur within minutes. However, when a muscle is repeatedly exercised over many days, it becomes stronger: the individual skeletal muscle cells enlarge. This is because high cytosolic calcium acts via NFAT (Medical Relevance 12.2 on page 214) to stimulate the transcription of genes coding for structural proteins. Furthermore, new blood vessels sprout and grow into the enlarging muscle. A growth factor called FGF is released by stimulated muscle. The receptor for FGF, like that for PDGF and VEGF (page 190), is found on endothelial and smooth muscle cells and is a tyrosine kinase that signals via Ras and MAP kinase to trigger cell division and hence the growth of new blood vessels. (FGF stands for fibroblast growth factor, but FGF is effective on a vast range of cell types, including both endothelial and smooth muscle cells).

IN DEPTH 11.2 FIGHT OR FLIGHT: SHOULD I STAY OR SHOULD I GO?

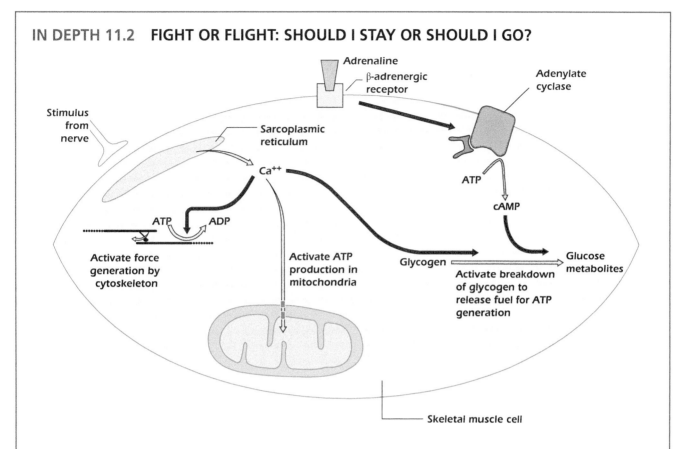

The **fight or flight response** is the body's response to stress. It is characterized by numerous physiological changes which prepare us for danger, likely tracing its roots back to ancient times. Adrenaline plays a key role in mediating flight or fight.

Adrenaline has multiple targets. In addition to relaxing smooth muscle cells to increase blood flow to skeletal muscle, it can also act directly on the skeletal muscle cells. For example, in the excitement before a race, the runner's adrenal glands release adrenaline into the blood. The adrenaline binds to β-adrenergic receptors on skeletal muscle cells, increasing cyclic AMP and activating protein kinase A (page 190), much in the same way as it does in smooth muscle. But in skeletal muscle, PKA phosphorylates and activates enzymes that in turn activate the breakdown of glycogen. This happens even when

cytosolic calcium is low. The end result is that even before the runner begins to run, the muscles are producing energy intermediates they will need once the race begins.

Adrenaline is also responsible for the increase in heart rate and the force of contraction during stress evident as "thumping." These effects are again mediated by β-adrenergic receptors. And again, they highlight how cyclic AMP and Ca^{2+} interact. In cardiac muscle, PKA phosphorylates numerous proteins involved in calcium signaling, including both voltage-gated calcium channels and ryanodine receptors (page 167), potentiating them and therefore causing larger calcium signals that in turn increase contraction.

If we are very frightened and too much adrenaline is released, so much blood is diverted to the muscles and away from the brain that we faint.

SUMMARY

1. G protein-coupled receptors are an important class of receptors that bind specific stimuli activating specific heterotrimeric G proteins resulting in the production of second messengers.

2. A second messenger is an intracellular solute whose concentration changes in response to cell stimulation; it activates or modulates a variety of cellular processes. The most important second messengers are IP_3, diacylglycerol, and cyclic adenosine monophosphate (cAMP).

3. IP_3 can be generated by the action of the β isoform of phospholipase C (PLCβ). Binding of extracellular chemical to a cell surface receptor activates the heterotrimeric G protein G_q, which in turn activates PLCβ.

4. IP_3 diffuses through the cytosol and opens a calcium channel in the membrane of the endoplasmic reticulum. Calcium then floods out, increasing the cytosolic calcium concentration.

5. A different set of G protein-coupled receptors activate G_s and hence adenylate cyclase, which makes cyclic AMP from ATP. Many of the actions of cAMP are mediated by the serine–threonine kinase cAMP-dependent protein kinase (protein kinase A).

6. Many receptor tyrosine kinases are caused to dimerize when their ligand binds. This allows each partner to phosphorylate the other upon tyrosine; this in turn recruits proteins containing SH2 domains.

7. Grb2 is an SH2-domain-containing protein that serves to bring together SOS and Ras, allowing activation of Ras and therefore activation of the MAP kinase pathway. This culminates in the transcription of genes necessary for DNA synthesis and cell division.

8. Phospholipase Cγ (PLCγ) and phosphoinositide 3-kinase (PI3K), two enzymes that act on the membrane lipid phosphatidylinositol bisphosphate (PIP_2), each have SH2 domains and are each recruited to, and phosphorylated by, receptor tyrosine kinases: this activates them. PLCγ generates IP_3 and therefore initiates a calcium signal, while PI3K generates PIP_3.

9. The serine–threonine kinase Akt is recruited to the plasma membrane by PIP_3; this allows it itself to be phosphorylated, which activates it.

10. In many cells, including fat and skeletal muscle, glucose carriers only appear at the plasma membrane when Akt is kept active through the activation of the insulin receptor.

11. Type 1 cytokine receptors become tyrosine phosphorylated when the transmitter binds, but unlike receptor tyrosine kinases the kinase is an independent protein called JAK.

12. Among the SH2 domain proteins recruited to activated type 1 cytokine receptors are STATs. When phosphorylated by JAK, STATs are active transcription factors.

13. Wnt receptors control the concentration of β catenin. In the absence of Wnt ligands, β catenin is phosphorylated and actively degraded by a destruction complex. In the presence of Wnt ligands, degradation is inhibited allowing β catenin to activate transcription of target genes.

14. Oxygen levels are sensed by the transcription factor, HIF. In the presence of oxygen, the alpha subunit of HIF is hydroxylated by oxygen-dependent enzymes and actively degraded. When oxygen levels decrease, hydroxylation and degradation of the alpha subunit is inhibited allowing it to associate with the beta subunit and activate transcription.

15. Guanylate cyclase is an intracellular receptor for nitric oxide. It produces the second messenger cyclic GMP from GTP. Cyclic GMP mediates many of its effects through cyclic GMP-dependent protein kinase. It is degraded by cyclic GMP phosphodiesterase.

16. Receptors for steroids are another class of intracellular receptors that act as transcription factors when they are bound by steroids. The glucocorticoid receptor is one example.

17. The control of muscle blood supply exemplifies many of the enzyme-mediated pathways covered and how the endothelium, muscle, and nerves work together.

The following text appears upside-down at the top of the page:

ing relaxation.

as the β-adrenergic receptor would. This would increase cyclic AMP and hence activate cAMP-dependent protein kinase, caus-

Consequently, the cytoplasmic parts of chimeric CCK/β-adrenergic receptor would bind and activate G_s much in the same way

In cells expressing plasmid D, CCK would be expected to cause relaxation. This is because β-adrenergic receptors signal via G_s.

way as the CCK receptor or the α-adrenergic receptor would.

Consequently, the cytoplasmic parts of the chimeric CCK/α-adrenergic receptor would bind and activate Gq much in the same

Similar results would be expected in cells expressing plasmid C. This is because α-adrenergic receptors also signal via Gq.

would allow CCK to activate Gq, produce IP_3 and increase the calcium concentration.

In cells expressing plasmid B, CCK would be expected to cause contraction. This is because the introduction of CCK receptors

smooth muscle cells do not normally express CCK receptors. Therefore, CCK does not mediate any biological effect.

Answer to Thought Question: In cells expressing plasmid A, CCK would have no effect on contractility. This is because

FURTHER READING

Berridge, M. Cell Signalling Biology (website). https://portlandpress.com/pages/cell_signalling_biology.

Marks, F., Klingsmüller, U., and Müller-Decker, K. (2017). *Cellular Signal Processing*, 2e. New York: Garland.

 REVIEW QUESTIONS

11.1 Theme: Processes downstream of receptor tyrosine kinases

A JAK
B phosphoinositide 3-kinase
C phospholipase Cβ
D phospholipase Cγ
E Akt
F SH2
G SOS
H STAT

From the above list of proteins or protein domains, select the one corresponding to each of the descriptions below.

1. a domain comprising a pocket with a positively charged arginine at the base. Proteins with this domain are recruited to phosphorylated tyrosine residues
2. a guanine nucleotide exchange factor for Ras
3. a hydrolytic enzyme that is activated when phosphorylated by receptor tyrosine kinases
4. a kinase that is activated when phosphorylated by receptor tyrosine kinases
5. an enzyme that is activated by Ras–GTP

11.2 Theme: Proteins activated by nucleotides

A calcium ATPase
B cAMP-gated channel
C cGMP-gated channel
D G_q
E G_s
F IP_3 receptor
G protein kinase A
H Akt = Protein kinase B
I protein kinase C
J Ras

From the above list of proteins, select the one corresponding to each of the descriptions below.

1. a protein kinase activated by cAMP
2. a protein responsible for generating electrical signals in photoreceptors
3. a protein that activates a phospholipase C when in the GTP bound state
4. a protein that activates adenylate cyclase when in the GTP bound state
5. a protein that activates MAP kinase kinase kinase when in the GTP bound state

11.3 Theme: Inositol compounds

A inositol
B inositol bisphosphate IP_2
C inositol trisphosphate IP_3
D inositol tetrakisphosphate IP_4
E inositol hexakisphosphate IP_6
F phosphatidylinositol bisphosphate PIP_2
G phosphatidylinositol trisphosphate PIP_3

From the above list of inositol compounds, select the one described by each of the descriptions below.

1. a ligand that binds to and opens a calcium channel
2. a lipid that recruits Akt to the plasma membrane
3. a substrate for phosphoinositide 3-kinase (PI3K)
4. a substrate for phospholipases C (PLC)
5. the product of phosphoinositide 3-kinase (PI3K)
6. a product of phospholipases C (PLC)

⬤ THOUGHT QUESTION

A molecular biologist has been given an expression plasmid (plasmid A). From this, a second plasmid encoding the CCK receptor (plasmid B) has been constructed. Two additional plasmids were generated where the sequence coding for the cytoplasmic parts of the CCK receptor are replaced with the corresponding sequences of either the α-adrenergic receptor (plasmid C) or the β-adrenergic receptor (plasmid D). Each of these plasmids was expressed in smooth muscle cells. What would you expect to happen to contractility when the cells expressing each of the plasmids are exposed to CCK? Explain these results with your knowledge of how each of the receptors signal.

SECTION 4

THE MECHANICS OF THE CELL

Eukaryotic cells are complex and structured, and that structure is not static but constantly in motion. In this section we will describe some of the aspects of this intracellular clockwork. We begin with trafficking: how cells move proteins from the location where they are made to their final destination, or take up proteins or other items from the environment and process them in specialized organelles. We then move to the cytoskeleton, the struts and ropes that hold the cell together, allow it to move and exert force, and upon which cell components are moved. Lastly, we turn to one of the most dramatic procedures carried out by the cytoskeleton: the sorting of chromosomes and pinching of the cell into two. We describe the control mechanisms that guide cell proliferation and cell death and discuss meiosis, the cell division that generates gametes and therefore underlies reproduction of not the cell but the whole organism.

12

INTRACELLULAR TRAFFICKING

Eukaryotic cells contain organelles (page 22). This compartmentation allows the spatial and temporal separation of different processes so that synthesis and degradation of molecules are segregated and materials destined for secretion are packaged separately from those taken up by the cell. For this to be possible, each type of organelle must have a specialized set of proteins. Proteins are synthesized by ribosomes so there must be systems to move proteins to their correct destination. In this chapter, we will discuss some of the ways in which a newly synthesized protein is precisely and actively moved (**translocated**) to the correct cellular compartment.

PRINCIPLES OF PROTEIN TRANSPORT

Newly synthesized proteins are delivered to their appropriate site of function within the cell after they are synthesized. The mechanisms and machinery that eukaryotic cells use to accomplish this are highly conserved from yeast to humans. Figure 12.1 depicts some ways how this is accomplished.

Proteins Enter Organelles in Different Ways

A common form of protein transport is **transmembrane translocation.** Here, unfolded polypeptide chains are threaded across one or more membranes to reach their final destination. Proteins that are synthesized by ribosomes on the rough endoplasmic reticulum are threaded across the membrane and into the interior of that organelle while they are being synthesized. Proteins destined for the interior of peroxisomes and mitochondria (and chloroplasts in plants) are synthesized on cytosolic ribosomes and may fold up completely or partially into their final form before being transported to and across the appropriate membrane. Transport into mitochondria involves the protein being unfolded and threaded through a hole in a transporter; import into peroxisomes is less well understood.

Once in an organelle, many proteins are transported between organelles in a process known as **vesicular trafficking.** Here, small closed bags made of membrane, **vesicles,** carry the protein. Transport of protein from the endoplasmic reticulum to the Golgi apparatus and from there to other compartments is by this means.

Some proteins fold into their final form as they are synthesized on free ribosomes in the cytosol and then move through an aqueous medium to their final destination, remaining folded all the way. Delivery of proteins to the nucleus follows this scheme; the proteins pass through nuclear pores into the nucleoplasm by a process called **gated transport.**

Cell Biology: A Short Course, Fourth Edition. Stephen Bolsover, Andrea Townsend-Nicholson, Greg FitzHarris, Elizabeth Shephard, Jeremy Hyams and Sandip Patel.
© 2022 John Wiley & Sons Ltd. Published 2022 by John Wiley & Sons Ltd.
Companion website: www.wiley.com/go/bolsover/cellbiology4

Vesicles Shuttle Proteins Around the Cell Through Fission and Fusion

Two main directions of vesicular traffic can be identified (Figure 12.1). The **exocytotic pathway** runs from the endoplasmic reticulum through the Golgi apparatus to the plasma membrane via two branches: constitutive and regulated secretion. The difference between the two is that constitutive secretion is always "on" (vesicles containing secretory proteins are presented for exocytosis continuously) while the regulated pathway is an intermittent one in which the vesicles containing the substance to be secreted accumulate in the cytoplasm until they receive a specific signal. This is usually an increase in the concentration of calcium ions in the surrounding cytosol, whereupon exocytosis proceeds rapidly. The **endocytotic pathway** runs from the plasma membrane to the lysosome. This is the route by which extracellular macromolecules can be taken up and processed. Bidirectional transport between the Golgi and lysosomes links the exocytic and endocytic pathways. If vesicles are to be moved over long distances, they are transported along cytoskeletal highways (page 222).

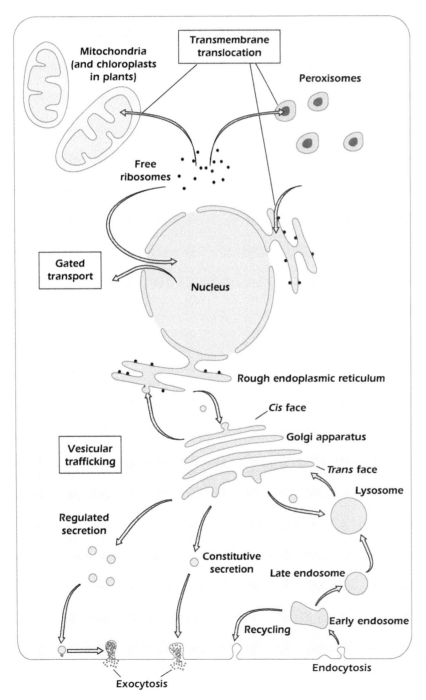

Figure 12.1. The three modes of intracellular protein transport.

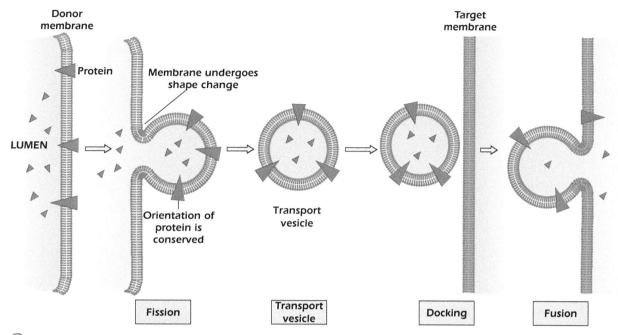

Figure 12.2. Vesicle fission and fusion.

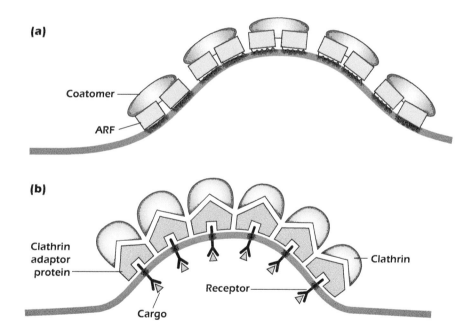

Figure 12.3. Generation of membrane buds by (a) coatomers and (b) clathrin.

Figure 12.2 illustrates how the budding of a vesicle from one organelle, followed by fusion with a second membrane, can transport both soluble proteins and integral membrane proteins to the new compartment. The process retains the "sidedness" of the membrane and the compartment it encloses: the side of an integral membrane protein that faced the lumen of the first compartment ends up facing the lumen of the second compartment, while soluble proteins in the lumen of the first compartment do not enter the cytosol but end up in the lumen of the second compartment, or in the extracellular medium if the target membrane is the plasma membrane.

Vesicles form by the shaping of the lipid membrane, with the help of cytosolic **coat proteins,** into a bud which is then pinched off in a process called **membrane fission.** The ordered assembly of cytosolic proteins into a coat over the surface of the newly forming vesicle is responsible for forcing the membrane into a curved shape (Figure 12.3). Coat proteins must be shed before fusion of the vesicle with its target membrane can occur.

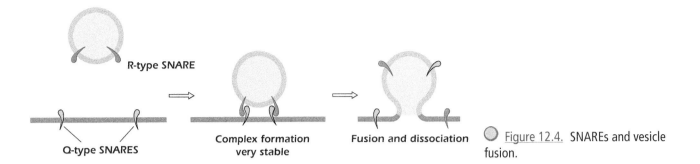

Figure 12.4. SNAREs and vesicle fusion.

Membrane fusion is the process by which a vesicle membrane incorporates its components into the target membrane and releases its cargo into the lumen of the organelle or, in the case of secretion, into the extracellular medium (Figure 12.4). Different steps in membrane fusion are distinguishable. First, the vesicle and the target membrane mutually identify each other. Then, proteins from both membranes interact with one another to form stable complexes and bring the two membranes into close apposition, resulting in the docking of the vesicle to the target membrane. Finally, the membranes then fuse.

SNARE proteins are critical for fusion. R-type SNARES located on the vesicles (also called v-SNARES) and Q-type SNARES on the target membranes (also called t-SNARES) interact to form a stable complex that holds the vesicle very close to the target membrane. Each type of vesicle must only dock with and fuse with the correct target membrane, otherwise the protein constituents of all the different organelles would become mixed with each other and with the plasma membrane. Not all R-SNARES can interact with all Q-SNARES, so SNARES provide specificity.

Example 12.1   SNARES, Food Poisoning, and Face-Lifts

Botulism, food poisoning caused by a toxin released from the anaerobic bacterium *Clostridium botulinum*, is fortunately rare. Botulinum toxin comprises a number of enzymes that specifically destroy those SNARE proteins required for regulated exocytosis in neurons. Without these proteins regulated exocytosis cannot occur, so the neurons cannot tell muscle cells to contract. This causes paralysis: most critically, paralysis of the muscles that

drive breathing. Death in victims of botulism results from respiratory failure.

Low concentrations of botulinum toxin (or "BoTox") can be injected close to muscles to paralyze them. For example, in a "chemical face-lift," botulinum toxin is used to paralyze facial muscles, producing an effect variously described as "youthful" and "zombie-like."

The Destination of a Protein Is Determined by Sorting Signals

When a protein is first made on the ribosome, it is simply a stretch of polypeptide. The initial sorting decisions are therefore made on the basis of particular amino acid sequences called **targeting sequences.** For proteins synthesized at ribosomes on the rough endoplasmic reticulum, additional sorting signals such as sugars and phosphate groups may be added by enzymes. In general, sorting operates by proteins containing a specific sorting signal binding to a receptor protein, which in turn binds to translocation machinery situated in the membrane of the appropriate compartment. Proteins without sorting signals, such as hemoglobin, are made on cytosolic ribosomes and remain in the cytosol.

Targeting sequences (also known as **localization sequences**) usually comprise a length of 3–80 amino acids

that are recognized by specific receptors that guide the protein to the correct site and make contact with the appropriate translocation machinery. Once the protein has been imported into the new location the targeting sequence may be removed by enzymes. Some targeting sequences have been characterized better than others. The targeting sequence encoding import into the endoplasmic reticulum consists of about 20 mostly hydrophobic amino acids at the N terminus of the protein called the **signal sequence.** The import signal for mitochondria is a stretch of 20–80 amino acids in which positively charged side chains stick out on one side of the helix and hydrophobic side chains stick out on the other, a so-called amphipathic helix. A cluster of about five positively charged amino acids located within the protein sequence targets a protein to the nucleus, while the best-known peroxisomal targeting sequence is the C-terminal tripeptide Ser-Lys-Leu-COOH.

Retention signals constitute another class of sorting signal. These do not activate the transport of a protein out of its present location but rather act as signals that the protein has reached its final destination and should not be moved. For example, proteins with the motif Lys-Asp-Glu-Leu-COOH (KDEL) at their C terminus are retained within the endoplasmic reticulum.

Proteins that are synthesized on the rough endoplasmic reticulum but otherwise have no particular sorting signal pass to the Golgi apparatus and then leave the cell by constitutive secretion, ending up in the extracellular medium. We can therefore regard this as the **default pathway** for ER-synthesized proteins.

Example 12.2   Holding Calcium Ions in the Endoplasmic Reticulum

One of the functions of the smooth endoplasmic reticulum is to hold calcium ions ready for release into the cytosol when the cell is stimulated. A protein called calreticulin (short for calcium-binding protein of the endoplasmic reticulum) helps hold the calcium ions. Its primary structure is

(NH_2) M L L S V P L L L G L L G L A V A E P A V Y F K E Q
F L D G D G W T S R W I E S K H K S D F G K F V L S S G K
F Y G D E E K D K G L Q T S Q D A R F Y A L S A S F E P F S
N K G Q T L V V Q F T V K H E Q N I D C G G G Y V K L F P N S
L D Q T D M H G D S E Y N I M F G P D I C G P G T K K
V H V I F N Y K G K N V L I N K D I R C K D D E
F T H L Y T L I V R P D N T Y E V K I D N S Q V E S G S
L E D D W D F L P P K K I K D P D A S K P E D W D E R
A K I D D P T D S K P E D W D K P E H I P D P D A K K P E D
W D E E M D G E W E P P V I Q N P E Y K G E W K
P R Q I D N P D Y K G T W I H P E I D N P E Y S
P D P S I Y A Y D N F G V L G L D L W Q V K
S G T I F D N F L I T N D E A Y A E E F G

N E T W G V T K A A E K Q M K D K Q D E E Q R L K E E E
E D K K R K E E E E A E D K E D D E D K D E D E E D E E D K
E E D E E E D V P G Q A K D E L (COOH)

The first 17 amino acids include 14 hydrophobic ones, shown in black: this is the signal sequence that triggers translocation of the protein into the endoplasmic reticulum. The last 4 amino acids, KDEL, include three hydrophilic amino acids, shown in green. This is the signal that ensures the protein remains in the endoplasmic reticulum. In between these two sorting signals is the functional core of the protein.

If you search for EF hand consensus sequences (see page 115) you will not find any! EF hands bind calcium with high affinity and can therefore bind calcium at the low concentrations found in the cytosol. Calreticulin binds calcium with much lower affinity, appropriate for an environment where the calcium concentration is hundreds of micromolar.

GTPases Are Master Regulators of Traffic

GTPases (page 90) that regulate and drive protein transport are numerous and distributed throughout the cell. Each seems to be found at one particular site where it regulates one specific transport event. For example, **Arf** and **dynamin** regulate fission in the ER and plasma membrane, respectively. The 60 or more members of the **Rab** family sit on vesicle membranes and regulate numerous aspects of trafficking to and from the Golgi apparatus and lysosome. **Ran** regulates the gated transport of proteins into and out of the nucleus. We will come across all of these GTPases in the following account of protein transport to various cellular locales.

 TRAFFICKING TO THE ENDOPLASMIC RETICULUM AND PLASMA MEMBRANE

Endoplasmic reticulum proteins enter the organelle during their synthesis. Plasma membrane proteins and proteins that are secreted begin their journey in the same way but traffic beyond the endoplasmic reticulum and through the Golgi in vesicles to reach their final destination.

Synthesis on the Rough Endoplasmic Reticulum

Proteins destined for import into the endoplasmic reticulum have a mainly hydrophobic signal sequence of about 20 amino acids at the amino terminal. Their synthesis starts on free ribosomes. When the growing polypeptide chain is about 20 amino acids long, the endoplasmic reticulum signal sequence is recognized by a **signal recognition particle** that is made up of a small RNA molecule and several proteins (Figure 12.5). When this particle binds to the signal sequence it stops protein synthesis from continuing. The complex of ribosome and signal recognition particle encounters a specific receptor called the signal recognition particle receptor (or "docking protein") on the endoplasmic reticulum. This interaction directs the polypeptide chain to a **protein translocator.** Once this has occurred the signal recognition particle and its receptor are no longer required

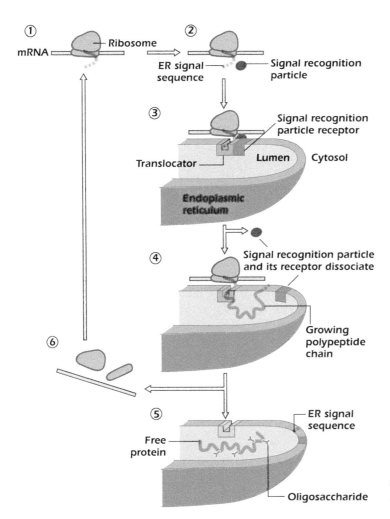

Figure 12.5. Transport of a growing protein across the membrane of the endoplasmic reticulum.

and are released. Protein synthesis now continues; and, as the polypeptide continues to grow, it threads its way through the membrane via the protein translocator, which acts as a channel allowing hydrophilic stretches of polypeptide chain to cross. Once the polypeptide chain has entered the lumen of the endoplasmic reticulum, the signal sequences may be cleaved off by an enzyme called signal peptidase. Some proteins do not undergo this step but instead retain their signal sequences.

The platelet-derived growth factor receptor is an example of an integral membrane protein. It contains a stretch of 22 hydrophobic amino acids that spans the plasma membrane (Figure 7.6 on page 110). The first part of the polypeptide to be synthesized is an endoplasmic reticulum signal sequence, so the polypeptide begins to be threaded into the lumen of the endoplasmic reticulum. This section will become the extracellular domain of the receptor. When the stretch of hydrophobic residues is synthesized, it is threaded into the translocator in the normal way but cannot leave at the other end because the amino acid residues do not associate with water. As synthesis continues, therefore, the newest length of polypeptide bulges into the cytosol. Once synthesis stops, this section is left as the cytosolic domain.

If a protein contains more than one hydrophobic stretch, then synthesis of the second stretch reinitiates translocation across the membrane, so that the protein ends up crossing the membrane more than once. Ion channels contain multiple transmembrane regions and are therefore synthesized in this way.

Glycosylation: The Endoplasmic Reticulum and Golgi System

Many polypeptides synthesized on the rough endoplasmic reticulum are glycosylated, that is they have sugar residues added to them, as soon as the growing polypeptide chain enters the lumen of the endoplasmic reticulum. In the process of N-glycosylation a premade chain of sugar residues (**oligosaccharide**) composed of two N-acetyl glucosamines, then nine mannoses, and then three glucoses is added to an asparagine residue by the enzyme oligosaccharide transferase. The three glucose residues are subsequently removed, marking the protein as ready for export from the endoplasmic reticulum to the Golgi apparatus.

Example 12.3 Cyclists and Glycosylation

Low oxygen in the kidney acts through HIF-α (page 189) to activate the production of the hormone erythropoietin. Erythropoietin in turn stimulates the formation of red blood cells in the bone marrow. The hormone has important medical applications in the treatment of anemias caused by chronic kidney disease or appearing as a side effect of cancer therapy. It is difficult to purify and medical use had to wait until recombinant DNA methods were developed (Chapter 8). Bacteria can be transfected so that they express human erythropoietin cDNA but they do not add the oligosaccharides that are critical in the protein's ability to bind to its receptor. Fortunately mammalian expression systems for the human gene were developed and these produce recombinant glycosylated erythropoietin that is active. Recombinant erythropoietin is now widely available for clinical use.

Soon erythropoietin found other uses among athletes taking part in endurance sports such as cycling as it boosted the oxygen-carrying capacity of the blood (although taking erythropoietin is not without risks). As recombinant erythropoietin has the same amino acid sequence as the regular endogenous human protein its presence is extremely hard to prove. However, the mammalian cells used to make commercial erythropoietin are not human and there are small differences in the exact glycosylation that do not affect the function. Work is being undertaken to develop tests that can discriminate between human erythropoietin and that made in nonhuman cells.

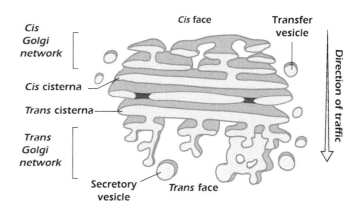

Figure 12.6. The Golgi apparatus.

The Golgi apparatus (Figure 12.6) is a stack of flattened membranous bags called **cisternae.** Each cisterna is characterized by a central flattened region where the luminal space, as well as the gap between adjacent cisternae, is uniform. The margin of each cisterna is often dilated and fenestrated (i.e. has holes through it). Small, spherical vesicles are always found in association with the Golgi apparatus, especially with the edges of the *cis* face. These are referred to as **transfer vesicles;** some of them carry proteins from the rough endoplasmic reticulum to the Golgi stacks, others transfer proteins between the stacks, that is, from *cis* to middle and from middle to *trans*. As proteins move through the Golgi apparatus, the oligosaccharides already attached to them are modified, and additional oligosaccharides can be added. As well as having important functions once the protein has reached its final destination, glycosylations play an important role in sorting decisions at the *trans* **Golgi** network.

Coatomer-Coated Vesicles

Transport along the default pathway uses coatomer-coated vesicles. This is the mechanism used in trafficking between the endoplasmic reticulum and Golgi, between the individual Golgi stacks, and in the budding of constitutive secretory vesicles from the *trans* Golgi. The coatomer coat consists of seven different proteins that assemble into a complex. The current model for coatomer coat formation at the Golgi is that a guanine nucleotide exchange factor (GEF) in the donor membrane exchanges GDP for GTP in the GTPase Arf. This causes Arf to adopt its active configuration, in which an α helix at the N terminus is exposed and embeds into the top of the donor membrane (Figure 12.3a). As well as anchoring Arf to the membrane, this also increases the area of the cytosolic face of the membrane, causing it to buckle outwards to form the bud. Membrane-bound Arf is the initiation site for coatomer assembly and coatomer-coated vesicle formation. The coat is only shed when the vesicle is docking to its target membrane. An Arf-GAP in the target membrane causes Arf to hydrolyze its GTP, and the resulting conformational change causes Arf to retract its hydrophobic N terminus and become cytosolic and this causes the coat to be shed.

Trans Golgi Network and Protein Secretion

At its *trans* face the Golgi apparatus breaks up into a complex system of tubes and sheets called the *trans* Golgi network (Figure 12.6). Although there is some final processing of proteins in the *trans* Golgi network, most of the proteins reaching this point have received all the modifications necessary to make them fully functional and to specify their final destination. Rather, the *trans* Golgi network is the place where proteins are sorted into the appropriate vesicles and sent down one of three major pathways: **constitutive secretion, regulated secretion,** or transport to lysosomes.

Vesicles for constitutive or regulated secretion, though functionally different, look very much alike and are directed to the cell surface (Figure 12.1). When the vesicle membrane comes into contact with the plasma membrane, it fuses with it. At the point of fusion the membrane is broken through, the contents of the vesicle are expelled to the extracellular space, and the vesicle membrane becomes a part of the plasma membrane. This process, by which the contents of a vesicle are delivered to the plasma membrane and following

membrane fusion and breakthrough are released to the outside, is called **exocytosis.**

Notice that in exocytosis the membrane of the vesicle becomes incorporated into the plasma membrane; consequently the integral proteins and lipids of the vesicle membrane become the integral proteins and lipids of the plasma membrane. This is the principal, if not the only, way that integral proteins made on the rough endoplasmic reticulum are added to the plasma membrane.

Major histocompatibility complex (MHC) **proteins** are a group of integral plasma membrane proteins that reach their final location by this route. Found only in vertebrates, the function of these proteins is to present short lengths of peptide to patrolling white blood cells. Figure 12.7a shows how this works. All the time, a selection of cytosolic proteins is being degraded into peptide fragments by the proteasome (page 98). These fragments are captured by a carrier located on the endoplasmic reticulum membrane called the TAP (for Transporter associated with Antigen Processing) and moved into the lumen, where they bind to a pocket in the MHC protein. From there the complex is processed together and passes via the Golgi complex to the plasma membrane.

One class of immune system cells called **T killer** patrols the body and kills any cell that is presenting a novel peptide (Figure 12.7b). In this way, cells infected by viruses or bacteria, or cells that have undergone somatic mutation, are detected and killed.

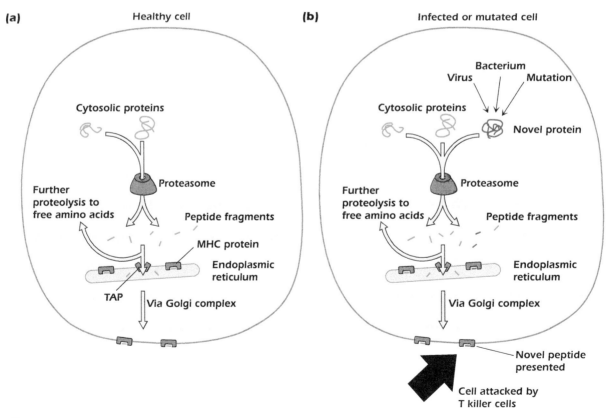

(a) Healthy cell   **(b)** Infected or mutated cell

Figure 12.7. Presentation of peptides by MHC proteins.

TRAFFICKING TO THE LYSOSOME

The function of the lysosome is to degrade unwanted materials. To carry out this function, lysosomes fuse with a vesicle containing the material to be digested. Vesicles with which lysosomes fuse are often bringing materials in from outside the cell or the cell surface. The lysosomal enzymes which degrade the material are trafficked to the lysosomes from the Golgi. Lysosomes also fuse with vesicles made by condensing a membrane around the cell's own cytoplasm. Vesicular trafficking thus features heavily. We will consider all of these processes.

Endocytosis Is a Gateway into the Cell

After secretion, vesicular membrane proteins are retrieved from the plasma membrane by **endocytosis** and transported to the **endosomes** (Figure 12.1). In a cell that is not growing in size, the amount of membrane area added to the plasma membrane by exocytosis is balanced over a period of minutes by endocytosis of the same area of plasma membrane. Soluble proteins and other cargoes in the extracellular fluid can also be internalized by this route. From the endosomes, vesicles are targeted to the lysosomes, back to the Golgi, or into the pool of regulated secretory vesicles.

Endosomes that are formed soon after pinching off from the plasma membrane are referred to as early endosomes. Here sorting decisions are made that determine the fate of the given protein. Those that are destined for degradation by the lysosomes are trafficked to late endosomes which then go on to fuse with lysosomes delivering the content. Others are recycled back to the plasma membrane in recycling endosomes. Many G-protein coupled receptors and growth-factor receptors (chapter 11) periodically move into the cell then back out to the surface, after binding their ligands, in this way.

IN DEPTH 12.1   SNEAKING INTO THE CELL

Enveloped viruses have their genomes encapsulated in a lipid coat derived from the host. Many hijack endocytosis to gain entry into cells. Filoviruses such as EBOLA and coronaviruses such as SARS CoV-2, which causes COVID-19, are notable examples. Enveloped viruses can traffic through the endocytic system along with other cargoes, thus evading the immune system. They then fuse with the host membrane, releasing their genome into the cytosol, allowing replication and the formation of new virions which are subsequently released from the cell. The cycle continues, resulting in more and more virus that propagates infection.

Clathrin-Coated Vesicles

In general, whereas vesicles coated with coatomer drive traffic out of the cells, it is **clathrin** that coats vesicles that traffic material in. For example, clathrin-coated vesicles carry proteins and lipids from the plasma membrane to endosomes. They also operate in other places where selective transport is required.

Figure 12.3b illustrates how clathrin generates a vesicle. The process starts when the cargo of interest binds to integral proteins of the donor membrane that are selective receptors for that cargo. **Clathrin adaptor proteins** then bind to cargo-loaded receptors and begin to associate, forming a complex. Lastly, clathrin molecules bind to this complex, forming the coat and bending the membrane into the bud shape.

Even though clathrin can force the membrane into a bud shape, it cannot force the bud to leave as an independent vesicle. One of the best-studied membrane fission events is endocytosis. Here, a GTPase called **dynamin** forms a ring around the neck of a budding vesicle. GTP hydrolysis then causes a change in dynamin's shape that mechanically pinches the vesicle off from its membrane of origin. Unlike coatomer, clathrin coats dissociate as soon as the vesicle is formed, leaving the vesicle ready to fuse with the target membrane.

Delivery of Enzymes to Lysosomes

One of the best-understood examples of sorting in the *trans* Golgi network is lysosomal targeting (Figure 12.8). Proteins that are destined for the lysosome are synthesized on the rough endoplasmic reticulum and therefore, like all proteins synthesized here, have a mannose-containing oligosaccharide added. Because they do not have an endoplasmic reticulum retention signal such as KDEL, they are transported to the Golgi apparatus. There proteins destined for the lysosome are modified by phosphorylation of some of their mannose residues to form mannose-6-phosphate. Once the proteins reach the *trans* Golgi network, specific receptors for mannose-6-phosphate recognize this sorting signal and cause the proteins to be packaged into vesicles that are transported to, and fuse with, the lysosome in a process guided by one of the Rab family of GTPases (Rab7). In the low pH (4–5) environment of the lysosome, the lysosomal protein can no longer bind to its receptor. The phosphate group is removed by a phosphatase. Vesicles containing the receptor bud off from the lysosome and deliver the mannose-6-phosphate receptors back to the *trans* Golgi network.

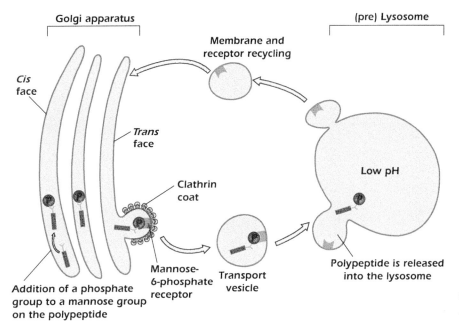

Figure 12.8. Targeting of protein to the lysosome.

Medical Relevance 12.1   Failure of the Lysosome-Targeting Signal

One severe lysosomal storage disease (page 25) is different in that, rather than a single enzyme, a whole group of hydrolytic enzymes is missing from the lysosomes. This is known as "I-cell disease" because of the large inclusions found in affected cells that can be easily seen with the light microscope. These inclusions are lysosomes swollen with undegraded materials. More curiously, the enzymes missing from the lysosomes are found instead in the extracellular medium, including the blood plasma! This mass failure of targeting occurs because the enzyme that adds the mannose containing oligosaccharide to proteins in the Golgi complex is missing or defective. This means that all the proteins that should go to the lysosome travel instead along the default pathway and are exocytosed.

Lysosomes Degrade Proteins from both Outside and Inside of the Cell: Autophagy

In addition to receiving cargo from the cell surface and the Golgi, lysosomes also receive cargo from the cytosol. Such cargoes include damaged proteins or worn-out organelles. This process is referred to as **autophagy** or "self-eating" (Figure 12.9).

Autophagy often involves the formation of a double membrane around portions of the cytosol. The resulting structure is known as an autophagosome. This then fuses with the lysosome, which eventually results in degradation of the content. The breakdown products are then exported out of the lysosomes to be reused.

Autophagy is activated when nutrients are scarce. Breakdown of cargo provides the building blocks to maintain cellular activity. In this way, autophagy matches the demands of a cell to the energy supply.

TRAFFICKING TO AND FROM THE NUCLEUS

In contrast to the situation in prokaryotes, RNA transcription and protein synthesis are separated in space and time in eukaryotes. Exchange of material between the nucleus (where transcription occurs) and the cytoplasm (where protein synthesis occurs) is essential for the basic functioning of these cells and must be tightly controlled. RNA and ribosomal subunits that are assembled in the nucleus have to enter the cytoplasm, where they are required for protein synthesis. On the other hand, proteins such as histones and transcription factors must enter the nucleus to carry out their functions. The nuclear pore complex mediates trafficking between the nucleus and the cytoplasm. The pore is made from a large number of proteins. Transport through the nuclear pore complex is mediated by signals and requires both energy and transporter proteins.

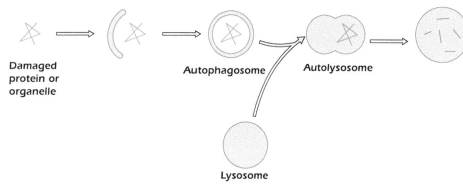

Figure 12.9. Autophagy allows cells to digest their own components.

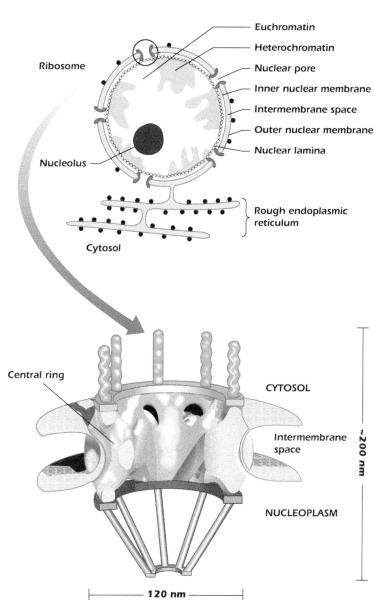

Figure 12.10. The nuclear pore.

The Nuclear Pore Complex

The nuclear pore complex is embedded in the double membrane of the nuclear envelope (Figure 12.10). It consists of over 30 different proteins called nucleoporins, plus some additional proteins. These form eight identical subunits arranged in a circle (see Figure 1.14 on page 16). Seen from the side, the pore complex comprises three rings. The central ring subunits have a transmembrane part that anchors the pore complex in the nuclear membrane and also project unstructured polypeptide chains inward to fill the hole of the pore. Small molecules can slip between these central tangled polypeptides but larger molecules are blocked and must be transported.

Gated Transport Through the Nuclear Pore

In general, a protein has to display a distinct signal for it to be transported through the nuclear pore. Proteins with a nuclear localization signal are transported in, while proteins with a nuclear export signal are transported out. Mobile transporter proteins, usually mediating either export or import, recognize the appropriate targeting sequence and then interact with the nuclear pore. The polypeptides tangled in the lumen of the pore have many glycine-phenylalanine repeats (GF repeats). Transporter proteins can bind these GF repeats and appear to hop from one to another through the pore, taking the protein to be transported with them. The precise mechanism by which large molecules move through the nuclear pore is not yet fully understood. The pore must expand or distort to allow movement of large assemblies such as ribosomal subunits. A typical nucleus will have 2000–4000 nuclear pores. Each pore can transport up to 500 macromolecules per second and allows transport in both directions at the same time. These complex structures vanish when the nucleus disassembles during mitosis and reform when the nuclei reform.

GTPases in Nuclear Transport

The discovery of the role played by the GTPase **Ran** in nuclear transport has given valuable insight into how directionality of transport into and out of the nucleus is achieved. GTP hydrolysis by Ran provides the energy for transport. The GEFs that operate on Ran are found in the nucleus and seem to be associated with chromatin, while Ran GAPs are attached to the cytosolic face of the nuclear pore complex (Figure 12.11). Thus nucleoplasmic Ran is predominantly in the GTP-bound state (Ran–GTP), while most cytosolic Ran is in the GDP-bound state (Ran–GDP).

Figure 12.12 shows how Ran regulates import of proteins into the nucleus. An import transporter binds the nuclear localization sequence on the protein. As long as the transporter remains on the cytosolic side of the nuclear envelope, its cargo will remain bound. However, once the transporter finds itself on the nucleoplasmic side, Ran in its active, GTP-bound state binds and causes the cargo to be released. Now, as long as the transporter remains on the nucleoplasmic side, it will have Ran–GTP bound and will be unable to bind cargo. However, once the transporter finds itself on the cytoplasmic side Ran GAPs will cause Ran to hydrolyze its bound GTP to GDP and the Ran–GDP will

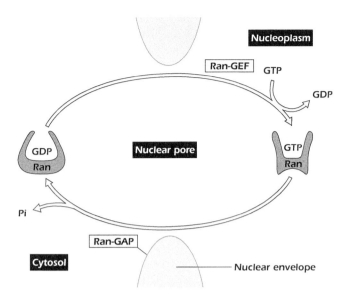

Figure 12.11. Ran GEF and GAP are localized to the nucleoplasm and cytosol, respectively.

dissociate from the transporter, which is then able to bind more cargo.

Much the same principle is used when a protein is to be exported from the nucleus (Figure 12.13). In this case the export receptor can only bind proteins with an export sequence when it is associated with Ran–GTP. As long as the transporter remains on the nucleoplasmic side of the nuclear envelope, its cargo will remain bound. However, once the transporter finds itself on the cytoplasmic side Ran GAPs will cause Ran to hydrolyze its bound GTP to GDP and both Ran–GDP and the cargo will dissociate from the transporter, which is then unable to bind more cargo until it moves back into the nucleoplasm where Ran–GTP is available.

Although Ran can drive both nuclear import and nuclear export, in each case it follows the standard rule of GTPases: it is the active GTP-bound form of the protein that binds to the target, in this case the transporter. The GDP-bound form is inactive and cannot bind the transporter.

Some proteins move back and forth between the cytosol and the nucleus by successively revealing and masking nuclear localization sequences. For example, the glucocorticoid receptor (page 79) only reveals its nuclear localization sequence when it has bound glucocorticoid.

TRAFFICKING TO OTHER ORGANELLES

Transport to Mitochondria

Mitochondria have their own DNA and manufacture a small number of their own proteins on their own ribosomes (In Depth 3.1 on page 36). However, the majority of mitochondrial proteins are coded for by nuclear genes. They are

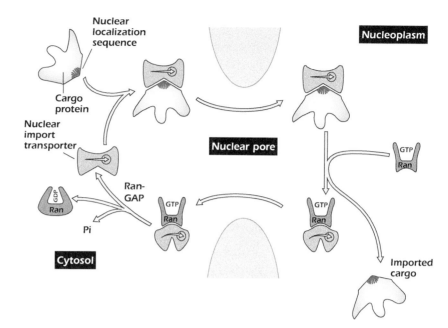

Figure 12.12. Nuclear import.

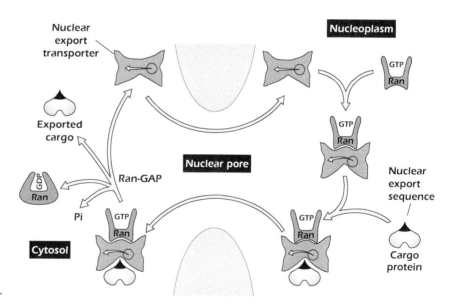

Figure 12.13. Nuclear export.

synthesized on cytoplasmic ribosomes and only imported into the mitochondrion post-translationally. For example, proteins destined for the mitochondrial matrix carry a targeting sequence at their N terminus and are recognized by a receptor protein in the outer membrane. This mitochondrial receptor makes contact with translocation complexes, which unfold the protein and move it across both outer and inner membrane simultaneously. Once the protein is translocated, the targeting sequence is cleaved off and the protein refolded.

Chaperones and Protein Folding. A correctly addressed protein may fail to be targeted to an organelle if it folds too soon into its final three-dimensional shape. For example,

movement of proteins into the mitochondrial matrix requires that a protein must move through channels through both the outer and inner membranes. These channels are just wide enough to allow an unfolded polypeptide to pass through. Our cells have proteins called **chaperones** which, as the name indicates, look after young proteins. Chaperones use energy derived from the hydrolysis of ATP to keep newly synthesized proteins destined for the mitochondrial matrix in an unfolded state. As soon as the protein moves through the channels and into the matrix, the matrix-targeting sequence is cleaved. The protein now folds into its correct shape. Some small proteins can fold without help. Larger proteins are

Medical Relevance 12.2   Blocking Calcineurin – How Immunosuppressants Work

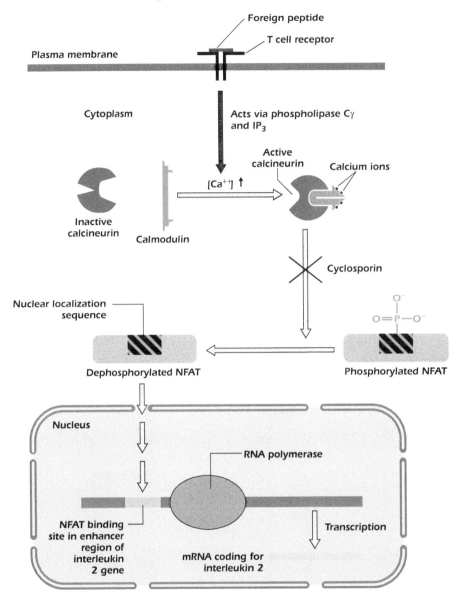

The drug cyclosporin A is invaluable in modern medicine because it suppresses the immune response that would otherwise cause the rejection of transplanted organs. It does this by blocking a critical stage in the activation of T lymphocytes, one of the cell types in the immune system. T lymphocytes signal to other cells of the immune system by synthesizing and releasing the cytokine interleukin 2 (page 187). Transcription of the interleukin 2 gene is activated by a transcription factor called NFAT.

NFAT has a sorting signal that would normally direct it to the nucleus, but in unstimulated cells this is masked by a phosphate group, so NFAT remains in the cytoplasm and interleukin 2 is not made. However, when major histocompatibility complex proteins present foreign peptides to the T lymphocyte, the concentration of calcium ions in the cytosol increases. Calcium activates the phosphatase calcineurin, which removes the phosphate group from many substrates including NFAT. NFAT then moves to the nucleus and activates interleukin 2 transcription. The released interleukin 2 activates other immune system cells that attack the foreign body. Cyclosporin blocks this process by inhibiting calcineurin, so that even though calcium rises in the cytoplasm of the T lymphocyte, NFAT remains phosphorylated and does not move to the nucleus.

helped to fold in the mitochondrial matrix by a chaperone protein called chaperonin, which provides a surface on which another protein can fold. Chaperones themselves do not change shape when helping another protein to fold.

Certain stresses that cells can experience, such as excessive heat, can cause proteins to denature (page 119). The cell responds by making a class of chaperone called **heat-shock proteins** in large amounts. The heat-shock proteins bind to misfolded proteins, usually to a hydrophobic region exposed by denaturation, and help the protein to refold. The heat-shock proteins are not themselves changed, but instead form a platform on which the denatured protein can refold itself. Heat-shock proteins are found in all cell compartments and also in bacteria.

Transport to Peroxisomes

Most organelles that are bound by a single membrane have their proteins made at the rough endoplasmic reticulum and transported to them in vesicles (Figure 12.1). Peroxisomes (page 25) are an exception: their proteins are synthesized on free ribosomes in the cytosol and then transported to their final destination. Peroxisomal targeting sequences on the protein bind to peroxisome import receptors in the cytosol. The complex of cargo and receptor docks onto the peroxisomal membrane and then crosses the membrane to enter the peroxisome. Here, the protein cargo is released and the import receptor is shuttled back into the cytosol.

SUMMARY

1. The basic mechanisms of intracellular protein trafficking are similar in all eukaryotic cells, from yeast cells to human neurons.

2. Vesicles shuttle between most of the single-membrane cellular organelles. Budding and then fission of vesicles from the donor membrane can be driven by coat proteins. Fusion is driven by SNARE proteins.

3. The final destination of a protein is defined by sorting signals that are recognized by specific receptors. Some sorting signals activate translocation of a protein to a new location, while others such as the endoplasmic reticulum retention signal KDEL cause the protein to be retained at its present location.

4. Protein transport is regulated by GTPases resident on specific compartments throughout the cell.

5. Proteins with an endoplasmic reticulum signal sequence are synthesized on the rough endoplasmic reticulum. The growing polypeptide chain is fed across the membrane as it is synthesized. The signal sequence may then be cleaved off.

6. Glycosylation in the endoplasmic reticulum and then in the Golgi apparatus has two functions: to produce the final, functional form of the protein and to add further sorting signals such as the lysosomal sorting signal mannose-6-phosphate.

7. The default pathway for proteins synthesized on the rough endoplasmic reticulum is to pass through the Golgi apparatus and be secreted from the cell via the constitutive pathway.

8. In vertebrates, peptide fragments of cytosolic proteins are presented at the cell surface on major histocompatibility complex proteins for inspection by patrolling T cells.

9. Endocytosis delivers proteins to the lysosomes via endosomes for degradation. Degradative enzymes are delivered to the lysosomes via the Golgi through the mannose-6-phosphate receptor. Defunct material in the cytosol is delivered to the lysosome through autophagy.

10. Nuclear proteins are synthesized on free ribosomes and carried through the nuclear pore by Ran-mediated gated transport. Other proteins with a nuclear export signal are carried the other way, again by a Ran-mediated process.

11. Peroxisomal proteins, together with that majority of mitochondrial and chloroplast proteins not coded for by mitochondrial or chloroplast genes, are synthesized on free ribosomes and then transported across the membrane(s) of the target organelle.

Answer to thought question: Peroxisomes, like mitochondria and chloroplasts, receive their proteins by transmembrane translocation. This and other features of peroxisomes have caused some scientists to wonder if they, like those other organelles, arose by the endosymbiosis of free-living bacteria. Since they have no genome of their own, the evidence is certainly weaker than for mitochondria and chloroplasts. Nevertheless, it is an appealing theory; indeed, one group of bacteria, the actinobacteria, has been proposed as the original symbiote.

Medical Relevance 12.3   How Protein Mistargeting Can Give You Kidney Stones

Primary hyperoxaluria type 1 is a rare genetic disease in which calcium oxalate "stones" accumulate in the kidney. Healthy people convert the metabolite glyoxylate to the useful amino acid glycine by the enzyme alanine glyoxylate aminotransferase (AGT). AGT is located in peroxisomes in liver cells. If glyoxylate cannot be converted to glycine, it is instead oxidized to oxalate and excreted by the kidney, where it tends to precipitate as hard lumps of insoluble calcium oxalate. Two-thirds of patients with primary hyperoxaluria type 1 have a mutant form of AGT that simply fails to work.

However, the other third have an AGT with a single amino acid change (G170R) that works reasonably well, at least in the test tube. This amino acid change is enough to make the mitochondrial import system believe that AGT is a mitochondrial protein and import it inappropriately, so that no AGT is available to be transported to the peroxisomes. For the clinician, the mistargeting of AGT in primary hyperoxaluria type 1 poses an unusual problem, namely, how to explain to a patient that the way to cure their kidney stones is to have a liver transplant!

FURTHER READING

Alberts, B., Johnson, A., Lewis, J. et al. (2014). *Molecular Biology of the Cell*, 6e. New York: Garland Science.

Donaldson, J. & Nava Segev, N. (2009) Regulation and coordination of intracellular trafficking: an overview. In the Madame Curie Bioscience Database at https://www.ncbi.nlm.nih.gov/books/NBK7285.

Duhita et al. (2010). The origin of peroxisomes: the possibility of an actinobacterial symbiosis. Gene 450: 18. https://doi.org/10.1016/j.gene.2009.09.014.

Karp, G.C. (2020). *Cell and Molecular Biology, Concepts and Experiments*, 9e. New York: Wiley.

Rothman, J.E. and Wieland, F.T. (1996). Protein sorting by transport vesicles. *Science* 272 (5259): 227–234. https://doi.org/10.1126/science.272.5259.227.

 REVIEW QUESTIONS

12.1 Theme: The three modes of intracellular transport

A gated transport
B remains in cytosol, no trafficking required
C transmembrane translocation
D vesicular trafficking

From the above list of transport modes, select the one experienced by each of the proteins below subsequent to its complete synthesis.

1. β globin
2. catalase (required inside peroxisomes)
3. glucocorticoid hormone receptor
4. histone H1
5. platelet-derived growth factor receptor
6. pyruvate dehydrogenase (required in the mitochondrial matrix)

12.2 Theme: Trafficking processes

A endocytosis
B forcing a membrane to buckle outward, forming a bud
C membrane fusion
D traffic through the nuclear pore
E transport into the endoplasmic reticulum

From the list of trafficking processes above, select a process with which each of the proteins below is associated.

1. Arf
2. dynamin
3. SNARES
4. Ran
5. signal recognition particle
6. TAP (transporter associated with antigen processing)

12.3 Theme: GTPases

A ADP
B ATP
C coatomer
D GDP
E GMP
F GTP
G GTPase-activating protein, GAP
H guanine nucleotide exchange factor, GEF
I nuclear import transporter

We describe the cycle of operation of a GTPase below. For each step, identify the molecule associated with that step.

1. GTPases have a binding site for a nucleotide. Identify the nucleotide present in the pocket when the GTPase is in its off state (e.g. Arf in the state that cannot associate with membranes).
2. GTPases are activated when the nucleotide in the binding pocket is replaced by a nucleotide present at higher concentration in the cytosol. Give the general name for the protein partner that catalyzes this switch.
3. GTPases turn off when the nucleotide in the binding pocket is hydrolyzed. Identify the product of this hydrolysis.
4. Hydrolysis of the nucleotide in the binding pocket is activated by a protein partner. Give the general name for the protein partner that accelerates the hydrolysis.

 THOUGHT QUESTION

Does the method by which peroxisomes acquire their proteins suggest their evolutionary origin?

13

CELLULAR SCAFFOLDING

Eukaryotic cells contain a dense network of structural filaments called the **cytoskeleton.** The cytoskeleton is multi-functional; it determines the cell's shape, and allows the cell to undergo further shape changes that can lead to cell movement; it is responsible for moving vesicles and organelles from place to place within the cell; and it plays key roles in cell division by separating chromosomes between two newly forming daughter cells. The cytoskeleton is composed of three distinct cytoplasmic filament networks: **microtubules, microfilaments,** and **intermediate filaments** (Figure 13.1). These core structural components of the cytoskeleton are accompanied by associated proteins, such as motor proteins that power the movement of cargoes along the filamentous networks. The term "cytoskeleton" implies a rigid and static frame within the cell, but nothing could be further from the truth. All three filament systems are highly dynamic and able to rapidly alter their organization in response to the needs of the cell at any given moment.

 MICROTUBULES

Microtubules possess a combination of physical properties that allows them to participate in multiple cellular functions. They can form bundles of rigid fibers to make structural scaffolds that serve an important role in the determination of

cell shape, and also provide tracks along which the directed movement of other cellular components such as vesicles and organelles can occur. Microtubules have an inherent structural polarity that enables the cell to control the directionality of this transport. Movement can be powered by enzymes called **microtubule motors** that move cargoes along the microtubule surface, or can occur by modification of the lengths of microtubules themselves. Microtubules can be rapidly formed and broken down, a property that allows the cell to respond to subtle environmental changes. Finally, they play a role in one of the most important and precise of all movements within the cell, the segregation of chromosomes at mitosis and meiosis into newly forming daughter cells (Chapter 14).

Microtubules are composed of the protein **tubulin,** which consists of two subunits designated α and β. These have been highly conserved throughout evolution and the α- and β- tubulins present in the cells of complex eukaryotes such as humans are much the same as those in a simple eukaryote such as a yeast. In the human genome there are eight α-tubulin genes and nine of β-tubulin. α-tubulin/β-tubulin dimers assemble into chains called **protofilaments,** 13 of which make up the microtubule wall (Figure 13.2). Within each protofilament the tubulin dimers are arranged in a "head-to-tail" manner, α/β, α/β and so on, which gives the microtubule a built-in molecular polarity. Tubulin subunits are added to, and lost from, one

Cell Biology: A Short Course, Fourth Edition. Stephen Bolsover, Andrea Townsend-Nicholson, Greg FitzHarris, Elizabeth Shephard, Jeremy Hyams and Sandip Patel.
© 2022 John Wiley & Sons Ltd. Published 2022 by John Wiley & Sons Ltd.
Companion website: www.wiley.com/go/bolsover/cellbiology4

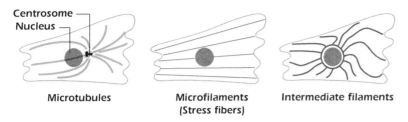

Figure 13.1. Typical spatial organization of microtubules, stress fibers (one form of microfilaments), and intermediate filaments.

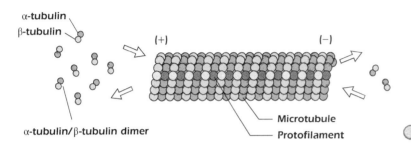

Figure 13.2. Microtubule structure.

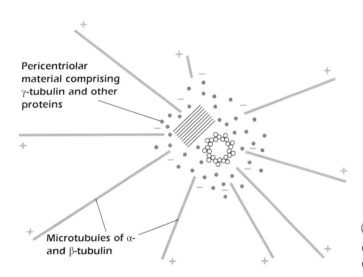

Figure 13.3. The microtubule organizing center or centrosome consists of amorphous material enclosing a pair of centrioles.

end much more rapidly than the other. By convention, the fast-growing end is referred to as the "plus" (+) end and the slow-growing end as the "minus" (−) end.

Microtubules are not scattered randomly within cells, but are arranged as a network. In most cells the major **microtubule organizing center** is the **centrosome,** a 1–2 μm structure that is usually tightly attached to the surface of the nucleus (Figure 1.2 on page 5). The centrosome comprises a pair of specialized microtubule tubes called **centrioles,** surrounded by a dense mass of material called the **pericentriolar material** (PCM). Centrioles have a characteristic ninefold symmetry that we will meet again later when discussing cilia and flagella. Found within the pericentriolar material is a third member of the tubulin superfamily, γ-tubulin, which plays a role in initiating microtubule assembly from the centrosome (Figure 13.3). The minus end of each microtubule is normally embedded in the

centrosome so that only the plus ends are free to grow or shrink. Individual microtubules undergo periods of slow growth followed by rapid shrinkage at their plus end, sometimes disappearing completely. This phenomenon of repeated rounds of shrinking and growing is referred to as **dynamic instability.** If by chance the growing ends of certain microtubules are captured by specialized sites at the plasma membrane or elsewhere in the cell, this can stabilize the microtubules so that they are protected from shrinkage. Having established a microtubule that reaches from the centrosome to the plasma membrane, its further growth can then influence the shape of the cell (Figure 13.4). Because microtubules are dynamic, the framework made up by multiple microtubules can be continually remodeled as the needs of the cell change. Figure 13.5 shows immunofluorescence images of two fibroblasts; in each the green microtubules radiate out from the microtubule-organizing center.

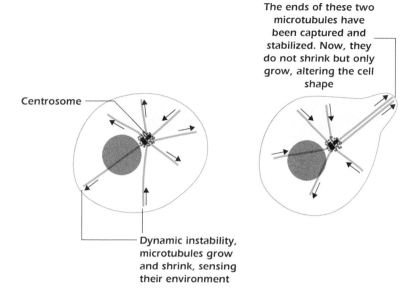

The ends of these two microtubules have been captured and stabilized. Now, they do not shrink but only grow, altering the cell shape

Centrosome

Dynamic instability, microtubules grow and shrink, sensing their environment

Figure 13.4. Microtubules show dynamic instability.

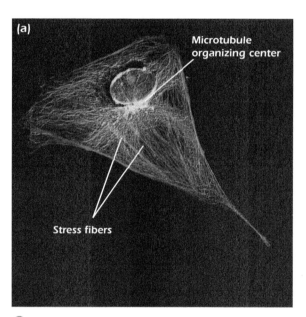

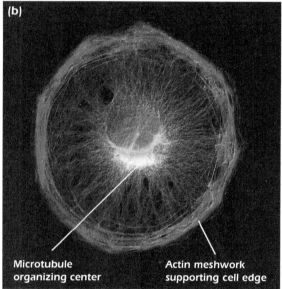

(a) Microtubule organizing center / Stress fibers

(b) Microtubule organizing center / Actin meshwork supporting cell edge

Figure 13.5. Microfilaments and microtubules in fibroblasts grown in culture. The green shows microtubules radiating out from the microtubule-organizing center while the red shows actin. (a) In this flattened cell the actin is organized as stress fibers. (b) In this rounded cell the actin is organized as a loose meshwork under the plasma membrane. The blue color is Hoechst staining and shows the nucleus. Source: Images by Professor David Becker, University College London; used with permission.

One of the most important tools in establishing microtubule function in cells has been the plant toxin colchicine. Extracted from the corms of the autumn crocus *Colchicum autumnale,* colchicine has been used since Roman times as a treatment for gout. Colchicine causes microtubules to become disassembled within cells. Cells exposed to colchicine therefore lose their shape and stop dividing and the movement of organelles within the cytoplasm is disrupted. When the drug is washed away, microtubules reassemble from the centrosome and normal functions can be resumed (Figure 13.6). Another drug, taxol, originally obtained from the bark of the Pacific yew, *Taxus brevifolia,* has the opposite effect, causing large numbers of very stable microtubules to form in the cell, an effect that is more difficult to reverse. Nowadays, taxol is synthesized chemically and is part of a widely used anti-cancer family of drugs, known as anti-mitotics, because it very effectively blocks cell division.

Example 13.1   Microtubule Dynamic Instability Helps Them "Search and Capture" Chromosomes in Mitosis

In mitosis the microtubules are reorganized into a characteristic spindle apparatus which is responsible for capturing the chromosomes and dispatching them to daughter cells during cell division (see Chapter 14). The dynamic instability of microtubules, wherein they undergo rounds of growth and shrinkage primarily at their plus ends, is essential for the spindle to perform this function. During mitosis microtubules grow from the centrosome into the cytoplasm. If the microtubule does not make contact with a chromosome, it subsequently undergoes catastrophe and shrinks back to the centrosome. However, if by chance a growing microtubule makes contact with the chromosome, then the microtubule becomes stabilized and docked onto that chromosome, and thus the chromosome becomes attached to the spindle. Through repeated rounds of growth and shrinkage of spindle microtubules all of the chromosomes are eventually captured by the spindle and can then be moved by the spindle into the daughter cells.

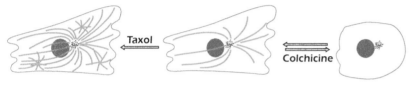

Figure 13.6. Effects of taxol and colchicine on microtubules.

FUNCTIONS OF MICROTUBULES

As mentioned above, the importance of the microtubule network is illustrated by the fact that disturbing it by using drugs brings many fundamental processes to a halt, including many types of cell movement, maintenance of the correct cell shape, and perhaps most notably the process of cell division that is covered in greater detail in Chapter 14. Here we discuss two important roles of the microtubule network that illustrate how the inherent properties and layout of microtubules described above can be harnessed to bring about essential cell functions, namely intracellular transport, and cell movement by **cilia** and **flagella.**

Intracellular Transport and Cellular Architecture

A major function of the microtubule network is to provide the tracks along which intracellular transport can take place. Cargoes that can be moved along the microtubule network include organelles, small vesicles, and even mRNA. These cargoes are moved by so-called **microtubule motor proteins** that use ATP as an energy source to move along microtubules like someone stepping up a ladder whilst carrying a load. Although usually referred to simply as motor proteins, many of them are in fact large complexes comprising several distinct subunits. Cargoes attach to these motor proteins either directly, or via adaptor molecules. The motor protein to which a cargo attaches determines the direction of its movement. One extreme but illustrative example of intracellular transport is the movement of cargoes along neuronal

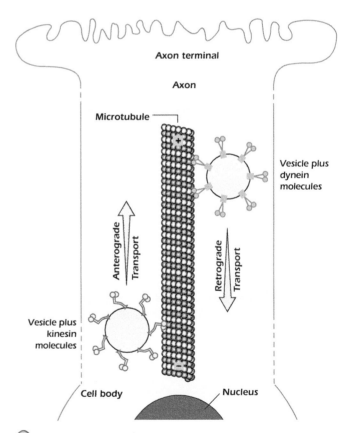

Figure 13.7. Axonal transport.

axons, which can extend up to 1 m from the cell body, referred to as **axonal transport** (Figure 13.7). Axons contain microtubules organized with their minus ends anchored

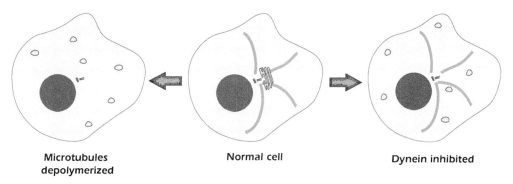

Microtubules depolymerized

Normal cell

Dynein inhibited

Figure 13.8. Dynein traffics the Golgi toward microtubule ends to position the Golgi next to the nucleus.

at the centrosome adjacent to the nucleus in the main cell body, and their plus ends oriented toward the terminal end of the axon. Movement of cargoes along the axon takes place in both directions, taking advantage of the inherent polarity of the microtubule. Dynein, sometimes referred to as **cytoplasmic dynein** to distinguish it from its relative in cilia and flagella, is a motor protein that hydrolyzes ATP to move along microtubules in a minus-end-directed manner. Dynein thus moves vesicles and organelles along the axon toward the cell body, termed **retrograde transport.** Many members of the **kinesin** family of motor proteins, on the other hand, move toward the plus end of microtubules, and are thus responsible for movement along microtubules away from the nucleus, in other words in the **anterograde** direction. Both dynein and kinesins consist of a tail that binds to the cargo to be transported and globular heads that interact with the surface of the microtubule, generating movement. Specificity is imparted to this process by having different adaptor proteins that link dynein and kinesins to specific types of cargo. The importance of kinesins can be inferred from the fact that the human genome contains 44 kinesin genes. These are structurally and functionally diverse but 18 members of this large superfamily have been shown to be associated with congenital abnormalities.

As well as motor proteins, some structural proteins associate with microtubules and modulate their function. One of the best studied is tau, which stabilizes microtubules in neurons. Displacement of tau, usually as the result of phosphorylation, results in microtubule destabilization and axon degeneration. The liberated tau forms highly insoluble neurofibrillary tangles which are diagnostic signatures of a number of neurodegenerative conditions such as Alzheimer's disease. So distinctive are these tangles that diseases that result in their formation are collectively referred to as tauopathies.

A knock-on effect of the role of the microtubule network in trafficking cargoes is that it effectively serves to determine the positioning and layout of intracellular organelles. An illustrative example of this is the positioning of the Golgi apparatus in the typical cell. Golgi components are continually trafficked by dynein toward the minus end of microtubules, and their continual trafficking in this direction causes the apparatus to accumulate at the cell center, close to the centrosome. As simple evidence of this, addition of a microtubule-disrupting drug such as colchicine causes the loss of the microtubule network, and as a result the Golgi becomes dispersed homogeneously throughout the cell (Figure 13.8).

Cell Movement by Cilia and Flagella

Cells move for a variety of reasons. Human spermatozoa in their millions swim frantically toward the egg; the soil amoeba, *Acanthamoeba* (said to be the most abundant eukaryote on Earth), crawls over and between soil particles, engulfing bacteria and small organic particles as it does so. Cells in the human embryo show similar crawling movement as they move into position to form tissues and organs; the invasive properties of some cancer cells is due to their reverting to this highly motile embryonic-like state. Both microtubules and microfilaments play important roles in these different types of cell motility. The roles of actin microfilaments in cell movement are discussed later in the chapter. Here we discuss two important examples of microtubule-based structures that enable cell motility, termed **cilia** and **flagella.**

Cilia and flagella appeared very early in the evolution of eukaryotic cells and have remained essentially unchanged to the present day. Both terms refer to fine protrusions of the plasma membrane containing microtubules that enable the protrusion to move. Generally, cilia are shorter than flagella (<10 μm compared to >40 μm) and are present on the surface of the cell in much greater numbers (ciliated cells often have hundreds of cilia but flagellated cells usually have only a single flagellum). The real difference, however, lies in the nature of their movement. **Cilia** row like oars. The movement is biphasic, consisting of an effective stroke in which the cilium is held rigid and bends only at its base ($1 \rightarrow 4$, Figure 13.9) and a recovery stroke ($4 \rightarrow 8 \rightarrow 1$, Figure 13.9) in which the bend formed at the base passes out to the tip.

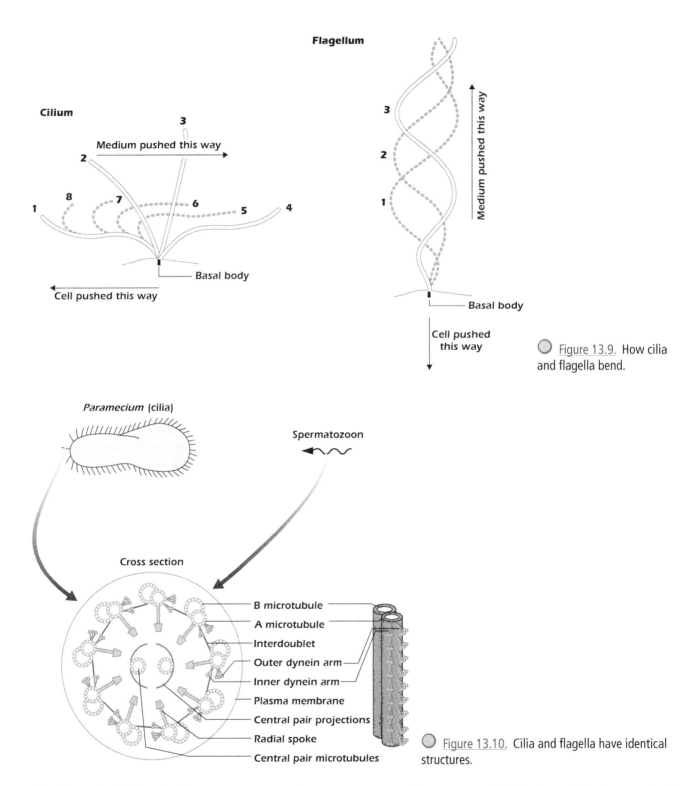

Figure 13.9. How cilia and flagella bend.

Figure 13.10. Cilia and flagella have identical structures.

Flagella wriggle like eels. They generate waves that pass along their length, usually from base to tip at constant amplitude. Thus the movement of water by a flagellum is parallel to its axis while a cilium moves water perpendicular to its axis and, hence, parallel to the surface of the cell.

Despite their different patterns of beating, cilia and flagella are structurally similar in animals (Figure 13.10). Both contain a ring of nine paired microtubules (the outer doublets)

surrounding two central single microtubules (the central pair). The overall structure is referred to as the **9+2 axoneme.** The axoneme is enclosed by an extension of the plasma membrane. Attached to the nine outer doublet microtubules are projections, or arms, composed of the motor protein **dynein.** Using ATP produced by mitochondria near the base of the cilium or flagellum as fuel, the dynein arms push on the adjacent outer doublets, forcing a sliding movement to occur between them.

Bacterial flagella use a fundamentally different mechanism. Like the propeller of a boat, the motion of the bacterial flagellum is entirely driven by the rotary motor at its base. The bacterial flagellum is made of one protein (flagellin) that has little similarity to tubulin or dynein.

Cilia are such a conspicuous feature of some unicellular eukaryotes that they are called ciliates. The swimming of a paramecium (Figure 13.10), for example, is generated by the coordinated motion of several thousand cilia on the cell surface. Cilia also play a number of important roles in the human body. The respiratory tract, for example, is lined with about $0.5\,m^2$ of ciliated epithelium, bearing in total some 10^{12} cilia (Figure 1.7 on page 10). The beating of these cilia moves a belt of mucus containing inhaled particles and microorganisms away from the lungs. This activity is paralyzed by cigarette smoke, with the result that mucus accumulates in the smoker's lung causing the typical cough.

Many mammalian cells possess a single primary cilium. This projects from the cell surface and acts as an antenna that receives signals and transmits them to the cell interior. Structurally, the primary cilium resembles the 9+2 structure of motile cilia, but the two central microtubules are missing; the so-called 9+0 structure. Many diseases result from defects in the function of the primary cilium. Collectively, these are known as ciliopathies and they include a variety of developmental abnormalities, kidney diseases, and defects in facial development.

MICROFILAMENTS

Microfilaments are fine fibers, about 7 nm in diameter, that are made up of subunits of the protein **actin.** The actin monomer is designated G-actin (for "globular"), while the filament that forms from actin when it polymerizes is referred to as F-actin. Each actin filament is composed of two chains of actin monomers twisted around one another like two strands of beads (Figure 13.11). In animal cells, actin is typically most concentrated within the **cortex** of the cell (i.e. at the cell periphery). When they are grown in plastic cell-culture dishes, non-motile cells have two main types of microfilaments: bundles of actin filaments called **stress fibers** run across the cell and help to anchor it to the dish, while under the plasma membrane can be seen a loose meshwork of filaments that give the cell cortex structural rigidity (Figures 13.1 and 13.5). Projections from the cell surface such as microvilli (page 8) are maintained by rigid bundles of actin filaments.

The equilibrium between G- and F-actin is affected by many factors, including **actin-binding proteins** (Figure 13.11). Filament nucleation proteins such as the Arp2/3 complex (Arp stands for **actin-related protein**) act as a base on which new filaments can form. Crosslinking proteins stick to two existing filaments, forming a mechanically strong lattice. Of these **villin,** found in microvilli, generates parallel bundles, while related proteins bind crisscrossing filaments together to form a viscous, three-dimensional cytoplasmic gel. In contrast, filament growth can be prevented by capping proteins that bind to F-actin filament ends. Cells are anchored to the extracellular matrix through transmembrane proteins such as **integrins.** These are dimeric proteins that have an extracellular domain that binds to collagen and other extracellular matrix proteins and an intracellular domain that attaches to actin microfilaments (Figure 13.12).

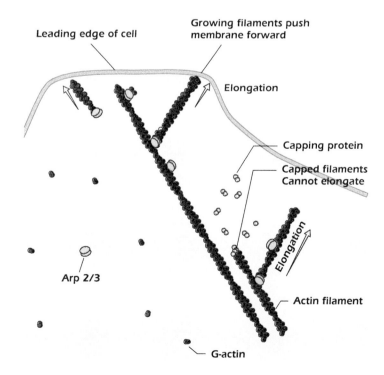

Figure 13.11. Actin polymerization is regulated by actin-binding proteins.

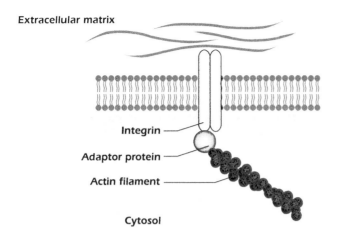

Extracellular matrix

Integrin

Adaptor protein

Actin filament

Cytosol

 Figure 13.12. Integrins anchor the actin cytoskeleton to the extracellular matrix.

The main class of motor protein associated with microfilaments is called myosin. The role of myosin in muscle cells is described in detail below. In non-muscle cells, several types of myosin are found. One of them, **myosin II,** is very similar to muscle myosin but does not assemble into filaments to the same extent, probably because the levels of force required within non-muscle cells is relatively small. The primary role of myosin II is in cell division (page 236). **Myosin V** on the other hand is responsible for carrying cargo (vesicles or organelles) along actin filaments. Unlike the microtubule-associated motors dynein and kinesin, which can make long journeys, myosin V can only make short excursions along an actin filament before falling off. Like microtubules, microfilaments have a defined orientation with plus and minus ends, and most myosins have their own direction of transport. For example, whereas myosin V moves "forward" toward the plus end of actin filaments, myosin VI moves in the reverse direction toward the minus end.

FUNCTIONS OF MICROFILAMENTS

Microfilaments made of polymerized actin play multiple indispensable roles in cells. As discussed above, they serve as tracks along which intracellular transport takes place and, as a major component of the cytoskeleton, contribute structural rigidity to the cell. As we will see in the next chapter, actin plays essential roles during cell division, first causing the dividing cell to become rounded in shape, and then serving with myosin as the force-generator that divides the cell into two new daughter cells. Here we will briefly consider two other well-studied cellular processes in which actin plays a central role; muscle contraction and cell movement.

Muscle Contraction

Striated muscle, the type of muscle found attached to bones or in our heart, gains its appearance because passing along the cell one encounters in turn regions of parallel microfilaments (which are thin and relatively transparent), then regions of thick filaments of a second protein called **myosin,** which is a microfilament-motor protein (Figure 13.13). The complete repeating unit is called a **sarcomere** and is delineated by the **Z disc,** which holds the microfilaments in a regular pattern, so that striated muscle has an almost crystalline appearance in transverse section (Figure 1.4 on page 8). The myosin motor has a distinctive structure, consisting of a tail and two globular heads. The thick filament is formed from a large number of myosin molecules arranged in a tail-to-tail manner. This layout means that the thick filament is a bipolar structure with myosin heads at both ends. Neuronal signals cause cytosolic calcium concentration to increase within the muscle fiber. Calcium subsequently binds to the protein **troponin,** which in turn changes shape to allow myosin access to the actin. The myosin motor then moves along the actin microfilaments, powered by the hydrolysis of ATP. Because

---

Medical Relevance 13.1　Some Bacteria Highjack the Cytoskeleton for Their Own Purposes

We are all familiar with the ways that bacteria and viruses spread from person to person by various forms of contact, but how does a bacterium spread through the cells of a human body? A number of bacteria, including the important pathogens *Listeria* (responsible for sepsis and meningitis in immunocompromised patients and for infections of the fetus during pregnancy) and *Shigella* (which causes dysentery), avoid contact with the human body's antibodies and white blood cells by remaining hidden within the cytoplasm of our cells as they spread through our tissues. The bacteria use actin to power their journey from cell to cell. *Listeria* has a protein called ActA at one end of its rod-shaped body. ActA activates the Arp2/3 complex to promote the formation of actin filaments. The force generated by the extending microfilaments pushes the bacterium through the cytoplasm in the opposite direction. Movement is random, but occasionally the bacterium will bump into the plasma membrane causing the formation of a finger-like projection from the cell surface. The membrane of the projection fuses with the membrane of the neighboring cell, transferring the bacterium in a membrane sac from which it quickly escapes to repeat the procedure.

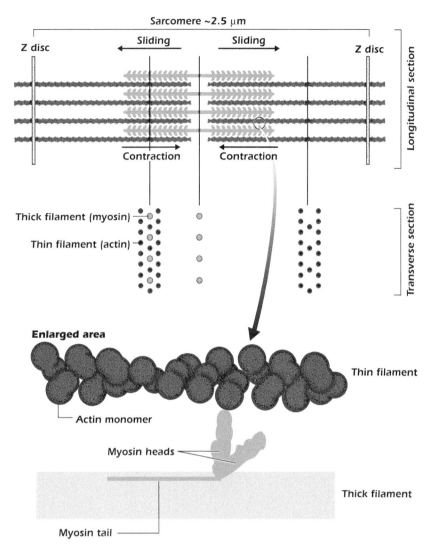

Figure 13.13. Muscle contraction.

of the layout of the myosin and actin, the two Z discs are pulled toward each other and the cell shortens, causing muscle contraction.

Microfilament-Based Cell Migration

Cell movement or **migration** is essential for development and for maintaining healthy tissues. For example cell migration is crucial in early development for establishing tissue lineages, and migration of white blood cells is critical for immune responses. Cell migration is also involved in pathological cell behavior, such as in cancer. Cell movement can occur by a number of distinct mechanisms. Among the best studied is so-called **mesenchymal cell migration,** in which dynamic remodeling of the actin cytoskeleton is key (Figure 13.14). This type of cell movement, which can be observed in two dimensions in the laboratory as cells moving along glass coverslips, depends upon the cell becoming polarized with a leading edge characterized by the GTPases Rac and CDC42. Rapid actin polymerization occurs at the leading edge, which enables the cell to generate protrusions

called **lamellipodia,** which reach forward in the direction of travel. The cell then makes strong attachment points to the extracellular matrix called **focal adhesions** made of integrins. Following maturation of the focal adhesions they link to thick actinomyosin **stress fibers**. Stress fibers are bundles of actin that form bands along much of the length of the cell, parallel to the direction of movement. Actinomyosin is then responsible for generating contraction forces along the stress fibers, which push the cytoplasmic contents of the cell forwards. Finally, as the cell contents move forward toward the leading edge, focal adhesions at the rear of the cell detach. In this manner, the cell trundles forward like a tank on caterpillar tracks. A second form of cell motility is also actin dependent. This form of cell motility, called **amoeboid migration,** occurs for cells that lack stress fibers and make less firm attachment to the extracellular matrix. In amoeboid migration, actin contraction causes the extension of cellular blebs that allow the cell to move. Notably, cells can switch back and forth between these two distinct modes of actin-dependent motility depending upon the extracellular environment.

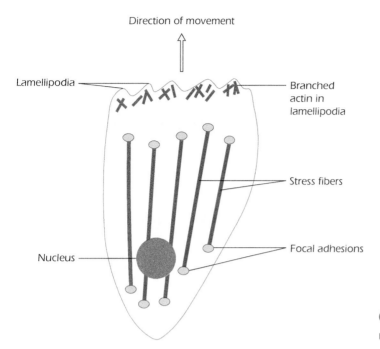

Figure 13.14. Mesenchymal cell migration depends upon actin microfilaments and focal adhesions.

INTERMEDIATE FILAMENTS

Intermediate filaments were so named because their diameter, 10 nm, lies between that of the thin and thick muscle filaments. They are the most stable (least dynamic) of the cytoskeletal filament systems. Although intermediate filaments from different mammalian tissues look much the same in the electron microscope, they are in fact composed of different protein monomers (and can therefore be distinguished by immunofluorescence, page 14). We have already met the **lamins,** which form filaments supporting the nuclear envelope (page 22). Neurons contain **neurofilaments,** muscle cells contain **desmin** filaments, fibroblasts contain **vimentin** filaments, and epithelial cells contain filaments composed of

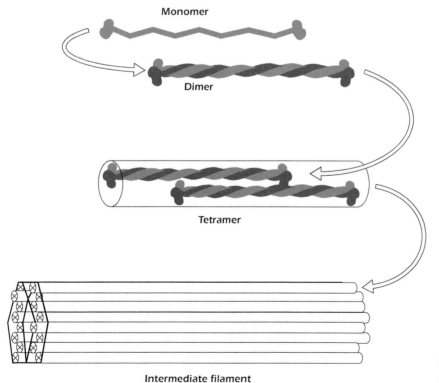

Figure 13.15. Intermediate filaments are formed from rod-shaped monomers.

IN DEPTH 13.1   CYTOSKELETAL PROTEINS IN PROKARYOTES

Until about 30 years ago, the absence of a cytoskeleton was one of the distinctive features of the prokaryotic cell (Table 1.1). This changed with the discovery in bacteria and archaea of a protein called FtsZ, which is a member of the tubulin superfamily. Although the primary amino acid sequences of the two proteins show weak homology, their crystal structures are virtually identical. FtsZ mutants are defective in cell division and the FtsZ protein forms a ring at the division site that constricts as cell fission proceeds. Although FtsZ does not form microtubules it does form filaments that resemble the protofilaments that make up the microtubule wall.

Like London buses, you wait for ages for one prokaryotic tubulin to appear, then two arrive together. TubZ forms a spindle-like structure as the cell divides. This has no role in the separation of the two daughter nucleoids but it does appear to be involved in the segregation of certain plasmids. Closely following the discovery of these prokaryotic tubulins, MreB was shown to be a bacterial actin. MreB is found in all non-coccoid bacteria and is involved in maintaining the rod-like shape. To complete the set, CreS, also known as Crescentin, forms coiled-coil filaments reminiscent of intermediate filaments. In the banana-shaped cells of *Caulobacter*, loss of function *cres* mutants become straight.

What is clear is that the cytoskeletal proteins evolved from prokaryotic ancestors that had already adapted to a wide range of cellular functions.

---

keratin, the protein that gives our skin its protective coating, forms our hair and fingernails, and makes the horns and hooves of our domestic animals and pets. These different proteins can generate a similar structure because all share the same basic design (Figure 13.15). This consists of a central α-helical rod connecting a head and tail. Most of the variation between intermediate filament proteins is in the head and tail, and these regions probably confer subtly different properties on different intermediate filament classes. The basic building block of intermediate filaments is a tetramer of pairs of subunit proteins joined by their central region to form coiled coils. Intermediate filaments tend to form wavy bundles that extend from the nucleus to the cell surface (Figure 13.1).

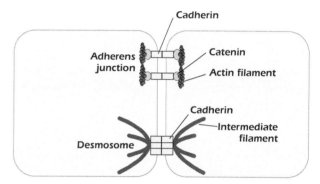

Figure 13.16. Anchoring junctions attach the cytoskeletons of adjacent cells.

FUNCTIONS OF INTERMEDIATE FILAMENTS

Anchoring Cell Junctions

The cells that form tissues in multicellular organisms are attached together via anchoring junctions (Figure 13.16). Integral membrane proteins called **cell adhesion molecules,** of which **cadherin** is an example, extend out from each cell and bind tightly together, while their cytosolic domains attach to the cytoskeleton. There are two basic types of anchoring junction. In **adherens junctions** the cell adhesion molecules are linked to actin microfilaments by linking proteins such as α and β catenin. In **desmosomes and hemidesmosomes,** however, the cell adhesion molecules are linked to intermediate filaments. Whereas desmosomes link cells to other cells, hemidesmosomes link cells to the extracellular matrix. Tissues that need to be mechanically strong, such as the epithelial cells of the gut (page 9) and cardiac muscle, have many anchoring junctions linking the cytoskeletons of the individual cells to each other and the

Answer to thought question: We have described how the activation of frizzled receptors by Wnt proteins promotes cell growth and division through β catenin acting to activate transcription of particular genes. β catenin locked up in adherens junctions is not available for this alternative role of promoting cell division. However, if many cells in a tissue are killed, the remainder will not be able to find partners with which to form adherens junctions, freeing the β catenin to promote cell division. Thus the dual role of catenin is part of the mechanism that allows the cells of an animal to divide when more cells are needed, but to stop dividing when there is no more space to be filled. Another aspect of this "contact inhibition" is described on page 241.

IN DEPTH 13.2   DYSTROPHIN: A STRONG MULTILINKER

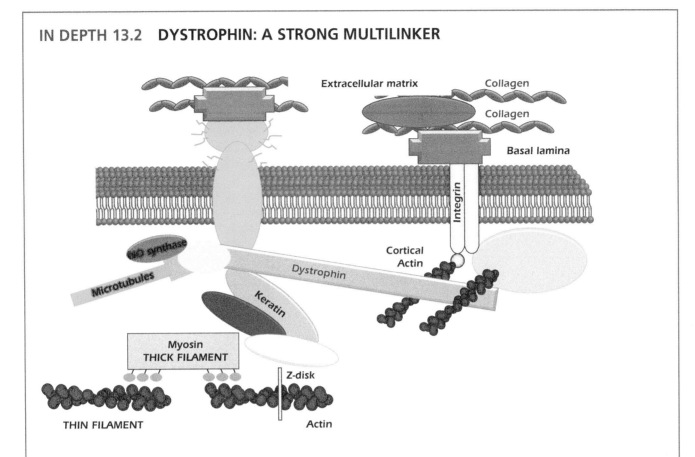

Dystrophin is an enormous protein that is encoded by the largest human gene. It is a strong structural protein that maintains the physical integrity of skeletal muscle cells as they contract and relax, but it also has binding sites for many other components of the cell so that each is placed in the correct location. The diagram above shows how dystrophin connects directly or through intermediates to all components of the cytoskeleton, to the extracellular matrix, and to signaling pathways via NO synthase. All the components shown are known proteins, but we label only those described elsewhere in this book. The loss of dystrophin in muscular dystrophy not only renders skeletal muscle cells physically weak but leaves the other components of the cell disorganized. In Medical Relevance 6.1 on page 94 we describe clinical strategies to alleviate the disease.

matrix. Anchoring junctions are one of the three types of cell junctions, the others being tight junctions (page 26) and gap junctions (page 26).

The Nuclear Lamina

The special class of intermediate filaments that form the nuclear lamina (Page 22) are called lamins. These are found only in animal cells and they have multiple roles in chromosome organization and gene expression. All animal cells have lamins but not all have cytoplasmic intermediate filaments. In mammalian cells there are three lamin proteins, A, B, and C. A and C are splice variants of the *LMNA* gene, whilst B is encoded by *LMNB*. The three lamins are structurally similar but have different patterns of expression; B is present in all cells, whilst A and C are only found in differentiated cells. In invertebrate cells that express only a single lamin it is always B.

Mutations in the lamin genes are responsible for a number of dystrophic human conditions, called laminopathies. Strikingly, alterations in *LMNA* give rise to a number of forms of progeria, or premature aging. This points to a role for lamins in the normal aging process.

Medical Relevance 13.2 Protected by the Dead

The epidermis of the skin is made up of a layer of living cells called keratinocytes covered by a protective layer of their dead bodies. Dead keratinocytes form a good protective layer because while alive they generate a dense internal cytoskeleton of the intermediate filament keratin, with adjacent cells being linked by desmosomes. When the cells die, the keratin fibers remain because intermediate filaments are stable. Since the intermediate filaments were joined by desmosomes, the resulting protective fibers do not stop at the edge of the now-dead cell, but are strongly connected with the fibers in the next cell, and the next, and so on, forming an extremely strong network of fibers.

Keratin mutants, collectively referred to as keratinopathies, predominantly give rise to blistering diseases of the skin. Mutations in the major skin keratins (K5 and K14), for example, give rise to epidermolysis bullosa simplex. In mild cases blistering is confined to the hands and feet, although in more severe forms it is widespread. Keratinocytes cultured from these patients show disorganized keratin filaments. This results in the epidermis becoming fragile and easily damaged.

SUMMARY

1. The cytoskeleton is made up of microtubules, microfilaments, and intermediate filaments.

2. Microtubules are composed of α and β tubulin. In animal cells the minus ends of microtubules are stabilized and anchored at the centrosome, while the plus ends are typically free to grow and shrink.

3. Cilia and flagella contain a 9 + 2 axoneme of microtubules plus the motor protein dynein.

4. Dynein is present elsewhere in cells where it transports cargo along microtubules in a retrograde (toward the minus end) manner. Cargo is moved along microtubules in the opposite (anterograde) direction by kinesins.

5. Microfilaments are composed of actin.

6. Actin-binding proteins control actin polymerization and the organization of actin filaments into networks.

7. The motor protein myosin is particularly prominent in muscle cells. When cytosolic calcium concentration rises myosin operates on actin microfilaments to drive muscle contraction.

8. Cell locomotion is driven by spatially distinct zones of actin polymerization and depolymerization, aided by myosin.

9. Intermediate filaments are more stable than microtubules and microfilaments and serve structural roles.

10. Anchoring cell junctions attach the cytoskeletons of adjacent cells together. They are divided into adherens junctions, which link to actin, and desmosomes/hemidesmosomes, which link to intermediate filaments.

FURTHER READING

Brouhard, G.J. and Rice, L.M. (2018). Microtubule dynamics: an interplay of biochemistry and mechanics. *Nature Reviews. Molecular Cell Biology* 19 (7): 451–463.

Lane, E.B. and McLean, W.H.I. (2004). Keratins and skin disorders. *Journal of Pathology* 204: 355–366.

Pollard, T.D. (2016). Actin and actin-binding proteins. *Cold Spring Harbor Perspectives in Biology* 8 (8): a018226.

Pollard, T.D. and Goldman, R.D. (2018). Overview of the cytoskeleton from an evolutionary perspective cold spring. *Harbour Perspectives in Biology* 10 (7): a030288.

Seetharaman, S. and Etienne-Manneville, S. (2020). Cytoskeletal crosstalk in cell migration. *Trends in Cell Biology* 30 (9): 720–734.

Svitkina, T.M. (2020). Actin cell cortex: structure and molecular organization. *Trends in Cell Biology* 30 (7): 556–565.

⬤ REVIEW QUESTIONS

13.1 Theme: Cytoskeletal structures

A   actin
B   intermediate filament proteins
C   tubulin

For each of the structures listed below, choose from the list above the relevant cytoskeletal protein that predominantly forms or supports the structure.

1. cilia
2. fingernails
3. flagella
4. microfilaments
5. microtubules
6. microvilli
7. stress fibers

13.2 Theme: Proteins of the cytoskeleton

A   actin
B   β tubulin
C   γ tubulin
D   keratin
E   kinesin
F   myosin
G   taxol

From the above list, select the protein corresponding to each of the descriptions below.

1. a building block of intermediate filaments
2. a building block of microfilaments
3. a building block of microtubules
4. a molecular motor that acts on microfilaments
5. a molecular motor that acts on microtubules
6. a protein that is concentrated at the centrosome

13.3 Theme: Fueling movement

A   actin
B   dynamin
C   dynein
D   troponin
E   kinesin
F   myosin
G   tubulin

From the above list of proteins, select the one described by each of the descriptions below.

1. the transport of Golgi components in toward the cell center is powered by ATP hydrolysis performed by this ATPase
2. the contraction of muscle is powered by ATP hydrolysis performed by this ATPase
3. the rowing motion of a cilium is powered by ATP hydrolysis performed by this ATPase
4. the transport of vesicles and organelles from the cell body to the tips of neuronal axons is powered by ATP hydrolysis performed by a member of this ATPase family
5. the wriggling motion of sperm tails is powered by ATP hydrolysis performed by this ATPase

⬤ THOUGHT QUESTION

We have already described how β catenin is a transcription factor in the Wnt pathway (page 188). Here it has a completely different function, as part of a cell–cell junction.

What might be the advantage to the organism of this dual role?

14

CONTROLLING CELL NUMBER

New cells arise by the division of an existing cell. The life of a cell from the time it is generated by the division of its parent cell to the time it, in turn, divides to generate two new daughter cells is called the cell division cycle or just the **cell cycle** for short. The duration of the cell cycle varies from only a few minutes in the rapidly dividing cells of a sea urchin embryo, to two to three hours in a single-celled organism like the budding yeast *Saccharomyces cerevisiae,* to around 24 hours in a human cell grown in a culture dish. During this period the cell duplicates its genome and then partitions it between two new daughter cells. In humans the cells of some tissues, such as the skin, the lining of the gut (see Chapter 1), and the bone marrow continue to divide throughout life. Others, such as the light-sensitive cells of the eye and skeletal muscle cells do not divide. Not only must the cell cycle be precisely controlled in terms of which cells divide when, but also the whole process has to know when to stop. A human is bigger than a mouse and smaller than an elephant. This is largely because humans are made up of more cells than a mouse and less than an elephant. How are these differences achieved? Why do humans not grow to be the size of elephants or whales? Perhaps more importantly, what happens if the cell cycle makes mistakes or gets out of control? As we will see in this chapter, tight control of the cell cycle is essential, as loss of control leads to rampant cell division that causes cancer.

The cell cycle is divided into phases (Figure 14.1). DNA replication occurs in **S-phase** (S for Synthesis), and the newly replicated chromosomes are dispatched into two newly forming daughter cells in **M-phase** (Mitosis). S-phase and M-phase are separated by phases termed G1 and G2 (Gap phases). G1, S, and G2 can be grouped together as **interphase,** and together often occupy as much as 90% of the time of the cell cycle, whilst M-phase is relatively short. Cells that have become quiescent and are no longer proliferating are said to enter G0. G0 cells can remain healthy for months or even years, and most of the cells in the human body are in fact in this nondividing state, and cells can only enter G0 from G1. Cells are said to be terminally differentiated if they are unable to return to the cell-division cycle. Neurons are one such example. In contrast, other differentiated cells such as glial cells (page 10) can recommence the cell cycle if they receive the correct signals from their neighbors. It is imperative that each phase of the cell cycle occurs in the correct order and without error since, as we will see in this chapter, mistakes can lead to cancer and developmental defects.

Since M-phase is the pivotal moment of the cell cycle when new cells are formed, effectively ending one cell cycle and beginning the next, we will begin our discussion of the cell cycle there. We then take a broader look at how the progression of the entire cell cycle is regulated. In each of

Cell Biology: A Short Course, Fourth Edition. Stephen Bolsover, Andrea Townsend-Nicholson, Greg FitzHarris, Elizabeth Shephard, Jeremy Hyams and Sandip Patel.
© 2022 John Wiley & Sons Ltd. Published 2022 by John Wiley & Sons Ltd.
Companion website: www.wiley.com/go/bolsover/cellbiology4

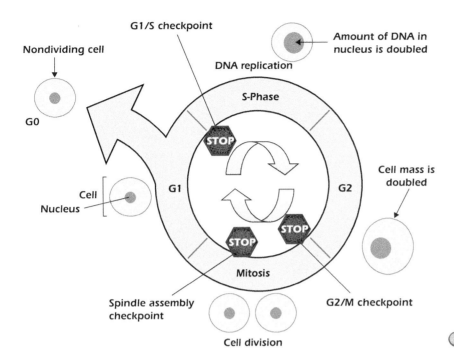

Figure 14.1. The cell division cycle.

IN DEPTH 14.1   STEM CELLS

As described in Chapter 1, stem cells are found throughout the body and divide to replace differentiated cells that die or are lost. The defining feature of a stem cell is that one or more cell divisions can generate both new, functionally identical stem cells and differentiated, specialized offspring, such as epithelial cells in the case of intestinal crypts (page 11). Stem cells often reside in specialized niches where they are encouraged to divide by factors such as Wnts (page 188) secreted by surrounding cells. Different stem cell populations differ in the range of specialized cells that they can generate. Intestinal stem cells will only differentiate into intestinal epithelial cells. In contrast the fertilized egg is **totipotent:** its daughter cells can contribute to not only all the tissues of the fetus but also to support structures including the placenta.

The ability of stem cells to replace damaged cells in the adult body has enormous medical potential. Initially research concentrated on embryonic stem cells which are **pluripotent,** that is, they can form a wide range of specialized cells. Techniques have been developed to induce differentiation into particular cell types or tissues. Use of human embryonic stem cells was always controversial, and now for both ethical and medical reasons has been overtaken by the ability to induce stem cell properties in differentiated adult cells which, if isolated from a patient, create no problems of immune rejection. In 2006 Shinya Yamanaka discovered that if only four transcription factors (Oct4, Sox2, Klf4, and c-Myc) were made active in differentiated cells a very small fraction – less than 1% – became induced pluripotent stem (**iPS**) cells with capabilities similar to cells isolated from early embryos.

As well as raising many exciting possibilities in medicine, this technique has become widespread in research since it allows generation in the laboratory of human cells that closely model adult differentiated cells. For example, mutations in a kinase called LRRK2 are known to predispose individuals to Parkinson's disease, which is characterized by the death of brain neurons. One could not take samples of brain neurons from patients, but one can take skin cells, induce them to become iPS cells, maintain them indefinitely in culture, and differentiate them into models of mature brain neurons for study.

these two subsections we will consider the implications of a failure to correctly regulate the cell cycle, and how this can lead to cancer. Next we will look at the highly specific situation of how the cell cycle is adapted to allow the egg and sperm to each provide only half of a genome at the time of fertilization, and again discuss how errors in this process can threaten the development of the newly formed embryo. Finally we will discuss the flip-side of how cell numbers are controlled, namely the process of how controlled cell death, termed apoptosis, serves to limit the overproliferation of cells within a tissue.

M-PHASE

M-phase is the critical moment of the cell cycle when chromosomes are dispatched into two newly forming daughter cells. This process comprises two separate series of events, **mitosis** and **cytokinesis.** Mitosis describes the process by which the chromosomes are organized and dispatched to the two new cells, whereas cytokinesis describes the physical partitioning of the cytoplasm into two to generate new daughter cells around the two sets of chromosomes. These two processes occur approximately simultaneously in most proliferating cells. However, they are very separate processes, as illustrated during early fly development where the absence of cytokinesis despite the continuation of the cell cycle causes thousands of nuclei to inhabit a single cytoplasm as a normal part of early development in that organism.

Mitosis

In mitosis, the chromosomes carry out a precisely choreographed sequence of movements that were first described well over a century ago. Classically, mitosis is divided into five stages, each of which is characterized by changes in the appearance of the chromosomes and their organization with respect to a new cellular structure called the mitotic **spindle** that is responsible for their segregation. The stages of mitosis are shown diagrammatically in Figure 14.2, and are as follows.

Prophase. The first evidence of mitosis in most cells is the compaction of the threads of chromatin that reside within the nucleus in interphase into chromosomes that are visible in the light microscope (upper cell in Figure 14.3). As the chromosomes condense, each can be seen to be composed of two sister **chromatids,** the result of each chromosome having been replicated in interphase (page 52). Chromosome condensation reduces the chance of long DNA molecules becoming tangled and broken. Each chromosome has a protein structure called the **kinetochore,** a structure that forms around a special region of the chromosome rich in satellite DNA (page 61) called the **centromere.** The kinetochore is the point of attachment of the chromosome to the spindle. At the same time as the chromosomes are condensing within the nucleus, two centrosomes, formed by replication of the single centrosome (page 220) during S phase, begin to separate to establish the mitotic spindle.

Prometaphase. At the breakdown of the nuclear envelope, the chromosomes become free to interact with the forming spindle. The spindle is a complex structure made from microtubules, described in the previous chapter. The newly separated centrosomes each form a focal point from which microtubules can grow. Microtubule assembly from the centrosomes is random and dynamic. As prometaphase progresses the growing plus ends of individual microtubules make chance contact with, and are captured by, the kinetochores. Because of the random nature of these events, the kinetochores of chromatid pairs are initially associated with different numbers of microtubules, and the forces acting upon each chromosome are unbalanced. Initially, therefore, the spindle is highly unstable and chromosomes make frequent excursions toward and away from the poles. Gradually, as microtubules cyclically make contact with and release kinetochores, chromosomes become aligned at the equator of the spindle, with the kinetochores of each member of a chromatid pair oriented toward opposite poles.

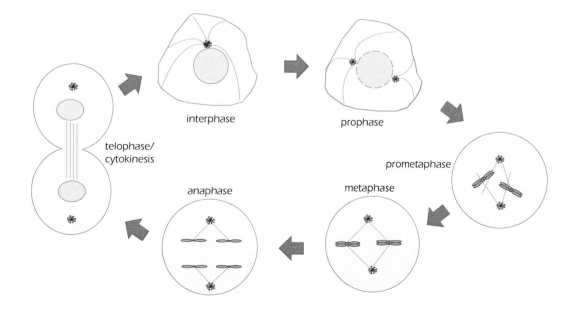

Figure 14.2. The stages of mitosis.

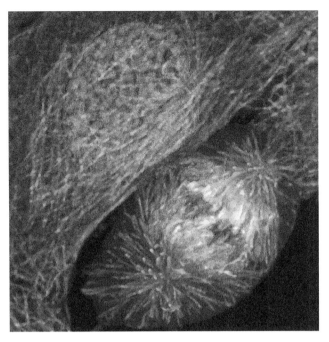

Figure 14.3. Mitosis in cultured breast cancer cells. The upper cell is in prophase. The chromosomes (shown orange) have compacted and are visible as independent structures. The microtubule cytoskeleton (shown green) has not yet reorganized into the mitotic spindle, and microtubules have not invaded the nucleus because the nuclear envelope (not visible in this image) is still intact. The lower cell is in anaphase. The microtubule cytoskeleton has been reorganized into the mitotic spindle. The two sets of chromosomes are being pulled to opposite spindle poles. Source: Image by Professor David Becker, University College London; used with permission.

Metaphase. Metaphase describes the period of time when chromosomes are fully aligned at the spindle equator ready to be separated. In many cells this period is fleeting, with the chromosomes being separated almost immediately after they are correctly aligned.

Anaphase. The trigger for the separation of the paired chromatids and the start of their journey to the spindle poles is the destruction of a protein complex called **cohesin,** which acts as the glue holding the pairs of chromatids together (Figure 14.4c). In **anaphase A** the microtubules holding the chromosomes shorten, pulling the chromosomes to the spindle poles (lower cell in Figure 14.3). By contrast, in **anaphase B** the microtubules that overlap at the spindle

equator lengthen and slide over one another, extending the distance between the poles.

Telophase. This stage sees the reversal of many of the events of prophase; the chromosomes decondense, the spindle disassembles, the nuclear envelope reforms, the Golgi apparatus and endoplasmic reticulum re-form and the nucleolus reappears. Each daughter nucleus now contains one complete copy of the parental cell genome.

Cytokinesis

Whereas the specific organization of microtubules into a highly specialized structure (the spindle) is a critical event in mitosis, it is a highly specialized reorganization of the actin cytoskeleton that causes the cell to be divided in two during **cytokinesis.** In animal cells, a **cleavage furrow** made of actin and its motor protein myosin II (page 226) constricts the middle of the cell. The positioning of the cleavage furrow is crucial and signals from both the center region of the spindle and the spindle poles ensure that it forms at the correct place on the cell cortex, to ensure constriction between the two newly forming daughter nuclei. Like mitosis, cytokinesis can be divided into specific stages, illustrated in Figure 14.5.

Positioning of the mitotic spindle leads to signaling between the spindle and the cortex that helps determine the position of the cleavage furrow. Specifically, signaling from the spindle midzone and spindle poles leads to the formation of a narrow band of active RhoA GTPase at the cell cortex, at the location of the future cleavage furrow. Active (GTP-bound) RhoA acts on effectors to locally promote actin polymerization and myosin II activation.

Furrow ingression is the process by which a constriction divides the cytoplasm into two around the two sets of chromosomes. The force necessary for this is generated by contraction of an actin/myosin ring in a process that is sometimes likened to the closing of a purse-string. At the end of furrow ingression the microtubules that previously formed the spindle remain and along with associated proteins form a relatively stable structure called the **midbody.**

Abscission describes the final cutting of the plasma membrane to form two new cells. This typically occurs ~1 hour after furrow ingression in mammals. It happens in an asymmetric manner such that the midbody remnant is inherited by only one of the two daughter cells. This subtle

Example 14.1   Taxol Stops Mitosis

During cell division, cytoskeletal microtubules break down and the tubulin monomers re-form as the mitotic spindle. Taxol stabilizes existing microtubules, making it impossible for the cell to form spindles and therefore preventing cell division. For this reason taxol, usually referred to by its generic name paclitaxel, is a valuable anti-cancer drug.

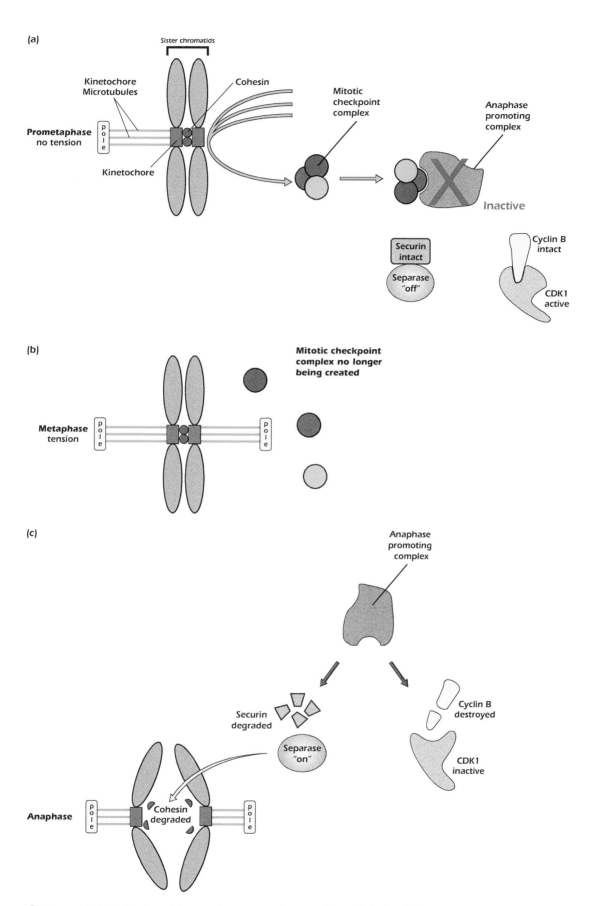

Figure 14.4. Activation of the anaphase-promoting complex and the breakdown of cohesin allows cells to pass the spindle assembly checkpoint.

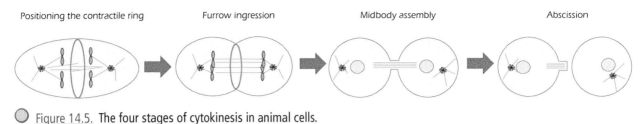

Positioning the contractile ring   Furrow ingression   Midbody assembly   Abscission

 Figure 14.5. The four stages of cytokinesis in animal cells.

Example 14.2   Counting Chromosomes

In interphase chromosomes are decondensed within the nucleus, and intertwined with each other, and are thus impossible to count using a microscope. However, at mitosis, when they condense, individual chromosomes can be seen relatively easily using a standard light microscope (Figure 3.6 on page 42). For many years so-called mitotic "chromosome spreads" have been used to count chromosomes to look for conditions associated with gains and losses of chromosomes such as trisomy 21 which causes Down's syndrome. Nowadays genome sequencing approaches such as next generation sequencing are considered more accurate and, importantly, do not require the chromosomes to be in mitosis to count them.

---

difference between the two daughter cells has been proposed to be a mechanism allowing daughter cells to have different behaviors and fates in some situations.

CONTROL OF THE CELL CYCLE

The Cell Cycle Is Driven by Kinase Activities

In dividing cells the orderly progression through the different phases of the cell cycle is ensured by a unique class of cell cycle enzymes, the **cyclin-dependent protein kinases** or **CDKs (CDK1, CDK2, etc.).** The amounts of each CDK remain largely constant, but each is active only when associated with regulatory subunits whose concentrations increase and decrease in phase with the cell cycle, hence their name, the **cyclins (cyclinA, cyclin B, etc.).** In simple eukaryotes such as the fission yeast, *Schizosaccharomyces pombe,* a single CDK-cyclin combination, **CDK1-cyclin B,** drives both S-phase and mitosis. At the G1/S transition the activity of the enzyme is weak but is sufficient to modify a small number of proteins at origins of DNA replication and initiate DNA synthesis. At G2/M enzyme activity is strong and a larger repertoire of substrates is phosphorylated. These induce chromosome condensation and spindle formation for mitosis together with the fragmentation of the membranes of the Golgi apparatus and endoplasmic reticulum to shut off endocytosis and secretion whilst mitosis proceeds. In multicellular organisms including humans, where cells must respond to a greater variety of extracellular and intracellular signals than yeast, there are multiple waves of CDK activity (Figure 14.6). **CDK4-cyclin D** and **CDK6-cyclin D** control early events in G1 whilst

CDK2-cyclin E controls G1/S itself. A wave of **CDK1-cyclin A** and **CDK2-cyclin A** ensures passage through S-phase and G2, whilst **CDK1-cyclin B** controls G2/M just as it does in yeast.

In addition to requiring a partner cyclin, CDK1 activity is modulated by phosphorylation. To illustrate this we will look in more detail at how CDK1 is regulated at G2/M (Figure 14.7). To ensure that CDK1 remains inactive through interphase, even after cyclin binding, it is itself phosphorylated by another protein kinase called Wee1 (named because yeast mutants, with unphosphorylated CDK1, divide too early, while they are still "wee" – Scottish for small!). Wee1 adds two phosphates to CDK1, one on a threonine residue at amino acid number 14, the other on a tyrosine residue at amino acid number 15. These two amino acids lie within the ATP-binding site of CDK1, and the presence of two phosphates prevents ATP binding, the first step of the phosphorylation reaction. The phosphates are removed at G2/M by a second enzyme, the protein phosphatase **Cdc25.** This leads to a burst of CDK1 activity and entry to M-phase. To complete mitosis it is then necessary to inactivate CDK1. This is achieved by the destruction of cyclin B and the rephosphorylation of CDKl by Wee1.

Once activated, CDK-cyclin complexes exert their effects by phosphorylating downstream proteins to initiate events that drive cells through the appropriate cell-cycle transition. In the case of M-phase entry, CDK1 activation leads to phosphorylation of nuclear lamina proteins to promote nuclear envelope breakdown, and CDK1 acts on multiple downstream pathways to cause cells to lose adherence to neighboring cells and become rounded in readiness for cell division. In the case of S-phase entry,

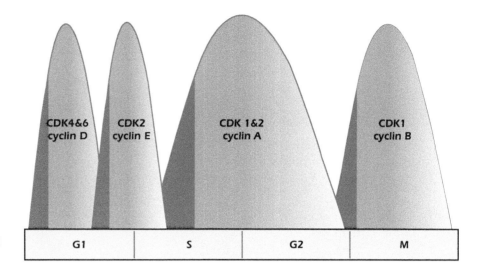

Figure 14.6. CDK activities through the cell cycle.

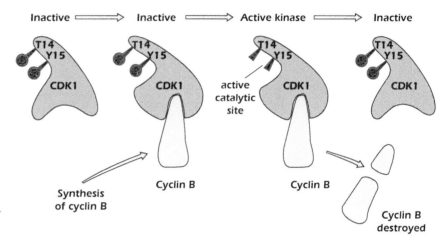

Figure 14.7. How CDK1 is controlled by cyclin B and phosphorylation in multicellular organisms.

the key target of the CDK4-cyclin D and CDK6-cyclin D enzymes is the protein **Rb** (Figure 14.8). Rb interacts with **E2F,** a transcription factor that is required to produce the enzymes required for DNA replication during S-phase. Among them is DNA polymerase α, the main polymerase involved in DNA replication in eukaryotes (page 52). By interacting with E2F Rb prevents this function. Phosphorylation of Rb by CDK4-cyclin D and CDK6-cyclin D causes Rb to release E2F and allow it to get on with its job. Mutation of Rb can generate forms that are permanently phosphorylated and hence cannot bind to and thus regulate E2F. When this happens, normal controls on DNA replication are removed and cells become cancerous (the role of the cell cycle in cancer is discussed further below).

Growth factors (page 184) promote cell division in target cells. One mechanism for this is the activation of MAP kinase which in turn stimulates the production of cyclin D

(page 238); cyclin D can then pair with and activate CDK4 and CDK6.

Checkpoints Tell the Cell Cycle When to Stop and When to Go

The cell cycle has built-in safety devices to ensure that each cell division results in a perfect copy of the parental cell's genome being passed on to its progeny. These were termed "checkpoints" by Leland Hartwell who, along with Paul Nurse and Tim Hunt, shared the 2001 Nobel Prize in Physiology or Medicine for their pioneering work on the cell cycle. A checkpoint pathway detects errors and passes a message down the line to stop or "arrest" the cell cycle until the problem has been attended to. Among the best-characterized cell-cycle checkpoints are those that respond to DNA damage or the failure to complete DNA replication. Both situations result in the activation of the protein kinase

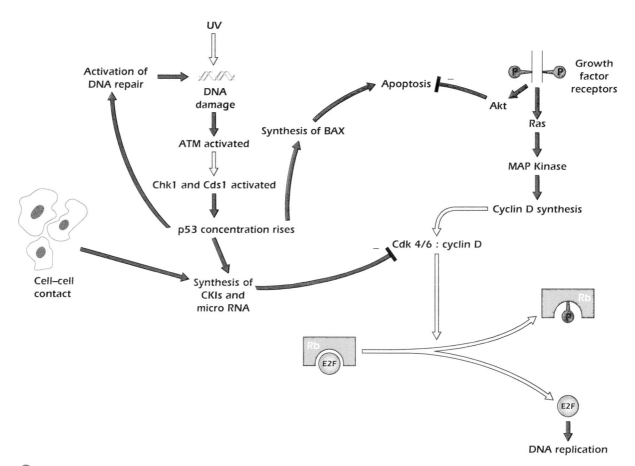

Figure 14.8. Retinoblastoma protein Rb sequesters E2F, the critical transcription factor controlling entry into S phase.

ATM. In response to DNA damage, for example by ultraviolet light (page 57), ATM activates two more protein kinases, Chk1 and Cds1 (Figure 14.9). In the case of incomplete DNA replication only Chk1 is activated. When active, these kinases prevent cells passing both the G2/M and the G1/S control points. To prevent entry into mitosis, Chk1 and Cds1 phosphorylate Cdc25. The phosphorylated form of Cdc25 is inactive, and without active Cdc25 cells are unable to activate CDK1-cyclin B and enter M-phase. This buys time for the cell to repair the damage to its DNA or to complete S-phase. To prevent entry into S-phase, Chk1 and Cds1 phosphorylate the transcription factor **p53** (Figure 14.8). p53 is produced all the time in cells but is broken down rapidly,

so its concentration is usually low. However, phosphorylated p53 is not destroyed so its concentration increases. In turn, the systems that repair DNA are activated (page 57) and genes whose products block progression through the cell cycle are upregulated. We have described one in Chapter 6: p53 upregulates expression of microRNA34a which then operates at the level of translation to block the production of Cdk4, Cdk6, and Cdc25 (page 238). p53 also upregulates the expression of p21[CIP1], a member of a class of proteins called **CKIs** for **cyclin-dependent kinase inhibitors.** These bind to and inactivate CDKs; in the case of p21 the target is CDK4-cyclin D. p21 binding therefore arrests the cell in G1, preventing it from entering another round of S-phase while

Medical Relevance 14.1   Retinoblastoma

Retinoblastoma is a cancer of the retina of the eye that generally occurs in children below the age of 5. It is usually caused by two independent somatic mutations in a photoreceptor cell lineage within one eye, each of which damages or destroys an *Rb* gene inherited from one parent. *Rb* encodes the Rb protein that sequesters E2F.

A rarer form, familial retinoblastoma, occurs in children who inherit one defective *Rb* gene. Every cell in the body therefore has only one working *Rb* gene, and a single somatic mutation that destroys this gene can generate a tumor. Such children also have an increased risk of other, non-optical cancers.

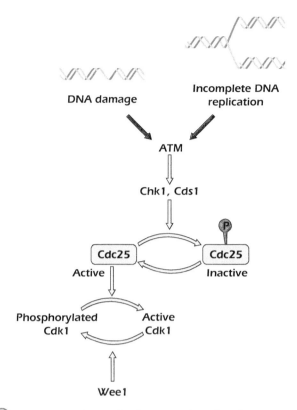

Figure 14.9. ATM activation stops the cell cycle.

the previous one is incomplete (Figure 14.8). There are many CKIs mediating a number of pathways all with the same outcome: stopping cell division.

One important example of the function of CKIs is in mediating an important phenomenon known as **contact inhibition** (Figure 14.8). Cells in a culture dish, or at the edge of a wound, divide until they touch each other. When they contact neighboring cells, they stop dividing because contact causes the production of two CKIs called p16[INK4a] and p27[KIP1]. These inhibit the G1 Cdks and therefore prevent DNA synthesis. Loss of contact inhibition is one of the first changes seen in the transformation of normal cells into cancer cells.

The Mitotic Checkpoint Determines When the Cell Cycle Ends

As described earlier, mitosis is the key moment in the cell cycle when chromosomes are segregated between the newly forming daughter cells, and errors would lead to the new daughter cells inheriting the wrong number of chromosomes, a hazardous situation termed aneuploidy. To avoid this, mitosis possesses its own specialized cell-cycle checkpoint, called the **spindle assembly checkpoint** or just the **mitotic checkpoint,** that ensures that anaphase is not initiated and chromosomes not segregated until all chromosomes are correctly aligned on the spindle. Figure 14.4 shows how this is achieved. During mitosis a group of proteins called

Spindle Assembly Checkpoint proteins bind to kinetochores and cause the formation of a diffusible protein complex called the **mitotic checkpoint complex (MCC).** The role of the MCC is to inhibit a second protein complex called the **anaphase promoting complex (APC/C)** which, as its name suggests, is responsible for initiating anaphase. When all the kinetochores are connected to the mitotic spindle the spindle assembly checkpoint proteins dissociate from the kinetochore, such that MCC is no longer produced, and so the APC/C is activated.

The APC is a large protein complex that catalyzes the attachment of a peptide sequence called ubiquitin to specific substrates and thus marks these for proteasome-mediated degradation (page 98). It initiates anaphase by activating two signaling pathways in parallel. First, APC/C activation causes the destruction of securin, which in turn releases the enzyme separase to cleave the cohesin molecules that held sister chromosomes together in mitosis – thus allowing them to be separated. Second, APC/C activation causes cyclin-B to be destroyed. Without cyclin-B, CDK1, which maintains the M-phase state, is no longer active and the cell enters G1.

Thus the spindle assembly checkpoint pathway confers upon the cell the ability to "wait" until all chromosomes are correctly aligned upon the spindle before anaphase takes place. Importantly, even a single kinetochore not correctly attached to the spindle can make enough MCC to prevent anaphase. This is illustrated beautifully in movies of chromosome segregation in live cells where anaphase can be seen to wait until all chromosomes are aligned, and then occur soon after the very last chromosome is aligned at the center of the spindle.

Cell Cycle Control and Cancer

Cancer occurs when cells evade the mechanisms that normally control the cell-division cycle, and thus cells overproliferate. Loss of cell-division control results in the formation of a tumor that can be either benign, meaning it remains in one place and is relatively easy for the surgeon to remove, or malignant, meaning it is able to invade other tissues and establish secondary cancers. Death from cancer is most frequently due to such metastases (from metastasis; the spread of a disease from one organ to another) and not the original tumor. There are many types of cancer, which are broadly named after the tissues in which they occur; carcinomas are cancers of the epithelial layers of the skin, gut, and breast; sarcomas are cancers of muscle, bone, and connective tissue; lymphomas are cancers of the immune system and leukemias are cancers of blood-forming tissues such as the bone marrow. Many of these are a result of mutations in key genes responsible for quality control during cell proliferation, whose failure leads to the overproliferation of the tissues.

Tumorigenesis (the formation of a cancer, also called oncogenesis or carcinogenesis) is a complex process that may take several decades. In a few cases the genetic changes leading to the development of the disease are well understood. Colon cancer is a good example. The first step is mutation of the gene adenomatous polyposis coli. This is referred to as a "gatekeeper" gene as colon cancer cannot develop without this crucial step. Mutation of adenomatous polyposis coli alone leads, at worst, to the formation of a small, benign growth that is often asymptomatic. Further progress requires mutations in a number of oncogenes (genes that can lead to cancer if mutated) including Ras and PI 3-kinase (pages 183, 185). Despite these multiple genetic changes the resulting tumor remains benign, forming only growths called adenomatous polyps that gave the gatekeeper gene its name. The crucial step in transforming a benign tumor into a malignant one is often the mutation of *p53* and therefore the elimination of the cell cycle checkpoints. p53, which is mutated in more than 50% of all human cancers, is therefore called a tumor-suppressor gene. The importance of p53 as a tumor suppressor is illustrated beautifully by the African elephant. Elephants are far less prone to cancer than humans, and this is explained at least in part by the fact that elephants have 20 copies of p53 in their genome, compared to the solitary copy found in humans.

Loss of cell-cycle control, such as is caused by p53 mutation, causes the over-proliferation of cells and thus tumors. But a defect in cell-cycle control can also be the initiating factor that triggers the path to cancer. For example, failure to correctly segregate chromosomes during anaphase of mitosis can cause aneuploidy, with daughter cells containing an incorrect number of chromosomes. Whilst in some cases this may lead to the death of the aneuploid cells, some may survive and become prone to further genetic errors that can lead to cancer. Alternatively, a failure of cytokinesis can lead to cells possessing twice the normal number of chromosomes, a state called tetraploidy. Tetraploid cells are highly likely to subsequently gain and lose chromosomes in future cell divisions, which also serves as a potential stepping-stone to cancer. Thus when it comes to cancer, failures in cell-cycle control and cell division are both the chicken and the egg.

MEIOSIS AND FERTILIZATION

Our genome is encoded in 23 chromosomes and is made up of 3×10^9 base pairs of DNA. Human cells contain 46 chromosomes, 23 inherited from each parent. Cells that contain two complete sets of chromosomes are said to be **diploid** and those containing a single set are **haploid.** Whereas the goal of most cell divisions (i.e. mitotic cell divisions) is to generate daughter cells with a full diploid array of chromosomes, at fertilization the goal is to bring together an egg and a sperm each with a haploid genome, to generate a diploid embryo containing chromosomes from each germ cell. In sexually reproducing organisms, the **germ cells** that give rise

to the eggs and sperm undergo a special type of cell division in which a single round of DNA synthesis is followed by two cell divisions known as **meiosis I** and **meiosis II,** to create haploid eggs and sperm (gametes). Fusion of an egg and sperm at fertilization thus restores the diploid state, and the development of the new embryo occurs by the recommencement of mitotic cell divisions.

Meiosis

Each meiotic division involves the formation of a meiotic spindle and the same sequence of prophase, prometaphase, metaphase, anaphase, and telophase that we saw earlier in mitosis. The steps are illustrated in Figure 14.10. Two chromosomes are shown; the paternal chromosome is shown in pink, whilst its maternal homolog (the matching chromosome from the mother) is gray. Because DNA synthesis has occurred in the preceding interphase, each chromosome enters meiosis as two sister chromatids. Mitosis is usually a rather brief process, but meiosis is often extended and in different organisms can last for months or even years. Most of this is occupied by a lengthy prophase of meiosis I, **prophase I,** during which the duplicated homologous chromosomes become aligned and then connected at specific points along their length to form a specific four-chromatid structure called a **bivalent.** We discuss the points of connection between the homologous chromosomes which are called **chiasmata** later (page 245). As the cell progresses to metaphase I, the maternal/paternal bivalents line up along the metaphase plate. At anaphase I, the homologous pairs separate but, unlike the situation in mitosis, the paired chromatids remain attached and journey together to the spindle pole. The two daughter nuclei formed during telophase I almost immediately enter meiosis II. Prophase II is often so brief as to be undetectable. Metaphase II and anaphase II resemble their mitotic counterparts, the chromatids finally separating to give haploid gametes (sperm or eggs). Since in Figure 14.10 we show only one pair of chromosomes the final panel in the figure shows each gamete containing one chromosome.

Although meiosis in male and female animals follows roughly the same program, there are some important distinctions. In males meiosis produces four equal-sized haploid cells called spermatids, each subsequently developing into a functional spermatozoon. In females, both meiotic divisions are asymmetric, resulting in only one large egg plus smaller cells called **polar bodies** that serve as the wastebasket of female meiosis, receiving chromosomes but then dying to make way for the egg (Figure 14.11). A second major difference is that in mammalian eggs the completion of meiosis II occurs only if the egg is successfully fertilized by a sperm. Prior to fertilization the egg is arrested at metaphase of meiosis II as a result of an oocyte-specific cell-cycle arrest activity called **cytostatic factor. Fertilization** triggers a series of increases of cytosolic calcium concentration (Chapter 10)

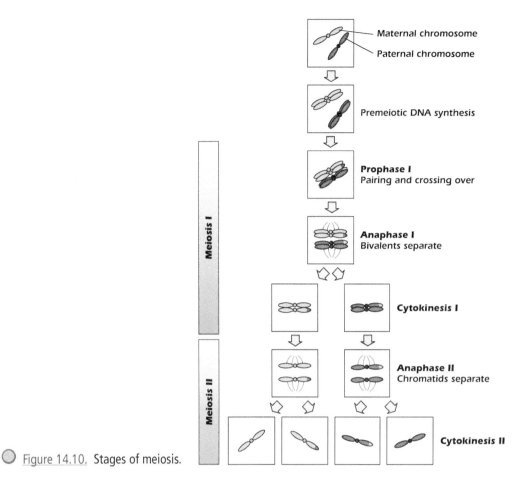

Maternal chromosome

Paternal chromosome

Premeiotic DNA synthesis

Prophase I
Pairing and crossing over

Anaphase I
Bivalents separate

Cytokinesis I

Anaphase II
Chromatids separate

Cytokinesis II

Meiosis I

Meiosis II

Figure 14.10. Stages of meiosis.

Medical Relevance 14.2   Down's Syndrome

Cell division by mitosis is a relatively error-free process, in part because of the spindle assembly checkpoint described on page 241. The cell-cycle checkpoints ensure that all of the cells of our body contain the correct number of chromosomes. Meiosis on the other hand is more prone to mistakes, particularly in females. This causes eggs that are aneuploid, having the wrong number of chromosomes. If fertilized, these aneuploid eggs would give rise to aneuploid embryos.

In most cases, fertilized embryos that contain abnormal chromosome numbers are eliminated by spontaneous abortion, often at an early stage of development (most first-trimester spontaneous abortions are associated with chromosome abnormalities). Some, however, survive, particularly when the additional chromosome is small. The most familiar example is Down's syndrome. First described by the British physician, John Langdon Down in 1866, Down's syndrome individuals are easily recognized by characteristic facial features. It was not until a century after Down's first description of the condition that affected individuals were shown to be trisomic for

(possess an additional copy of) chromosome 21. This is the smallest human chromosome other than the tiny, male-specific Y and Down's syndrome is now often referred to as Trisomy 21.

The likelihood of oocyte aneuploidy increases with maternal age. Thus children with Downs syndrome are more likely to be born of women of older age, and since most embryo aneuploidy leads to miscarriage, this is also the reason for the decrease in fertility that happens typically in a woman's 40s. The explanation for this increase in oocyte aneuploidy with age probably lies in the fact that all of the eggs that a woman will produce throughout her lifetime are produced months before her own birth, and remain at prophase of meiosis-I until ovulated during adulthood. Oocytes ovulated when a woman is 40 have therefore been sitting in the ovary for that long. It is thought that oocytes lack a strong spindle assembly checkpoint, and that the cohesin molecules that hold chromosomes together gradually deteriorate during that time, causing chromosomes to be less likely to align properly on the spindle in meiosis, thus contributing to aneuploidy.

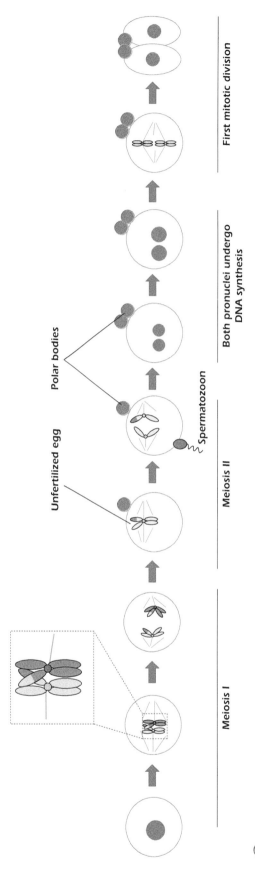

Meiosis I   Meiosis II   Both pronuclei undergo   First mitotic division
DNA synthesis

Unfertilized egg   Polar bodies

Spermatozoon

Figure 14.11. Meiosis and fertilization.

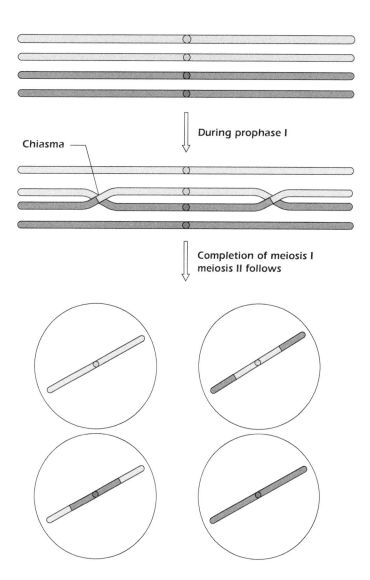

○ Figure 14.12. Chiasmata allow the crossing over of genetic material during prophase I of meiosis.

that cause the inactivation of cytostatic factor, and thus completion of meiosis II leading to the expulsion of one set of maternal sister chromatids a few hours after fertilization. Thus, perhaps paradoxically, the mammalian egg is never truly haploid, and is briefly triploid just after fertilization, but overall the process of meiosis allows the egg to contribute a haploid genome to the new developing individual.

Crossing Over and Linkage

The biological advantage of sexual reproduction is that it allows organisms to possess a random selection of the genes from their ancestors. Those individuals with a complement of genes that makes them better suited to their environment tend to do better, allowing evolution by natural selection of the individuals possessing the better genes. This ability to generate genetic diversity during reproduction is achieved as a result of meiotic **crossing over.**

Prophase I lasts such a long time because the chromosomes tie themselves in knots and then untangle (using topoisomerase II, In Depth 3.2 on page 41). Figure 14.12 illustrates the process. At the top we see the paternal and maternal chromosomes, each composed of two chromatids, lined up side by side. As before, we show the chromosome that originated from the father in pink and the maternal one in gray. During prophase of meiosis I (prophase I) the chromosomes are cut and resealed at points called **chiasmata** (singular **chiasma**) so that lengths of paternal chromosome are transferred to a maternal one and vice versa. This process is called crossing over. The rest of meiosis I proceeds, followed by meiosis II, and the end result is that some gametes contain chromosomes that are neither completely paternal nor completely maternal but are a **recombination** of the two. The phenomenon is called **homologous recombination.**

Without crossing over, genes that are located on the same chromosome would remain **linked** down the generations, greatly reducing the number of gene permutations possible at each generation. Crossing over allows a child to inherit, for example, his grandmother's red hair without also inheriting thalassaemia caused by a defective α globulin gene (page 60), although both genes are on chromosome 16. Even with crossing over, genes on the same chromosome are inherited together more than they would be if they were on different chromosomes. The closer the genes, the less likely it is that a chiasma will form between them, and therefore the greater the probability that they will be inherited together. This phenomenon of genetic linkage is used to help identify the genes responsible for specific diseases such as cystic fibrosis (Chapter 15).

 ## CELL DEATH

An adult human is made up of about 30 trillion (3×10^{13}) cells, all of which originate from a single fertilized egg. If this first cell divides into two, the two daughter cells into four, and so on, it would take only about 45 rounds of division to produce the number of cells required to make an adult human. Moreover, cell division does not stop when we have finished growing. In fact, when one counts up all the cell divisions of active stem cells (page 10) that occur in the adult body one finds that 3×10^{13} new cells are created every two weeks. The reason that we do not double in size every two weeks is that the proliferation of cells is balanced by cell death. Cells can die for two quite different reasons. One is accidental, the result of mechanical trauma or exposure to some kind of toxic agent, and often referred to as **necrosis.** The other type of death is more deliberate, the result of an in-built controlled suicide mechanism known as **apoptosis.** The two types of cell death are quite distinct.

In cases of necrosis, cell injury causes ATP concentrations to fall so low that the Na+/K+ ATPase can no longer operate, and therefore ion concentrations are no longer controlled. This causes the cells to swell and then burst. The cell contents then leak out, which can cause the surrounding tissues to become inflamed. Cells that die by apoptosis, on the other hand, shrink, causing the cytoplasm to appear dense, and the cellular contents are packaged into small membrane-bound packets called apoptotic bodies. The nuclear DNA becomes chopped up into small fragments, each of which becomes enclosed in a portion of the nuclear envelope. The dying cell modifies its plasma membrane, signaling to macrophages (page 10) which respond by engulfing the blebs and the remaining cell fragments and by secreting cytokines that inhibit inflammation. The changes that occur during apoptosis are the result of the activation of a family of proteases called **caspases** (short for cysteine-containing, cleaving at aspartate). Most of the cells of our body contain caspases, but they are normally kept in an inactive form by an integral inhibitory domain of the protein. Proteolysis cleaves the inhibitory domain off, releasing the active caspases. The advantage to the cell of this strategy is that no protein synthesis is required to activate the apoptotic pathway – all the components are already present. Thus, for example, if a virus infects a cell and takes over all protein synthesis, the cell can still commit suicide and hence prevent viral replication. Apoptosis can be broadly defined as occurring by two separate pathways, the **intrinsic pathway** (or **mitochondrial pathway**) which is activated as a result of cellular stress, or the **extrinsic pathway** in which cells are commanded to initiate apoptosis by external signals acting upon so-called death domain receptors. Figure 14.13 summarizes these control systems that regulate the decision to die or survive.

Cell Stress Activates the Intrinsic Apoptotic Pathway

If a unicellular organism is damaged, it will try to repair itself since the alternative is the death of the organism. However, if a cell of a multicellular organism is stressed or damaged, it may be more efficient to allow the cell to undergo a quick suicide and replace it by cell division of a healthy neighbor. There are therefore a number of mechanisms that trigger apoptosis in response to cell stress. One mechanism, termed the intrinsic apoptotic pathway, operates via mitochondria. When mitochondria are stressed, they can spontaneously release pro-apoptotic proteins such as cytochrome c into the cytoplasm (Medical Relevance 9.1 on page 148). Although cytochrome c plays a vital role in allowing mitochondria to generate an H+ gradient and hence ATP, once in the cytosol it is deadly. It binds to the **Apoptotic**

Example 14.3   Sunburn, Cell Death, and Skin Cancer

Without knowing it, we have all observed apoptosis in action. All of us at some time or other have been out in the sun without proper protection. The ultraviolet (UV) light causes damage to the DNA of the skin cells and activates p53. If the DNA damage is minor, p53 simply arrests the cell cycle until the DNA repair machinery has had time to repair the lesions. However, if the damage was more severe, the skin cells activate the apoptotic pathway so that the sunburned skin dies and sloughs off.

The reason why this is a long-term danger is that in areas of skin where sunburn has caused cell death there has been strong selection for any cells in which an earlier somatic mutation has inactivated the p53 system. The new skin becomes enriched in these mutant cells, so that any further mutations in oncogenes risks generating not a benign tumor but a malignant one.

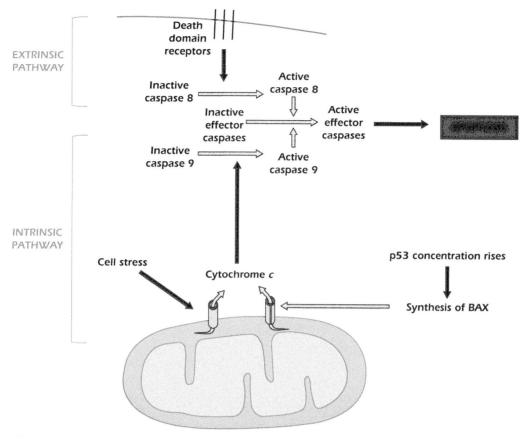

Figure 14.13. The extrinsic and intrinsic pathways for activating apoptosis.

Protease Activating Factor 1 (Apaf-1), which in turn activates pro-caspase-9. Activated caspase-9 cleaves pro-caspase-3 to generate caspase-3, a so-called effector caspase that ultimately triggers the subsequent events of apoptosis, including DNA degradation and apoptotic body formation. The unwanted activation of this pathway can be a major medical problem if it occurs in tissues that are unable to rebuild by cell division: for example in hearts during heart attacks and in brain neurons during a stroke.

The leakage of cytochrome *c* into the cytoplasm can occur via several pathways. One such pathway operates through the transcription factor p53, which we have already met in the context of cell-cycle control (Figure 14.8). p53 concentrations increase when DNA is damaged. DNA repair mechanisms are activated, but so is transcription of the *BAX* gene and synthesis of a member of the **bcl-2** protein family called **Bax.** If the DNA is not repaired in time, the Bax migrates to the outer mitochondrial membrane and dimerizes to form a channel big enough to allow cytochrome *c* to escape into the cytosol (Figure 14.13). Without this mechanism, cancers, which result from damage to and ultimately changes in the DNA sequence of our cells, would be far more abundant. Figure 14.14 shows live cells that have been transfected to express a chimera of cytochrome *c* and green fluorescent protein. Fluorescence imaging reveals the location of the chimera. The cells have been treated

with a drug that evokes apoptotic cell death through activation of BAX. The appearance of some cells (e.g. cell 1) has not yet changed: the cytochrome *c* chimera is localized to mitochondria, as it is in control cells. In some cells (e.g. cell 2) all the cytochrome *c* has been lost from the mitochondria and has distributed evenly throughout the cell, including the nucleus. Cell 3 is intermediate: the signal from the nucleus shows that some cytochrome *c* has been lost from mitochondria, but the mitochondria can still be seen as bright green organelles. With time, all the cells will come to look like cell 2, and will then show the other signs of apoptosis such as shrinking and blebbing.

Since caspases can self-activate by mutual proteolysis, there is a steady, slow rate of activation even in a healthy cell. To prevent this triggering apoptosis, cells produce **apoptosis inhibitor proteins,** which block caspase action. If the extent of caspase activation exceeds the capacity of the inhibitory proteins, death results.

Communication with the External Environment Can Activate the Extrinsic Apoptotic Pathway

Whereas the intrinsic pathway is initiated by individual cells as a result of stress, the extrinsic pathway provides a mechanism for cells to initiate apoptosis when instructed by signals

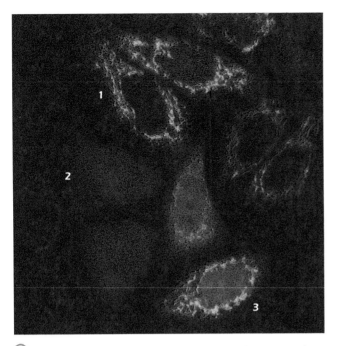

A green fluorescent protein chimera revels cytochrome *c* translocation in cells. Live human cells (HeLa) were transfected to express a chimera of cytochrome *c* and GFP then treated with staurosporine, a drug that evokes apoptotic cell death. Source: Data of Choon Hong Tan and Professor Michael Duchen, University College London; used with permission.

from other cells. Cells can be instructed to die when an extracellular ligand binds to one of a family of cell surface receptors containing so-called **death domains.** For example, cytokines of the tumor necrosis factors family (**TNF**) frequently serve as the extracellular ligand, and bind to the TNF receptor known as Fas. This activation of the Fas receptor leads to the assembly of a so-called Death Inducing Signaling Complex (**DISC**), causing caspase 8 to be activated (Figure 14.13). In turn, caspase 8 can hydrolyze and hence activate the **effector caspases** that begin the processes of cell destruction.

Default Death: Apoptosis as a Result of Absence of Growth Factors

Cells that are not required by the organism die. To make sure that this occurs, death is the default option for the cells of a multicellular organism: only if a cell receives growth factors from other cells will it survive. We have already described the first part of this pathway (page 185). Active receptor tyrosine kinases activate phosphatidylinositol 3-kinase, which generates the highly charged membrane lipid phosphoinositide trisphosphate. This causes Akt to localize to the plasma membrane, where it is itself phosphorylated and hence activated. Akt phosphorylates BAX. Phosphorylated BAX is inactive. However, if Akt is ever allowed to stop working, BAX loses its phosphates and is then able to release cytochrome *c* from mitochondria (Figure 11.9 on page 186).

Example 14.4   Neurotrophin Trafficking

During fetal development motor neurons die unless they succeed in growing their axons all the way to an appropriate muscle. Those that do find the target are bathed in **neurotrophin** 3, a growth factor released by the muscle cells. Neurotrophin 3 binds to its receptor tyrosine kinase, Trk C, on the neuron surface. However, the activated kinase then has to activate Akt in the cell body in order to prevent apoptosis, and the cell body may be many millimeters away. To achieve this, the neurotrophin-receptor complex is endocytosed using the clathrin mechanism (page 209). The endocytotic vesicles are then transported back to the cell body by dynein (page 223). Once in the cell body, Akt is activated and apoptosis prevented.

The nerve that contains motor neurons also contains pain-receptor axons. Instead of Trk C these express a related receptor, Trk A, which requires a different neurotrophin (number 1): Trk A cannot be activated by neurotrophin 3. However, both motor neurons and pain receptors express a second, unrelated receptor called the p75 neurotrophin receptor, which binds both neurotrophin 1 and neurotrophin 3. p75 is a death-domain receptor. The overall result is that when an axon arrives at a target, it will automatically receive a signal to die. However, if it has arrived at the correct location, it will receive a countermanding signal to survive.

Answer to thought question: In normal cells PIP_3 is constantly destroyed by PTEN, so in order to keep protein kinase B active and hence avoid apoptotic death cells must be constantly bathed in growth factors that act via receptor tyrosine kinases to cause the activation of PI 3-kinase. Cells in which PTEN is inactive no longer need so much growth factor, because once PIP_3 is made it persists so that Akt in turn remains active. This allows the cells to proliferate even though the relevant growth factors are in short supply.

SUMMARY

1. The cell cycle comprises S, G1, M, and G2 phases. G2, S, and G1 phases together constitute interphase. DNA is replicated in S phase, so that each chromosome becomes a pair of identical chromatids. The cell physically divides in M phase.

2. Cells that no longer divide enter the G0 resting state. Many are able to re-enter the cell cycle if they receive the correct signals from neighboring cells. Some terminally differentiated cells cannot re-enter the cell cycle.

3. Mitosis comprises prophase, prometaphase, metaphase, anaphase, and telophase. The end result of mitosis plus cytokinesis is two diploid cells whose chromosome complement is the same as that of the original cell before it underwent S phase.

4. While the chromosomes are being segregated, the cell divides in two by a process termed cytokinesis, which is driven by the contraction of an actinomyosin ring.

5. The cell cycle has in-built control points that act as stop/go switches. The main checkpoints are at G1/S, G2/M, and at the metaphase/anaphase transition of M phase.

6. Cells enter mitosis when cyclin-dependent kinase 1 is active. This in turn requires that cyclin B be present in high enough concentration and that cyclin-dependent kinase 1 be dephosphorylated by Cdc25. Once mitosis is initiated, cyclin-dependent kinase 1 is rapidly turned off through phosphorylation by Wee1 in parallel with the proteolytic destruction of cyclin B.

7. The entry of cells into S phase is a complex decision involving cyclin-dependent kinases 2, 4, and 6. The main effect of active cyclin-dependent kinase 4 is to phosphorylate Rb, causing it to release the transcription factor E2F and hence allowing the synthesis of proteins required for DNA synthesis. Important components of the decision are a raised concentration of cyclin D as a result of MAP kinase activity and a low concentration of CKIs. Cell to cell contact upregulates CKIs so that when an organ has filled the space available it stops growing.

8. p53 is continually produced in cells but is as quickly destroyed. An increase in the concentration of p53 follows DNA damage or other cell stress and has three main effects: (i) activation of DNA repair mechanisms, (ii) synthesis of CKIs, preventing cell division, and (iii) activation of apoptosis.

9. Cell cycle checkpoints stop the cycle if DNA is damaged or if DNA synthesis (S phase) is incomplete. A key checkpoint occurs during mitosis, called the spindle assembly checkpoint, which ensures that all chromosomes are properly aligned on the spindle before anaphase begins.

10. Cells become cancerous when oncogenes are activated or tumor suppressor genes are inactivated.

11. Errors in cell division such as chromosome missegregation and cytokinesis failure can contribute to tumorigenesis by generating cells with abnormal numbers of chromosomes.

12. Haploid cells contain only one copy of each chromosome. Diploid cells contain two copies, one from the organism's father, one from the mother.

13. Meiosis generates haploid germ cells: in vertebrates, eggs and sperm. Like mitosis, it follows an S phase but comprises two cycles of cell division, so that the end result is four cells whose chromosome complement is only half that of the original cell before it underwent S phase.

14. During meiosis I homologous chromosomes undergo recombination, a physical re-splicing of homologous chromosomes that allows information on chromosomes originating from mother and father to be mixed.

15. In contrast to necrosis, which causes inflammation, apoptosis is a regulated mechanism of cell suicide that has little effect on the surrounding tissue. The final effectors of apoptosis are a family of proteases called caspases.

16. Apoptosis can be triggered in three ways: (i) cell stress activates the intrinsic or mitochondrial apoptotic pathway (ii) binding of ligand to death domain receptors, termed the extrinsic pathway (iii) denial of growth factors.

FURTHER READING

Clift, D. and Schuh, M. (2013). Restarting life: fertilization and the transition from meiosis to mitosis. *Nature Reviews. Molecular Cell Biology* 14 (9): 549–562.

Green, R.A., Paluch, E., and Oegema, K. (2012). Cytokinesis in animal cells. *Annual Review of Cell and Developmental Biology* 28: 29–58.

Hunter, N. (2015). Meiotic recombination: the essence of heredity. *Cold Spring Harbor Perspectives in Biology* 7: 12.

Morgan, D.O. (2007). *The Cell Cycle: Principles of Control*. London: New Science Press.

O'Connor, C. (2008). Cell division: stages of mitosis. *Nature Education* 1 (1): 188.

Pollard, T.D. (2017). Nine unanswered questions about cytokinesis. *Journal of Cell Biology* 216 (10): 3007–3016.

Walczak, C.E., Cai, S., and Khodjakov, A. (2010). Mechanisms of chromosome behavior during mitosis. *Nature Reviews. Molecular Cell Biology* 11 (2): 91–102.

 ## REVIEW QUESTIONS

14.1   Theme: Cell division

A   anaphase
B   cytokinesis
C   metaphase
D   prometaphase
E   prophase
F   telophase

From the stages of cell division listed above, select the one in which each of the events described below occur.

1. chromosome condensation occurs in the nucleus and spindle formation begins in the cytoplasm
2. the nuclear envelope breaks down and chromosomes become associated with the spindle
3. the chromosomes are aligned on the spindle and no longer make individual excursions toward and away from the spindle poles
4. paired chromatids separate and begin to move toward the spindle poles
5. chromosomes decondense and the nuclear envelope re-forms
6. physical separation into two cells

14.2   Theme: Checkpoints in the cell cycle

A   Cdc25
B   cohesin
C   cyclin A
D   cyclin B
E   cyclin D
F   mitotic checkpoint complex
G   p53
H   Rb
I   securin
J   separase
K   Wee1

From the above list of proteins, select the one described by each of the descriptions below.

1. For animal cells to begin DNA replication, many conditions must be met. One is that the transcription factor E2F must be released from a ligand that holds it in an inactive dimer. What is this ligand?
2. If the DNA is damaged, it must be repaired before it is replicated. DNA damage activates two kinases, Chk1 and Cds1. These phosphorylate a transcription factor that acts to upregulate cyclin-dependent kinase inhibitors, CKIs. Phosphorylation of the transcription factor allows its concentration to increase in cells. What is the identity of this anti-division transcription factor?
3. Once DNA is replicated and the cell is large enough, it can enter mitosis. Passing the G2/M checkpoint requires activity of cyclin-dependent kinase 1, CDK1. CDK1 is regulated by phosphorylation. When phosphorylated on threonine 14 and tyrosine 15 it is inactive. What kinase is responsible for phosphorylating and inactivating CDK1?
4. What phosphatase is responsible for dephosphorylating CDK1, readying it for action?
5. CDK1 is only active at the G2/M boundary when dimerized with a protein partner; identify this essential partner.
6. During prometaphase the chromosomes line up on the metaphase plate. The sister chromatids are joined together by which protein complex?
7. Kinetochores that are not attached to spindle fibers allow generation of this factor that prevents anaphase.
8. When all the kinetochores are under tension the anaphase promoting complex targets securin for destruction, allowing activation of an enzyme that digests the link between the chromatids, permitting the separation of the chromatids in anaphase. What is that link-destroying enzyme?

14.3   Theme: Life and death

A   bcl-2
B   caspases

C Cdk1
D cytochrome *c*
E Fas
F p53
G protein kinase A
H Akt

From the above list of compounds, select the one described by each of the descriptions below.

1. In animals, cells are kept alive by the action of other cells that supply growth factors. Growth factor receptors activate PI 3-kinase, which generates PIP_3 at the plasma membrane. PIP_3 recruits a critical survival-promoting kinase to the plasma membrane; name that kinase.

2. If PIP_3 disappears from the plasma membrane and the kinase described above becomes inactive, BAX is activated, allowing a protein to escape from the mitochondria. Name the released mitochondrial protein.

3. White blood cells can kill target cells by secreting tumor necrosis factors which bind to and activate this cell surface death domain receptor.

4. Cells also die if their DNA is damaged so badly that it cannot be repaired in a reasonable time. What is the transcription factor whose concentration increases after DNA damage and which upregulates the synthesis of BAX?

5. All the death initiation pathways described above converge on the activation of this family of cytosolic proteases.

⬤ THOUGHT QUESTION

Medical Relevance 7.1 on page 108 describes how Ras mutations are often found in cancers, while in this chapter we describe how P53 is mutated in over 50% of human cancers. Another protein that is often mutated in cancers is PTEN, an enzyme that hydrolyzes PIP_3, generating PIP_2. Mutations that inactivate PTEN are found in 75% of gliomas. Suggest why inactivating mutations of PTEN are selected for in cell lineages leading to cancer.

SECTION 5

Only by understanding the whole of cell biology, from the code in DNA to membrane voltage and the effects of phosphorylation, can we understand how cells and tissues function and how they can fail. In this final chapter we discuss the most common serious genetic disease, cystic fibrosis, and how all the aspects of cell biology described in this book come together to allow us to understand what goes wrong in cystic fibrosis and how this knowledge leads to medical innovation.

Chapter 15: Case Study: Cystic Fibrosis

15

CASE STUDY: CYSTIC FIBROSIS

In the final chapter of the book we describe the biochemical basis of the disease **cystic fibrosis (CF),** the search for the gene, the mutations that cause CF, the prospects for CF therapy, and prenatal diagnosis. We have selected this disease as a case study to show how the combined efforts of biochemistry, genetics, molecular and cell biology, and physiology were needed to find the cause of CF and the function of the protein encoded by the cystic fibrosis transmembrane conductance regulator *(CFTR)* gene. The basic principles of many of the techniques used in achieving this knowledge are described in Chapters 1–14.

 CYSTIC FIBROSIS IS A SEVERE GENETIC DISEASE

Among Caucasians (white, non-Jewish) about 1 child in 2500 is born with CF. Inheritance is simple: when both parents are carriers, there is a one in four chance that a pregnancy will result in a child with the disease. The disease is a distressing one. Most of its symptoms arise from faults in the way the body moves liquids, leading in particular to a buildup of inadequately hydrated, sticky mucus in various parts of the body. In the lungs this leads to difficulty in breathing, a persistent cough, and a greatly increased risk of infection. A bacterium, *Pseudomonas aerigunosa*, thrives in the lung mucus and is particularly recalcitrant to antibiotic treatment. The pancreas, which provides a digestive secretion that flows to the intestine, is also affected and may be badly damaged (which explains the disease's full name, cystic fibrosis of the pancreas). Often, there are digestive problems because the damaged pancreas is not producing enzymes. The reproductive system is also harmed and most males who survive to adolescence are infertile. Even though there are so many varied symptoms, the simple Mendelian pattern of inheritance led researchers to believe that the cause was an abnormality or absence of a single protein.

In the 1960s, the life expectancy for a child with cystic fibrosis was under three years. Today, about half the people with CF will live past the age of 40 years. This remarkable improvement has resulted from intensive therapy designed to reverse individual symptoms. Digestive enzymes are taken by mouth to replace those proteins the pancreas fails to produce, while physiotherapy – helping patients to cough up the mucus in their lungs by slapping their backs – reduces the severity of the lung disease. Nevertheless CF patients still die tragically young. Today, the situation is once again changing dramatically as treatments based on an understanding of the underlying molecular and cell biology are coming onstream. In addition, more adults are being diagnosed with CF due to the increased recognition of milder presentations of the condition. The oldest person diagnosed with CF for the first time in the US was 82, in Ireland was 76, and in the UK was 71.

Cell Biology: A Short Course, Fourth Edition. Stephen Bolsover, Andrea Townsend-Nicholson, Greg FitzHarris, Elizabeth Shephard, Jeremy Hyams and Sandip Patel.
© 2022 John Wiley & Sons Ltd. Published 2022 by John Wiley & Sons Ltd.
Companion website: www.wiley.com/go/bolsover/cellbiology4

THE FUNDAMENTAL LESION IN CYSTIC FIBROSIS LIES IN CHLORIDE TRANSPORT

According to the Almanac of Children's Songs and Games from Switzerland, "the child will soon die whose brow tastes salty when kissed." Cystic fibrosis has long been recognized as a fatal disease of children. The link between a salty brow and a disease known for its effects on the pancreas and lungs was not, however, made until 1951. During a heatwave in New York that summer, Paul Di Sant' Agnese and coworkers noticed that babies with cystic fibrosis were more likely than others to suffer from heat prostration. This prompted them to test the sweat, and they found that it contained much more salt than the sweat of unaffected babies.

Sweat glands have two regions that perform different jobs (Figure 15.1). The secretory region deep in the skin produces a fluid that has an ionic composition similar to that of the extracellular medium; that is, it is rich in sodium and chloride. If the sweat glands were simply to pour this liquid onto the surface of the skin, they would do a good job of cooling but the body would lose large amounts of sodium and chloride. A reabsorptive region closer to the surface removes ions from the sweat, leaving the water (plus a small amount of sodium chloride) to flow out of the end of the gland. CF patients produce normal volumes of sweat, but this contains lots of sodium and chloride, implying that the secretory region is working fine but that the reabsorptive region has failed. The pathways by which sodium and chloride ions are removed are distinct. Which one has failed in CF? A simple electrical test gave the answer. A normal sweat gland has a small voltage difference across the gland epithelium (Figure 15.2). In CF patients the inside of the gland is much more negative than usual. This result tells us at once that it is the chloride transport (and not sodium transport) that has failed. In CF sweat glands the

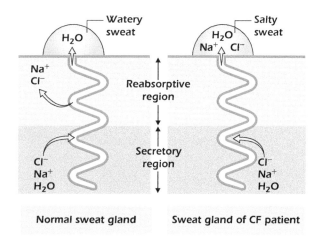

Figure 15.1. Sodium, chloride, and water transport in the sweat gland.

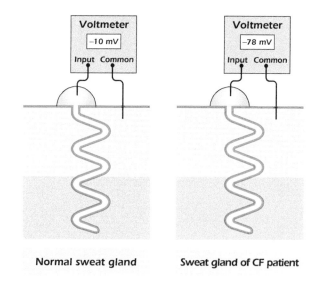

Figure 15.2. The transepithelial voltage of CF sweat glands is much more negative.

reabsorptive region has the transport systems to move sodium, but the chloride ions remain behind in the sweat, giving it a negative voltage. The sodium transport system cannot continue to move sodium ions out of the sweat in the face of this larger electrical force pulling them in, so sodium movement stops too, and sodium chloride is lost in the sweat, which then tastes salty.

The hypothesis that chloride transport had failed in the sweat gland of CF patients widened over the next decades to the view that chloride transport was abnormal in a number of locations, and then to the understanding in the 1980s that all the symptoms of CF were the result of a failure of chloride transport across membranes.

CLONING THE *CFTR* GENE

The first family studies were carried out more than 50 years ago and family pedigrees showing CF as a classic case of recessive inheritance were published in 1946. However, there was no obvious way to identify the millions of carriers who carry a single copy of the defective CF gene. There were many attempts to find a test for CF. For a time it seemed that a simple stain might do the job but this was abandoned. Other tests were more eccentric. For example, there was a claim that an extract from the blood of CF patients, and perhaps even from that of their parents, who must be carriers, slowed the beating of the cilia of oysters.

Most other claims that particular gene products were peculiar to CF carriers also failed to stand up or proved to be symptoms rather than cause. For instance, trypsinogen, a precursor of the digestive enzyme trypsin, is found in the fluid bathing CF fetuses and not in others – but this proves to be because the pancreas is already dying at this time.

This happens because although the acinar cells can secrete proteolytic enzymes (page 181), the duct cells cannot generate the movements of sodium and chloride and therefore the osmotic water movements (page 23) to carry those enzymes away from the acini. In the 1980s there was hope that this result could be used to diagnose affected pregnancies but the approach was quickly overtaken by developments in studying the gene itself. After much time-consuming work, in 1985 the research group led by Lap-Chee Tsui published data indicating that the gene for CF lay on chromosome 7. Researchers then used linkage measurements (page 245) to try to pinpoint the region of chromosome 7 containing the CF gene. An international scientific collaboration analyzed DNA from 200 families. The results indicated that the CF gene must lie between two marker regions on chromosome 7, named met and D7S8. The distance between the markers was 2 million base pairs. It was now possible to think about isolating the CF gene. Sheer hard work won the day when it came to cloning the CF gene and in 1989 scientists published the sequence of a cDNA isolated from a sweat gland library which corresponded to the entire CF mRNA. The cDNA was used to screen a genomic library to find the rest of the CF gene. Alignment of the cDNA sequence with the genome sequence obtained from the Human Genome Project (In Depth 4.2, page 62), now reveals that the gene with all of its transcriptional control elements is 216 700 bp long and has 27 exons.

THE *CFTR* GENE CODES FOR A CHLORIDE ION CHANNEL

Once CF cDNA had been sequenced, the nucleotides could be translated into amino acids using the genetic code (page 44), revealing that CF codes for a protein of 1480 amino acids. Although we knew that CF was caused by a defect in chloride ion transport, it was not immediately clear that the product of the CF gene was itself a chloride channel. Although hydropathy plots (In Depth 7.1 on page 106) indicated that the CF protein was a transmembrane protein, investigators had to consider the possibility that it was, for example, a cell surface receptor whose activation triggered expression of the proteins mediating chloride transport. The protein was therefore given the catch-all name **"cystic fibrosis transmembrane regulator" (CFTR)** and its gene was named *CFTR*. However, for once things were simple: in 1991 Michael Welsh and coworkers showed that CFTR is indeed itself the chloride ion channel in the plasma membrane. Figure 15.3 shows our current understanding of how the CFTR works. Two **nucleotide-binding domains** lie within the cytosol but are prevented from dimerizing by a regulatory domain that lies between them (left-hand sketch). If the regulatory domain is phosphorylated by protein kinase A (page 190) it is repelled out of the way, allowing the nucleotide-binding domains to dimerize. This strains the transmembrane part of the protein which flips into an open state, allowing chloride ions to pass.

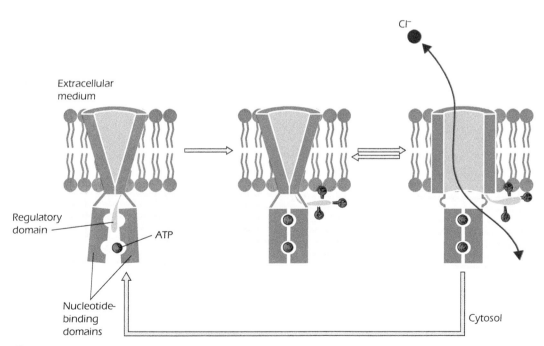

Figure 15.3. The CFTR is a chloride channel gated by phosphorylation and ATP binding.

Brain Box 15.1   Dorothy Hansine Anderson, Lap-Chee Tsui, and Michael Welsh

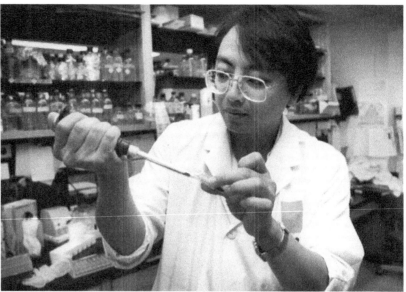

Dorothy Hansine Anderson, Lap-Chee Tsui, and Michael Welsh. Sources: Octavio L https://commons.wikimedia.org/wiki/File:Dorothy_Hansine_Andersen.jpg; Boris Spremo/Getty Images; Howard Hughes Medical Institute.

Three pivotal milestones in our understanding of cystic fibrosis have enabled us to understand the molecular mechanisms by which different mutations in *CFTR* lead to the disease and have aided in the development of increasingly effective treatments for it.

The first key discovery was the identification of cystic fibrosis in 1938 by Dorothy Hansine Anderson, a clinician with a passion for medical research, whose further research in the area led to the establishment of the sweat test – the gold standard for diagnosing CF. As a tribute to her identification of CF, Andersen was posthumously inducted into the National Women's Hall of Fame for her contributions to medical science. The second key discovery, in 1989, was the identification of *cystic fibrosis transmembrane conductance regulator (CFTR)*, the gene that causes CF – a tour de force of molecular biology and a breakthrough in medical genetics – from a team led by the Chinese Canadian molecular biologist Lap-Chee Tsui. The third key discovery came from Michael Welsh, a Howard Hughes Medical Institute investigator, and colleagues who showed that *CFTR* encodes a chloride channel.

IN DEPTH 15.1 LIPID BILAYER VOLTAGE CLAMP

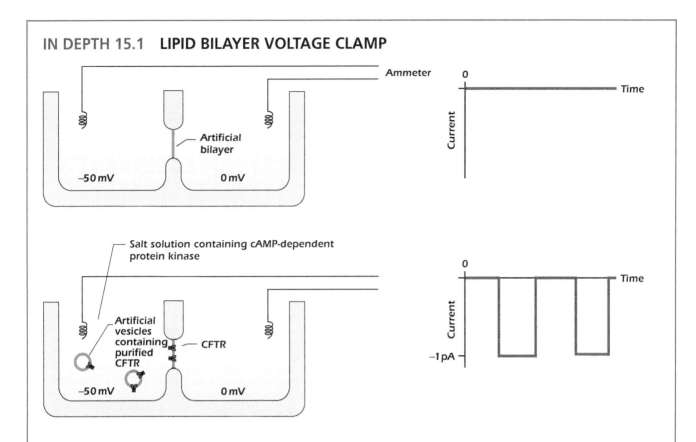

In 1992 Christine Bear and coworkers published an unequivocal demonstration that the CFTR is a chloride ion channel. They used a technique called the lipid bilayer voltage clamp, which is illustrated in the accompanying figure. A lipid bilayer is constructed from phospholipid and used to plug a hole between two chambers filled with a salt solution. Since a lipid bilayer is a barrier to ion movement, no current is recorded when a voltage is imposed across the bilayer. However, when artificial vesicles containing integral membrane proteins are added to one bath, the vesicles fuse with the artificial membrane so that the proteins now span the barrier

bilayer. When Bear and her coworkers did this with purified recombinant CFTR, they recorded currents with the direction and amplitude expected of a chloride-selective channel. In particular, no current flowed when the voltage difference across the membrane was equal to the calculated chloride equilibrium voltage (page 155). No other integral membrane proteins were present in this experiment, so the CFTR must itself be a chloride channel.

Since then this and similar techniques have been used to reveal the gating and permeability mechanisms of the channel and how these fail in specific mutant forms.

⬤ REPLACING OR REPAIRING THE GENE

Perhaps the most obvious approach to treating CF would be to introduce the non-mutant gene so that the patient's cells can produce functional CFTR. However, **gene therapy** is a difficult procedure and one that is not without risk to the patient. To date, it has not provided the wonder cure that was hoped for.

In 1992 the National Institutes of Health in the United States granted permission for a *CFTR* gene therapy trial. Because viruses are so good at infecting our cells, they are useful tools for transferring DNA into somatic cells. Viral vectors for gene therapy have been engineered to remove

their harmful genes. Using many of the recombinant DNA techniques described in Chapter 8, the gene to be transferred into the patient is inserted into the modified viral genome. A modified adenovirus vector containing the *CFTR* cDNA was introduced into the lung cells of four patients. Each of the patients made CFTR protein, but only for a short time. Some patients showed adverse reactions to the therapy, probably due to the adenovirus itself. Over the subsequent decades this idea has been tweaked to try and improve it but has to date not had much success. A recent double-blind study on 140 patients in London and Edinburgh used lipid droplets, inhaled into the lung once every month for a year, to introduce a plasmid encoding CFTR. Since no virus was used there was less chance of adverse reaction. At the end of the year one measure of lung function, forced expiratory volume FEV1, was significantly greater in the treatment group, but the difference was not large: about 4%. Other measures of lung function such as mucus buildup and severity of lung disease did not show significant differences.

If introducing the gene does not work very well, could one correct the gene in the lungs of the patient? So far, this has only been tried in a dish, where CRISPR (page 139) has been used successfully to repair CFTR nonsense mutations in human adult stem cells and airway cells. These new methodologies have great potential but have a long way to go before they can be tried in humans.

Currently, attempts are also being made to find alternatives to gene therapy. A number of new and promising nucleic acid-based therapies are being investigated. mRNA therapeutics that can be translated into normal CFTR protein inside a patient's cells and tRNAs that fool the ribosome into reading through nonsense codons are being developed.

Medical Relevance 15.1 Gene Therapy for Leber Congenital Amaurosis

Massive effort has gone into trying to find more effective treatments for cystic fibrosis because it is the commonest severe genetic disease. However, genetic lesions that affect individual organs may well be more amenable to genetic intervention. Among these, the eye has received considerable attention, in part because the gene vectors can be injected directly to the site where they are needed, and in part because even a partial rescue of sight gives a dramatic improvement in patients' quality of life.

RPE65 is a retinal enzyme that makes 11-*cis*-retinal, the prosthetic group of rhodopsin. Babies with defective RPE65 lose all vision within the first year of life in a condition known as Leber congenital amaurosis (**LCA**). Three groups, at University College London and the universities of Pennsylvania and Florida, carried out preliminary trials in which a vector encoding RPE65 was injected into the eye of human patients. The results were promising and further work led to the development of Luxturna, a viral vector that delivers RPE65 to the retina. In 2017, Luxturna was approved by the US Food and Drug Administration, becoming the first directly administered gene therapy approved for use in the USA. Luxturna is now in clinical use in many regions, including Canada, Australia, and Europe. This example shows that gene therapy does indeed hold out the hope of treatment of severe genetic disorders, although the pace at which our basic knowledge is resulting in clinical gains is painfully slow.

TAILORING TREATMENT TO THE PATIENT'S LESION

New drugs to treat CF are now appearing which address specific problems with particular CFTR mutations. Before we describe these drugs, we will describe the different functional types of mutation so that we can then see how each might be treated.

Class I mutations result in the absence of CFTR protein. One relatively common class I mutation is the nonsense mutation G542X, in which a GGA coding for glycine at position 542 is mutated to UGA, a STOP codon. About 2% of mutant genes are this mutation.

Class II mutations generate proteins with a slightly different amino acid sequence that nevertheless can function as chloride channels, albeit often at lower efficiency, but which never arrive at the plasma membrane. The changed primary structure disrupts folding and the cell detects the misfolded protein and sends it to the proteasome (page 98). This class of mutations includes the most common CF mutation, ΔF508 (Figure 15.4), which represents about 74% of the mutant genes in the UK population.

In contrast, class III and IV mutations generate proteins that are delivered to the plasma membrane but which do not function correctly. In class III gating is impaired, so that the channel rarely switches to the open state, while in class IV the pore itself is distorted and does not allow chloride through so readily. G551D, which represents about 3% of mutant genes, is a class III mutant. A glycine at position 551, within the first nucleotide-binding domain, is mutated to a charged aspartate. The protein folds correctly and is trafficked to the plasma membrane, but the mutant nucleotide-binding domain cannot dimerize with its partner so the CFTR remains shut even when the regulatory domain is phosphorylated (Figure 15.5a).

Class V and VI mutants are relatively rare. In these the protein is functional, but because of stability or problems with transcription or translation it is present at low levels in the cell. Because some functional protein is present, the disease tends to be less severe in these cases.

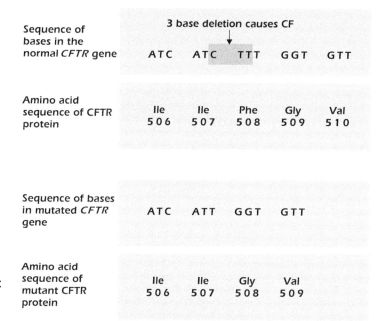

	3 base deletion causes CF			
Sequence of bases in the normal *CFTR* gene	ATC	ATC TTT	GGT	GTT

Amino acid sequence of CFTR protein	Ile 506	Ile 507	Phe 508	Gly 509	Val 510

Sequence of bases in mutated *CFTR* gene	ATC	ATT	GGT	GTT

Amino acid sequence of mutant CFTR protein	Ile 506	Ile 507	Gly 508	Val 509

Figure 15.4. The mutation seen in 70% of CF patients: a deletion of three nucleotides from the CFTR gene has removed a phenylalanine at position 508 (ΔF508).

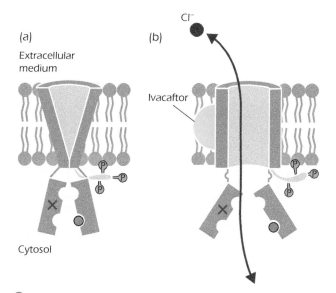

Figure 15.5. Ivacaftor causes CFTR to open even in class III mutants such as G551D.

NEW TREATMENTS FOR CF

Discussion of the new drugs for CF is easiest if we consider the mutation classes in the reverse order.

Classes III through VI have some CFTR at the plasma membrane but insufficient chloride channel function. The drug Ivacaftor, marketed as Kalydeco^R, is now licensed for patients with these mutations. Ivacaftor binds to the transmembrane part of the CFTR and causes this to switch to the open state. This "potentiation" effect works even without nucleotide-binding domain dimerization (Figure 15.5b) and is therefore helpful for class III mutations such as G551D.

Over half of CF patients have two copies of the most common mutant gene, ΔF508, which has a class II malfunction. Ivacaftor alone is without effect in these patients, since the mutant CFTR does not get trafficked to the plasma membrane. However, a number of drugs have been developed that act to stabilize or "correct" the mutant CFTR protein so that some reaches the plasma membrane. Lumacaftor, Tezacaftor, and Elexacaftor are three such licensed drugs. They are given in combination with the potentiator Ivacaftor in therapies marketed as Orkambi^R, Symdeko^R, and Trikafta™, with the potentiator present to ensure that the relatively small amount of CFTR that does reach the plasma membrane spends as much time as possible open.

Patients with the nonsense mutations that cause class I disease will not benefit from these therapies since there is no full-length CFTR protein synthesized for the drugs to act on. In Medical Relevance 6.1 on page 94 we described how Ataluren (Translarna™) allows ribosomes to read through a STOP codon. It was therefore hoped that the drug might be efficacious for class I CF patients. However, clinical trials, reported in 2014, failed to show significant improvement in either lung function or the sweat abnormality and Ataluren is no longer being developed as a potential treatment for CF.

DIAGNOSTIC TESTS FOR CF

CF is one of the genetic conditions for which carrier screening is recommended and screening for CF is now a well-established public health policy across the globe. Because 70% of CF patients in the USA and Europe carry at least one copy of the commonest mutation, ΔF508, one of the first screening tests for CF was developed based on the detection of this mutation using PCR (page 132). This test required a few cells from the

1. PCR amplify section of *CFTR* gene
spanning codon for Phe 508

2. Dot DNA onto membrane filters

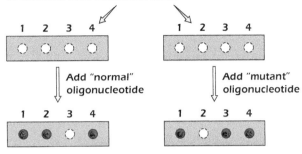

Individual 1: CF carrier
Individual 2: Both copies of gene are normal
Individual 3: Both copies of gene are mutant: CF patient
Individual 4: CF carrier

 Figure 15.6. A diagnostic test for the most common CFTR gene mutation.

inside of the mouth and DNA primers to amplify a stretch of DNA that included the region encoding amino acid 508. It also required two detection oligonucleotides, one containing the sequence encoding ΔF508 and the other lacking the codon for this mutation (Figure 15.6). By comparing the patterns of hybridization, carriers of CF could be detected.

Guidelines have now defined a panel of 23 different mutations and variants to be tested in population screening for CF carriers. However, these guidelines were established before the completion of the Human Genome Project and the recent advances in NGS sequencing (page 134). As costs continue to decrease significantly, NGS sequencing has become more affordable and is now being recommended as a first-line methodology for identifying carriers of disease-causing CF mutations.

Tests to identify individuals with CF can be carried out prenatally or after birth. PCR is currently being used, together with NGS sequencing, in non-invasive prenatal testing (**NIPT**) for CF. Increased levels of trypsinogen are associated with CF and blood samples obtained from a heel prick are used to determine the levels of immunoreactive

trypsinogen (**IRT**) present in newborns. It is only a matter of time before the whole genome sequence of every individual can be obtained, perhaps even before birth, and disease-causing mutations identified as a matter of routine.

Prenatal implantation diagnosis for CF

In 1992 the first preimplantation diagnosis for CF was carried out. Three sets of parents took part in the trial. Each parent was a carrier for the ΔF508 mutation. Oocytes were removed from the women and fertilized with sperm from their partners. Success was reported for only one couple. Six of their oocytes developed into fertilized embryos. A single cell was tested from each embryo, using PCR, for the presence of the defective *CFTR* gene. Five embryos were characterized. Two embryos each had two copies of the normal *CFTR* gene, one on each of their chromosome 7s. Two embryos each had two copies of the F508 deletion. One embryo tested positive for both the normal gene and the F508 deletion and was therefore a carrier. One normal/normal embryo and the carrier embryo were implanted into the mother. A baby girl was born. Her DNA showed that both copies of her *CFTR* gene were normal. Since this initial success a slow but steady stream of babies have been born following preimplantation diagnosis of a selected embryo and the procedure is now offered by a number of clinics worldwide.

● CONCLUSION

Cystic fibrosis remains a very serious disease. The new treatments offer enormous hope but are unlikely to completely reverse the symptoms. Furthermore, the cost of drug development has been high, making it difficult to ensure that patients worldwide have access to these new therapies.

The study of cystic fibrosis shows how all of the disciplines covered in this book, from electrophysiology to protein folding, contributed to our understanding of how the protein works normally and in disease. We hope that we have conveyed both some of the excitement of understanding how cells work and how this knowledge is leading to better medical care.

Answer to thought question 1: It is in the Golgi apparatus that additional oligosaccharides are added to proteins destined for secretion (page 207). Sugar residues, being rich in hydroxyl groups, are strongly hydrophilic, so that a heavily glycosylated protein adsorbs lots of water and forms a hydrated gel. The proteins that form airway mucus are extremely heavily glycosylated, 70% of their total mass being carbohydrate. They therefore form the thin, watery gel that lubricates the airways and other internal spaces of the body. Barasch and coworkers suggested that an abnormal pH in the Golgi lumen, resulting from a failure to transport Cl⁻ as the counter ion to H⁺ (in the same way that Na⁺ transport fails in the sweat gland because Cl⁻ cannot move to neutralize the positive charge on Na⁺) impairs the glycosylation of mucus proteins. The question is difficult to study because the lungs of even symptom-free CF patients contain many more bacteria than those of a non-CF subject, and bacterial enzymes digest polysaccharides. It is difficult to differentiate the primary effects of the mutation from the secondary effects resulting from bacterial invasion.

Answer to thought question 2: By affecting splicing. Both mutations generate novel splice sites which upon processing generate mRNAs that contain premature STOPs. A genome-editing strategy based on CRISPR (page 139) has been developed that, *in vitro*, corrects the splicing defects and restores physiological levels of CFTR expression and function for each of these two mutations. So far clinical trials of this therapy have not been reported.

SUMMARY

1. Cystic fibrosis is the most common serious single-gene inherited disease in the Western world. Many organs are affected. Sticky mucus builds up in the reproductive tract and in the lungs. The pancreas is always affected and usually fails completely.

2. Electrical measurements show that the basic problem is in chloride transport.

3. CF is a classic recessive disorder. The gene is on chromosome 7.

4. A combination of hard work and novel techniques helped isolate the normal *CFTR* cDNA from a sweat gland clone library.

5. From the cDNA it was possible to identify the gene and infer its amino acid sequence in both normal and mutated forms. The gene codes for a chloride channel called CFTR.

6. Thousands of different *CFTR* mutations have now been found. Tests have been devised for prenatal diagnosis and to detect carriers of the commonest mutations.

7. The focus of current research is to generate therapies geared to the specific genetic lesion. For the most common genetic error, ΔF508, this aims to block the mechanisms that normally prevent misfolded proteins from being trafficked to the plasma membrane. Therapies that improve chloride transport have also been developed.

FURTHER READING

Anderson, M.P., Gregory, R.J., Thompson, S. et al. (1991). Demonstration that CFTR is a chloride channel by alteration of its anion selectivity. *Science* 253 (5016): 202–205.

Bear, C.E., Li, C.H., Kartner, N. et al. (1992). Purification and functional reconstitution of the cystic fibrosis transmembrane conductance regulator (CFTR). *Cell* 68: 809–818.

Bell, S.C., Mall, M.A., Gutierrez, H. et al. (2019). The future of cystic fibrosis care: a global perspective. *The Lancet Respiratory Medicine* 2020 8 (1): 65–124. Erratum in: Lancet Respir Med. 2019 7 (12): e40.

Bragonzi, A. and Conese, M. (2002). Non-viral approach toward gene therapy of cystic fibrosis lung disease. *Cuff. Gene Therapy* 2: 295–305.

Handyside, A.H., Lesko, J.G., Tarin, J.J. et al. (1992). Birth of a normal girl after *in vitro* fertilization and preimplantation diagnostic testing for cystic fibrosis. *The New England Journal of Medicine* 327: 905–909.

Kerem, B., Rommens, J.M., Buchanan, J.A. et al. (1989). Identification of the cystic fibrosis gene: genetic analysis. *Science* 245: 1073–1080.

Pearson, H. (2009). Human genetics: one gene, twenty years. *Nature* 460: 164–169.

Rich, D.P., Anderson, M.P., Gregory, R.J. et al. (1990). Expression of cystic fibrosis trans-membrane conductance regulator corrects defective chloride channel regulation in cystic fibrosis airway epithelial cells. *Nature* 347: 358–363.

Riordan, J.R., Rommens, J.M., Kerem, B. et al. (1989). Identification of the cystic fibrosis gene: cloning and characterization of complementary DNA. *Science* 245: 1066–1073.

Tsui, L.C. and Dorfman, R. (2013). The cystic fibrosis gene: a molecular genetic perspective. *Cold Spring Harbor Perspectives in Medicine* 3 (2): a009472.

Welsh, M.J. and Smith, A.E. (1995). Cystic fibrosis. *Scientific American* 273: 52–59.

REVIEW QUESTIONS

15.1 Regulation and operation of CFTR

A calcium
B cAMP
C cGMP
D chloride
E nucleotide-binding domains
F regulatory domain
G sodium
H transmembrane domains

From the domains and solutes above, choose the appropriate answer for each of the questions or statements below.

1. Activation of β-adrenergic receptors in the airway epithelium increases the cytosolic concentration of which intracellular messenger, which in turn acts to activate CFTR?
2. CFTR is primed to open by phosphorylation on the . . .
3. Opening of the channel is then triggered by dimerization of the . . .
4. To increase the time that CFTR spends open the drug Ivacaftor binds to the . . .
5. When open, CFTR allows which ion to pass?

15.2 Diagnosis from the sweat

A −80 mV, lumen with respect to interstitial fluid
B −10 mV, lumen with respect to interstitial fluid
C +10 mV, lumen with respect to interstitial fluid
D +80 mV, lumen with respect to interstitial fluid
E greater or equal to 100 mmoles liter^{-1}
F less than 60 mmoles liter^{-1}

From the list of value ranges above, choose the range in which the values indicated below might be found.

1. In a non-CF patient, sweat Na^+ concentration would be expected to be:
2. In a non-CF patient, sweat Cl^- concentration would be expected to be:
3. In a CF patient, sweat Na^+ concentration would be expected to be:
4. In a CF patient, sweat Cl^- concentration would be expected to be:
5. In a non-CF patient, a typical value of transepithelial voltage might be:
6. In a CF patient, a typical value of transepithelial voltage might be:

15.3 Mutations in CF

A nonsense mutation
B missense mutation
C frameshift mutation

From the list of mutation types above, select the one that describes each of the known CFTR mutations below. Please note that each group of three nucleotides represents a CFTR codon. You should refer to Figure 3.9 on page 44 while answering these questions.

1. the insertion of a T after position c.3773,
   changing GCT TTT TTG AGA CTA to GCT TTT TT<u>T</u> GAG ACT A
2. a single base substitution in which a G at position c.350 is converted to A,
   changing GAG GAA C<u>G</u>C TCT ATC to GAG GAA C<u>A</u>C TCT ATC
3. a single base substitution in which a C at position c.1657 is converted to T,
   changing GGT CAA <u>C</u>GA GCA AGA to GGT CAA <u>T</u>GA GCA AGA
4. a single base substitution in which a T at position c.1705 is converted to G,
   changing GAT TTG <u>T</u>AT TTA TTA to GAT TTG <u>G</u>AT TTA TTA
5. a single base substitution in which a G at position c.3846 is converted to A,
   changing CAA CAG TG<u>G</u> AGG AAA to CAA CAG TG<u>A</u> AGG AAA
6. the deletion of A at position c.313,
   changing GGA AGA <u>A</u>TC ATA GCT . . . to GGA AGA TCA TAG . . .

⬤ THOUGHT QUESTIONS

---

1. A hypothesis put forward by Barasch and co-workers in 1991 proposed that the mucus abnormalities seen in CF are in part the result of an unusual pH in the Golgi complex. What processes that occur in the Golgi complex might affect the ability of mucus proteins to form a hydrated gel?
2. The mutations 3272-26A>G and 3849+10kbC>T, which are single base substitutions within introns (the terminology is complicated and we will not explain it here!), generate mRNAs that contain a premature STOP. The first mutation converts an A to a G, the second converts a C to a T, in both cases deep with an intron. How can a mutation in an intron affect the mRNA exported to the cytosol?

ANSWERS TO REVIEW QUESTIONS

1.1 Theme: Dimensions in cell biology

General comment: $1\,m = 10^3\,mm = 10^6\,\mu m = 10^9\,nm$.

1. E. 2000 nm = 2 μm. Most bacteria are in the range 1–2 μm.
2. F. 20 000 nm = 20 μm. Most eukaryotic cells are in the range 5–100 μm.
3. I. 1 000 000 000 nm = 1 m. Neurons supplying the fingers and toes are about this length.
4. D. 250 nm. The wavelength of green light is 500 nm; transmission light microscopes can achieve a resolution of about half this.
5. B. 0.2 nm. This allows the electron microscope to reveal structures that are invisible in the light microscope.

1.2 Theme: Types of cell

1. C = fibroblast. Fibroblasts secrete collagen and other components of the connective tissue extracellular matrix.
2. E = glial cell. Glial cells and neurons are the two main types found in nervous tissue.
3. B = epithelial cell. An epithelium is a sheet of cells.
4. D = macrophage. Macrophage means "big eater" in Latin; macrophages take up unwanted material and digest it.
5. A = bacterium. Prokaryotic cells have no nuclear envelope; the genetic material is free in the cytosol.

6. F = skeletal muscle cell. Many precursor cells, each with one nucleus, fuse to form the large cells that form our muscles.
7. G = stem cell.

1.3 Theme: Revealing cell organization

1. B = Fluorescence microscopy using a DNA stain such as Hoechst. Hoechst and similar stains are quick, cheap, extremely bright, and require little preparation of the tissue.
2. C = Fluorescence microscopy using a GFP chimera. Creating cells that express the chimera takes time (see Chapter 8), but once they are created the assay is quick and easy and can be scaled up to assay many samples simultaneously.
3. D = Fluorescence microscopy using a specific antibody. The tissue is fixed and an antibody against tubulin used to label these structures specifically. Figure 14.3 on page 000 shows an example.
4. E = Phase contrast light microscopy. Most animal cells have little color and do not show up in bright-field microscopy but are readily seen under phase.
5. F = Scanning electron microscopy. Cilia are too small to see in a light microscope. A transmission electron microscope shows a single thin plane of the specimen; a

Cell Biology: A Short Course, Fourth Edition. Stephen Bolsover, Andrea Townsend-Nicholson, Greg FitzHarris, Elizabeth Shephard, Jeremy Hyams and Sandip Patel.
© 2022 John Wiley & Sons Ltd. Published 2022 by John Wiley & Sons Ltd.
Companion website: www.wiley.com/go/bolsover/cellbiology4

scanning electron microscope image (such as Figure 1.7) shows the whole surface, allowing the density of cilia to be estimated.

6. G = Transmission electron microscopy. This is the technique that reveals the detail of individual planes through a cell; a plane through the nucleus will reveal ribosomes on its surface (as we have shown in Figure 1.2).

2.1   Theme: Membranes

1. D. Mitochondria, the nucleus and, in plant cells, chloroplasts, are surrounded by a double membrane envelope.
2. A
3. B
4. F
5. A

2.2   Theme: Organelles in eukaryotic cells

1. A
2. C
3. D. Only the nucleus, mitochondria and (in plants) chloroplasts contain DNA.
4. F
5. E. Chromatin, so called because it stains strongly with colored dyes, is a complex of DNA and histones.
6. B
7. D
8. B. The cell center is a specific region, close to the nuclear envelope, where the Golgi complex and the centrosome are found.

2.3   Theme: Transport across membranes

1. C. RNA molecules are hydrophilic because they bear multiple charges. They cannot therefore cross lipid bilayers by simple diffusion. Most are also too large to go through gap junctions. The Mr of this particular RNA was 10 000, and gap junctions do not allow molecules of Mr > 1000 to pass.
2. B. Inositol trisphosphate is highly hydrophilic since it bears six negative charges. It cannot therefore cross lipid bilayers by simple diffusion. However, inositol trisphosphate is small enough to pass through gap junctions. This is thought to be an important mechanism of cell:cell signaling – we will learn some of the signaling functions of inositol trisphosphate in Chapter 11.
3. B. As a small ion, K⁺ is highly hydrophilic and cannot cross lipid bilayers by simple diffusion. However K⁺ is readily permeable through gap junctions. Passage of electrical current, by the movement of K⁺ and other ions, allows heart cells to contract in concert.
4. A. NO is a small uncharged molecule and has sufficient solubility in hydrophobic solutes to be able to cross lipid bilayers by simple diffusion. In Chapter 11 we will

describe how the movement of NO from one cell to its neighbor in this way is critical in allowing the blood supply to respond to the needs of the tissues.

Review question covering chapters 1 and 2

1. D
2. C
3. F
4. B
5. A

3.1   Theme: Mutations

1. C. UAA is a STOP codon, so translation into a polypeptide chain will terminate prematurely.
2. D. CUA, like the original UUA, codes for leucine, so the encoded polypeptide sequence is unchanged. This type of mutation is known as a synonymous mutation.
3. B. The AGC now codes for serine in place of the original isoleucine.
4. C. The deletion has generated the STOP codon UGA, so translation into a polypeptide chain will terminate prematurely.
5. A. From the deletion onwards, the sequence will be read using the wrong reading frame, creating a different polypeptide sequence.

3.2   Theme: Bases and amino acids

1. J. Uracil replaces thymine in RNA.
2. C. Both positively charged amino acids, arginine and lysine, are present in large amounts in histones, the proteins around which DNA wraps to form the nucleosome.
3. F. The notation G5E means that the 5th amino acid in the protein is usually glycine (G) but is glutamate (E) in the mutant.
4. E
5. A

3.3   Theme: Structures associated with DNA

1. E. Heterochromatin is the form adopted by DNA that is not being transcribed into RNA. In Chapter 2 we described how heterochromatin is located at the edge of the nucleus.
2. F. This is the organization found in prokaryotes.
3. G
4. C

4.1   Theme: Normal or damaging chemical change in DNA

1. B
2. D. Unlike the other processes in the list, this is not a damaging change but a normal one that follows a short while after synthesis of a daughter DNA strand.

3. A

4. A

5. B

6. F

4.2 Theme: DNA replication

1. B. DNA ligase connects adjacent deoxyribonucleotides within an otherwise complete double-stranded DNA molecule, both during normal DNA synthesis, as here, and during DNA repair.

2. F. DNA polymerase I is responsible for building the DNA chain in those regions previously occupied by the RNA primer.

3. E. DNA polymerase III runs nonstop to create the leading strand. It also synthesizes most of the lagging strand, but is incapable of running through the sections already occupied by RNA primers; DNA polymerase I is required in those sections.

4. D. We have described how during mismatch repair exonuclease I runs in a 3′ to 5′ direction destroying the DNA strand.

5. D. We have described how during mismatch repair exonuclease VII runs in a 5′ to 3′ direction destroying the DNA strand.

6. A. As the first step in DNA replication, helicase splits the DNA double helix into two single strands.

7. C. This occurs repeatedly during synthesis of the lagging strand. As we will describe in Chapter 5 other enzymes (the RNA polymerases) synthesize RNA during transcription, but here we are concerned with DNA replication.

8. G. In prokaryotes DNA polymerase performs this function and then goes on to synthesize the complimentary DNA strand; however, in eukaryotes there is a specialized standalone enzyme to do the job.

4.3 Theme: Regions within eukaryotic chromosomes

1. A. In eukaryotic genes, only the exons encode the polypeptide sequence of a protein.

2. E

3. C

4. G. Examples are the genes that encode ribosomal RNA, transfer RNA, and histone proteins. Note that the answer "gene family" is wrong – members of a gene family encode significantly different, albeit strongly similar, proteins.

5. F

6. F. Note that the answer "long interspersed nuclear element" is wrong – the copy number of LINEs is in the thousands, not in the millions.

5.1 Theme: Codes within the base sequence

1. E. TATA binding protein, a component of transcription factor IID, binds to the TATA box on the DNA, allowing recruitment of the other transcription factors and then RNA polymerase II.

2. H. The resulting RNA strand forms a hairpin loop in the GC-rich region, reducing the size of the transcription bubble, while the attachment of the following uracils to the string of adenines is weak since there are only two hydrogen bonds in each UA pair.

3. E. This is an alternative transcription initiator used in many eukaryotic genes. A protein called Sp1 binds to the GC box and recruits TATA binding protein, allowing recruitment of the other transcription factors and then RNA polymerase II.

4. F. The −10 box is part of the prokaryotic promoter sequence, which recruits the σ factor, allowing recruitment of the other subunits of RNA polymerase.

5. D. All eukaryotic introns conform to the sequence GU. . ..AG, and although the process in incompletely understood, these bases are involved in the process by which the spliceosome recognizes the length of RNA to be removed.

6. C. This sequence is found close to the 3′ end of most mRNAs and is thought to be a signal for poly(A) polymerase to attach and add a string of A residues to the 3′ end of the mRNA.

5.2 Theme: The control of transcription

1. C. In the presence of cyclic AMP, catabolite activator protein binds to a regulatory site of the *lac* operon of *E. coli* and increases transcription. In turn cAMP is only high if glucose is in low supply.

2. C. In the presence of glucocorticoid hormone, the glucocorticoid hormone receptor binds to an enhancer site of various mammalian genes and increases transcription.

3. B. If the *lac* repressor protein binds a β-galactoside sugar, it adopts a shape that can no longer bind to the operator region of the operon. Thus, the enzymes that allow utilization of β-galactoside sugars are only made if these sugars are present.

4. D. When the aporepressor protein binds tryptophan, it can then bind to the operator region of the *trp* operon and inhibit transcription. Thus the enzymes that synthesize tryptophan are only made when tryptophan is lacking.

5.3 Theme: Events that occur after transcription in eukaryotes

1. D. A long stretch of adenosine residues is added to the 3′ end of the mRNA transcript.

2. A. The methylated guanosine is added by a 5′ to 5′ phosphodiester link unlike the 3′ to 5′ links formed by RNA polymerase.

3. E. Alternative splicing allows one primary mRNA transcript to give rise to two processed mRNAs that share some exons but differ in others. It is distinct from the

phenomenon of polycistronic mRNA seen in prokaryotes, where sequential lengths of mRNA on the same molecule are translated to generate completely different proteins.
4. E. Splicing removes the introns to leave only exonic, coding RNA. We specified "prior to its subsequent translation into protein." Answer B, digestion by nucleases, is therefore wrong because after digestion the mRNA cannot be translated into protein.

6.1 Theme: Translation initiation

1. H
2. M
3. E. The Kozak sequence includes the universal start codon AUG, so even if you don't remember the rest of the consensus sequence you should have been able to identify this answer as the Kozak sequence.
4. D
5. C

6.2 Theme: Translation elongation and termination

1. A
2. A
3. D
4. B
5. I
6. F

6.3 Theme: Wobble

1. A. If you answered G = 5′ UAC 3′ you forgot that nucleic acid strands pair up in an antiparallel fashion.
2. D. The wobble phenomenon is operating: the G at the 5′ end of the anticodon can pair with either U or C at the 3′ end of the codon.
3. B. Once again the wobble phenomenon is operating: the G at the 5′ end of the anticodon can pair with either U or C at the 3′ end of the codon.
4. F. The inosine at the 5′ end of the anticodon can pair with any of U, C, or A at the 3′ end of the codon, allowing a tRNA with the anticodon 5′ ICC 3′ to pair with any of 5′ GGU 3′, 5′ GGC 3′, or 5′ GGA 3′ in the codon. Other tRNAs would be needed when the codon is 5′ GGG 3′.

7.1 Theme: Amino acids

1. E, glutamate
2. E, glutamate
3. R, arginine
4. P, proline
5. C, cysteine

7.2 Theme: Terms used to describe proteins

1. D
2. A
3. E
4. C
5. F
6. G

7.3 Theme: Specific binding partners

1. C
2. A. Catabolite activator protein also binds cAMP, but does not directly bind either glucose or β-galactoside sugars. Rather, low glucose concentrations cause the concentration of cAMP to rise, and it is cAMP that activates catabolite activator protein, so it can bind to its specific base sequence on DNA.
3. D. Connexins on one cell bind a compatible connexin on a neighboring cell to form a gap junction channel.
4. A. The glucocorticoid receptor also binds glucocorticoid hormones, and can only adopt a shape that binds to the DNA when it has the steroid bound.

8.1 Theme: A mammalian expression plasmid

1. D. The multiple cloning site comprises the recognition sites of a number of restriction enzymes, allowing the experimenters to choose a site also present at the appropriate places on the DNA encoding the cytochrome c, cut both pieces of DNA with the same enzyme, and then assemble the recombinant plasmid.
2. E. The bacteria are grown on agar containing the toxic antibiotic. Only bacteria containing the resistance gene survive.
3. A. The origin of replication is the point where the host enzyme DnaA attaches, allowing formation of the open replication complex.
4. B. The cytomegalovirus promoter is used by the virus to drive transcription of its genes in preference to those of the host. This promoter is very commonly used by experimenters to drive expression from plasmids introduced into mammalian cells. This promoter is not recognized by the bacterial RNA polymerase and therefore the chimeric mRNA (and hence protein) will not be made in bacterial cells.

8.2 Theme: Choosing an oligonucleotide for a specific task

1. B. The oligonuclotide, 5′ TGCCTACTGCAGCGTCTGCA 3′ can be used to copy the top strand of the sequence shown in the 5′ to 3′ direction.

2. E. The sequence the enzyme will recognize is simply added to the 5′ end of the oligonucleotide.

3. A. Most eukaryotic mRNAs have at their 3′ end a poly(A) tail. The primer 5′ TTTTTTTTTTTTTTTT 3′ will hydrogen bond with the tail and can then be extended by reverse transcriptase.

4. F. Both strands of DNA are present on a Southern blot so the sequence 5′ GTGCATCTGACTCCTGTGGAGAA GTCT 3′ will hydrogen bond with its complimentary sequence.

5. G. The oligonucleotide must be complimentary in sequence to the mRNA. Oligonucleotides are made as DNA and therefore contain Ts instead of Us.

8.3 Theme: Uses of cDNA clones

1. A. The strands are separated at high temperature and then the primers (which have been chosen such that once attached the 3′ ends point toward each other) allowed to anneal to their respective strands. Thermostable DNA polymerase then generates the complementary strands so that for each cycle of operation there is a replication of the DNA between the primer two attachment points. At present polymerase chain reaction amplification of lengths of up to 4 kb are routine; some users amplify lengths of up to 8 kb.

2. D. DNA polymerase will attach to the 3′ end of the annealed oligonucleotide and then run along copying the remainder of the DNA sequence until a dideoxynucleotide is incorporated, at which point replication of that molecule stops. Note that this is automated sequencing, in which detection of the various products is by fluorescence of labeled dideoxynucleotides. The oligonucleotide does not need to be radiolabeled (and both cost and safety considerations mean that a radioactive reagent would never be used when a nonradioactive one will suffice).

3. C. The oligonucleotide will hybridize to the specific DNA sequence and can then be detected by autoradiography.

4. C. The oligonucleotide will hybridize to the specific RNA sequence and can then be detected by autoradiography.

9.1 Theme: Cytosolic and extracellular concentrations of important ions

1. H. A typical value for cytosolic calcium is 100 nmoles liter^{-1}.

2. D

3. C

4. A

5. H. A typical value for cytosolic H$^+$ is 60 nmoles liter^{-1}, so the response "less than or equal to 100 nmoles liter^{-1}" is correct.

6. H. A typical value for extracellular H$^+$ is 40 nmoles liter^{-1}, so the response "less than or equal to 100 nmoles liter^{-1}" is correct.

7. A. A typical value for cytosolic potassium is 140 mmoles liter^{-1}, so the response "greater than or equal to 100 mmoles liter^{-1}" is correct.

8. C

9. A. A typical value for extracellular sodium is 140 mmoles liter^{-1}, so the response "greater than or equal to 100 mmoles liter^{-1}" is correct.

9.2 Theme: Pathways for solute movement across the plasma membrane

1. A. The sodium/potassium ATPase hydrolyzes ATP to ADP + Pi; this hydrolysis provides the energy that drives sodium out of the cytosol and brings potassium in.

2. E

3. B. The molecular weight of glucose is 180, well below the 1000 or so limit for a connexon. Note that C is not correct because the glucose carrier is not a channel.

4. D

5. F

6. G

9.3 Theme: Ion fluxes in a neuron

1. B. H$^+$ has a crucial role in mitochondria, but the concentration gradient across the plasma membrane is small, with the concentration in the cytosol being about 1.5 times that in the extracellular medium.

2. D

3. C

4. A. When the cytosol becomes less negative, the voltage force pulling the positive ions H$^+$, K$^+$, and Na$^+$ will get smaller. Only a negative ion such as Cl$^-$ will experience a greater inward electrochemical gradient when the cell is depolarized.

5. A

10.1 Theme: The role of calcium at the neuromuscular junction

1. E

2. E. Skeletal muscle cells express a specific voltage-gated calcium channel that links physically to ryanodine receptors on the sarcoplasmic reticulum.

3. A. A specific calcium ATPase (called SERCA for Sarcoplasmic and Endoplasmic Reticulum Calcium ATPase) is responsible for pumping calcium into the endoplasmic reticulum and sarcoplasmic reticulum.

4. C

5. D. We will return to muscle contraction in Chapter 13.

10.2 Theme: Pathways for calcium movement

1. E
2. B
3. C
4. A. A specific calcium ATPase (called SERCA for Sarcoplasmic and Endoplasmic Reticulum Calcium ATPase) is responsible for pumping calcium into the endoplasmic reticulum and sarcoplasmic reticulum.
5. B. Cardiac muscle cells lack the physical link between voltage-gated calcium channels and ryanodine receptors that is found in skeletal muscle cells. Rather, the ryanodine receptors are triggered to open by the calcium ions entering the cell through voltage-gated calcium channels on the plasma membrane.

10.3 Theme: Synapses

1. D. Since the postsynaptic cell is already depolarized, the opening of GABA receptors will cause chloride ions to move in, carrying negative charge and reducing the depolarization of the postsynaptic cell.
2. B. Note that the stimulus was painful. For the stimulus to be perceived as painful by the subject, the dorsal horn neuron must have been depolarized to threshold.
3. B. The synapse between motoneurones and skeletal muscle cells is unusual in that one presynaptic action potential evokes a postsynaptic depolarization large enough to evoke an action potential.
4. E. Chloride is close to equilibrium across the plasma membrane of an otherwise unperturbed cell.
5. A

11.1 Theme: Processes downstream of receptor tyrosine kinases

1. F
2. G
3. D
4. B. A kinase is an enzyme that transfers the γ phosphate of ATP to another molecule; most of the kinases described in this book are protein kinases that phosphorylate serine, threonine, or tyrosine residues, but phosphoinositide 3-kinase phosphorylates the lipid PIP_2. Note that neither JAK nor Akt are correct answers. JAK is a kinase that is activated when phosphorylated on tyrosine, but the kinase that does the phosphorylation is JAK itself, not the cytokine receptor. Akt is indeed activated by being phosphorylated, but this phosphorylation is not performed by receptor tyrosine kinases. Furthermore the phosphorylation of Akt is on serine and threonine, not on tyrosine.
5. B. As well as activating MAPKKK, Ras can also activate PI 3-kinase, as shown in Figure 11.13.

11.2 Theme: Proteins activated by nucleotides

1. G
2. C
3. D
4. E
5. J

11.3 Theme: Inositol compounds

1. C
2. G
3. F
4. F
5. G
6. C

General comment: Unphosphorylated inositol, IP_2 and IP_4 play important roles in mammalian cells but are not covered in this book. IP_6 is found in plants but is relatively unimportant in animals.

12.1 Theme: The Three Modes of Intracellular Transport

1. B. β globin is used to assemble hemoglobin, which remains in the cytosol.
2. C. All other organelles bound by a single membrane receive the majority of their proteins by vesicular trafficking, but proteins destined for peroxisomes are synthesized on cytosolic ribosomes and only then imported into peroxisomes.
3. A. If the glucocorticoid hormone receptor binds its steroid hormone, it is imported into the nucleus.
4. A. Histones are synthesized in the cytosol, then imported into the nucleus.
5. D. If you answered C, transmembrane translocation: yes, integral membrane proteins are threaded through the protein translocator during synthesis – but this stops once the transmembrane part of the protein has been synthesized. We specified transport *subsequent to* complete synthesis.
6. C. Proteins destined for mitochondria are synthesized on cytosolic ribosomes and only then imported into mitochondria.

12.2 Theme: Trafficking Processes

1. B
2. A
3. C
4. D
5. E
6. E

12.3 Theme: GTPases

1. D
2. H
3. D
4. G

General comment: The term GTPase-activating protein is somewhat confusing. These proteins activate the GTPase catalytic activity of the GTPase, and therefore speed the process whereby the GTPase turns itself off.

13.1 Theme: Cytoskeletal structures

1. C
2. B
3. C
4. A
5. C
6. A
7. A

13.2 Theme: Proteins of the cytoskeleton

1. D
2. A
3. B
4. F
5. E
6. C

13.3 Theme: Fueling movement

1. C
2. F
3. C
4. E
5. C

14.1 Theme: Cell division

1. E
2. D
3. C
4. A
5. F
6. B

14.2 Theme: Checkpoints in the cell cycle

1. H
2. G
3. K
4. A
5. D
6. B

7. F
8. J

14.3 Theme: Life and death

1. H
2. D
3. E
4. F
5. B

15.1 Regulation and operation of CFTR

1. B, cAMP. We did not specifically describe the activation of CFTR in airway epithelium, but you should know that β-adrenergic receptors signal via Gs to adenylate cyclase, and that protein kinase A is activated by cAMP.
2. F
3. E
4. H
5. D

15.2 Diagnosis from the sweat

1. F. The sodium concentration in normal sweat is considerably lower than the value in interstitial fluid (140 mmoles liter^{-1}, Table 9.1 on page 000). Typically sweat values are in the range 20–50 mmoles liter^{-1}.
2. F. The chloride concentration in normal sweat is considerably lower than the value in interstitial fluid (100 mmoles liter^{-1}, Table 9.1 on page 000). Typically sweat values are in the range 15–40 mmoles liter^{-1}.
3. E. CF patients cannot significantly reduce the sodium concentration of the sweat below the concentration in interstitial fluid. Sodium concentrations in CF sweat are typically over 100 mmoles liter^{-1}.
4. E. CF patients cannot significantly reduce the chloride concentration of the sweat below the concentration in interstitial fluid. Chloride concentrations in CF sweat are typically over 100 mmoles liter^{-1}. A sweat test for CF is taken as positive if the chloride concentration is 60 mmoles liter^{-1} or more.
5. B. The lumen is negative with respect to the interstitial fluid because it is sodium that is actively moved out, leaving the lumen negative, with chloride following passively. Nevertheless because chloride can move easily, the net voltage difference is small.
6. A. The lumen is strongly negative because sodium is actively moved out, leaving the lumen negative, and chloride cannot follow to even up the charge movement.

15.3 Mutations in CF

1. C: Frameshift mutation. This region of the gene normally encodes Ala Phe Leu Arg Leu. . .; the mutant gene encodes Ala Phe Phe Glu Thr. . . This mutation is referred to in CF databases as 3905insT or L1258Ffs (frameshift starting with mutation of Leu1258 to Phe). It creates a STOP codon further on down the gene and severe disease in homozygotes.

2. B: Missense mutation. Arginine in the normal protein is changed to histidine in the mutant. This mutation is referred to in CF databases as R117H. It is a class III and class IV mutant: the channel opens less frequently, and is less permeable to chloride when open.

3. A: Nonsense mutation. The mutation has created the STOP codon TGA (UGA in RNA). This mutation is referred to in CF databases as R553* or R553X – the former is the IUPAC terminology but the second is more commonly used. It is a class I mutation: no CFTR protein is produced.

4. B: Missense mutation. Tyrosine in the normal protein is changed to aspartate in the mutant. This mutation is referred to in CF databases as Y569D. It is a class III mutation: the channel opens less frequently.

5. A: Nonsense mutation. The mutation has created the STOP codon TGA (UGA in RNA). This mutation is referred to in CF databases as W1282* or W1282X – the former is the IUPAC terminology but the second is more commonly used. It creates severe disease in a homozygote.

6. C: Frameshift mutation. This mutation is referred to in CF databases as CF444delA or I105Sfs (frameshift starting with mutation of Ile105 to Ser). It has created the STOP codon TAG (UAG in RNA) and hence creates severe disease in homozygotes.

GLOSSARY

7-methyl guanosine cap modified guanosine found at the 5′ terminus of eukaryotic mRNA. A guanosine is attached to the mRNA by a 5′ – 5′ -phosphodiester link and is subsequently methylated on atom number 7 of the guanine.

9 + 2 axoneme structure of cilium or flagellum; describes the arrangement of nine peripheral microtubules surrounding two central microtubules.

α-amino acid amino acid in which the carboxyl and amino groups are attached to the same carbon.

A (adenine) one of the bases present in DNA and RNA – adenine is a purine.

α helix a common secondary structure in proteins, in which the polypeptide chain is coiled, each turn of the helix taking 3.6 amino acid residues. The nitrogen atom in each peptide bond forms a hydrogen bond with the oxygen four residues ahead of it in the polypeptide chain.

A site (aminoacyl site) site on a ribosome occupied by an incoming tRNA and its linked amino acid.

abscission final step of cytokinesis by which the plasma membrane is cut to cause the two new daughter cells to become separate.

acetylcholine

$$CH_3\text{-}\underset{\underset{O}{\|}}{C}\text{-}O\text{-}CH_2\text{-}CH_2\text{-}\underset{\underset{CH_3}{|}}{\overset{\overset{CH_3}{|}}{\overset{+}{N}}}\text{-}CH_3$$

A transmitter released by various neurons including motor neurons.

acetylcholine receptor integral membrane protein that binds acetylcholine. There are two types: the nicotinic acetylcholine receptor is ionotropic while the muscarinic acetylcholine receptor is linked via G_q to phospholipase Cβ.

acid a molecule that readily gives H⁺ to water. Most organic acids are compounds containing a carboxyl group, although thiols (-SH) are also weakly acidic. The word is also used as an adjective to mean solutions of low pH, that is, with a high concentration of H⁺ (really H_3O^+) ions.

actin subunit protein of microfilaments. G-actin is the monomeric form while microfilaments are formed of F-actin.

actin-binding proteins proteins that bind to and modulate the function of G-actin or F-actin.

Cell Biology: A Short Course, Fourth Edition. Stephen Bolsover, Andrea Townsend-Nicholson, Greg FitzHarris, Elizabeth Shephard, Jeremy Hyams and Sandip Patel.
© 2022 John Wiley & Sons Ltd. Published 2022 by John Wiley & Sons Ltd.
Companion website: www.wiley.com/go/bolsover/cellbiology4

actin-related protein (Arp) an actin nucleation protein. New actin filaments can grow out from an Arp base.

action potential an explosive depolarization of the plasma membrane.

active site the region of an enzyme where the substrate binds and the reaction occurs.

acyl group
a group having the general formula:

$$C_nH_m - \overset{\displaystyle}{\underset{\displaystyle O}{C}} -$$

Acyl groups are formed when fatty acids are attached to other compounds by ester bonds.

adenine one of the bases present in DNA and RNA – adenine is a purine.

adenosine adenine linked to the sugar ribose. Adenosine is a nucleoside.

adenosine diphosphate (ADP) adenosine with two phosphates attached to the 5′ carbon of ribose. ADP is a nucleotide.

adenosine monophosphate (AMP) adenosine with one phosphate attached to the 5′ carbon of ribose. AMP is a nucleotide.

adenosine triphosphate (ATP) adenosine with three phosphates attached to the 5′ carbon of ribose. ATP is a nucleotide, and one of the cell's energy currencies.

adenylate cyclase (adenyl cyclase) an enzyme that converts ATP to the intracellular messenger cyclic AMP (cAMP).

adherens junctions type of anchoring junction in which the cell adhesion molecules are linked to actin microfilaments.

adipose tissue type of fatty connective tissue.

ADP adenosine diphosphate – adenosine with two phosphates attached to the 5′ carbon of ribose. ADP is a nucleotide.

adrenaline a hormone released into the blood when an individual is under stress. Adrenaline acts at β-adrenergic receptors to activate G_s and hence adenylate cyclase.

adrenergic receptor a receptor for the related chemicals adrenaline and noradrenaline. There are two isoforms, α and β. To a first approximation, noradrenaline acts mainly on α receptors linked to G_q and therefore generates a calcium signal, while adrenaline acts mainly on β receptors linked to G_s and therefore generates a cAMP signal.

α helix a common secondary structure in proteins, in which the polypeptide chain is coiled, each turn of the helix taking 3.6 amino acid residues. The nitrogen atom in each peptide bond forms a hydrogen bond with the oxygen four residues ahead of it in the polypeptide chain.

Akt a protein kinase that is activated when it is itself phosphorylated; this in turn only occurs when Akt is recruited to the plasma membrane by phosphatidylinositol trisphosphate (PIP_3). Akt phosphorylates proteins on serine and threonine residues (for example, the bcl-2 family protein BAX). An alternative name for Akt is protein kinase B.

alkali a strong base (that will take H^+ from water) such as sodium hydroxide or potassium hydroxide.

alkaline of a solution: one with a low concentration of H^+ (really H_3O^+) so that the pH is greater than 7.0.

allolactose a disaccharide sugar. Lactose is converted to allolactose by the enzyme β-galactosidase. Allolactose is an inducer of *lac* operon transcription.

alternative splicing phenomenon in which a single eukaryotic primary mRNA transcript can be processed to yield a number of different processed mRNAs and can therefore generate a number of different proteins.

α-amino acid amino acid in which the carboxyl and amino groups are attached to the same carbon.

amino acid a chemical that has both a carboxyl group and an amino group. In an α-amino acid, both the carboxyl and amino groups are attached to the same carbon atom. All proteins are generated using the genetically encoded palette of 19 α-amino acids plus proline.

α-amino acid amino acid in which the carboxyl and amino groups are attached to the same carbon.

aminoacyl site (A-site) site on a ribosome occupied by an incoming tRNA and its linked amino acid.

aminoacyl tRNA tRNA attached to an amino acid via an ester bond.

aminoacyl tRNA synthases family of enzymes, each of which attaches an amino acid to the appropriate tRNA.

γ-amino butyric acid a γ-amino acid that acts to open chloride channels in the plasma membrane of sensitive neurons. Usually called γ-amino butyric acid rather than γ-amino butyrate, even though the latter (NH_3^+-CH_2-CH_2-CH_2-COO^-) is the form in which it is found at neutral pH.

amino group the $-NH_2$ group. Amino groups are often basic, accepting a proton at normal body pH to become the positively changed $-NH_3^+$ group.

amino terminus (N terminus) the end of a peptide or polypeptide which has a free α-amino group.

amoeboid migration mode of cell migration in cells that lack stress fibers in which actin contraction causes the extension of cellular blebs.

amorphous without form. Not a specifically scientific word.

AMP adenosine monophosphate – adenosine with one phosphate attached to the $5'$ carbon of ribose. AMP is a nucleotide.

amphipathic "hating both" – a molecule with a hydrophobic region and a hydrophilic region is said to be amphipathic.

amphiphilic "loving both" – a synonym for amphipathic, that is, a molecule with a hydrophobic region and a hydrophilic region.

anaerobic without air. Obligate anaerobes are poisoned by oxygen and therefore can only function anaerobically. Other cells, such as yeast and skeletal muscle, can switch to using anaerobic glycolysis when they are denied oxygen.

anaphase the period of mitosis or meiosis during which sister chromatids or homologous chromosome pairs separate; consists of anaphase A and anaphase B.

anaphase A the part of anaphase in which the chromosomes move to the spindle poles.

anaphase B the part of anaphase in which the spindle poles are separated.

anaphase I anaphase of the first meiotic division (meiosis I).

anaphase II anaphase of the second meiotic division (meiosis II).

anaphase promoting complex (APC/C) protein complex that becomes active when the spindle assembly checkpoint is inactivated. The APC/C ubiquitinates cyclin-B and securin, causing them to be destroyed by the proteasome. The last C in APC/C stands for cyclosome, an alternative name for the anaphase-promoting complex.

anchoring junction class of cell junction that attaches the cytoskeleton of one cell to the cytoskeleton of its neighbor, forming a physically strong connection. There are two types, desmosomes which connect to intermediate filaments and adherens junctions which connect to actin microfilaments.

aneuploidy a cell is aneuploid if it has a few more or less than the normal number of chromosomes.

anion a negatively charged ion, e.g. chloride, Cl^-, or phosphate, HPO_4^{2-}.

anode a positively charged electrode, for example in a gel electrophoresis apparatus used for SDS-PAGE.

anterograde forward movement; when applied to axonal transport it means away from the cell body.

antibiotic a chemical that is produced by one type of organism and kills others – often by inhibiting protein synthesis. The most useful antibiotics to man are those that are selective for prokaryotes.

antibody a protein formed by the immune system that binds to and helps eliminate another chemical. Antibodies can be extremely selective for their ligand and are useful in many aspects of cell biology, such as immunofluorescence microscopy and western blotting.

anticodon the three bases on a tRNA molecule that hydrogen-bond to the codon on an mRNA molecule.

antigen any molecule that can be bound with high affinity by an antibody or (when presented by an MHC complex protein) by a T cell receptor.

antiparallel β sheet β sheet in which alternate parallel polypeptide chains run in opposite directions.

antisense oligonucleotide a short length of RNA complementary to a target mRNA.

Apaf-1 (apoptotic protease activating factor) apoptosis-pathway protein that promotes caspase activation.

APC/C (anaphase promoting complex) protein complex that becomes active when the spindle assembly checkpoint is inactivated. The APC/C ubiquitinates cyclin-B and securin, causing them to be destroyed by the proteasome. The last C in APC/C stands for cyclosome, an alternative name for the anaphase-promoting complex.

AP endonuclease a DNA repair enzyme that cleaves the phosphodiester links on either side of a depurinated or depyrimidinated sugar residue. AP stands for apurinic/apyrimidinic.

apoptosis a process in which a cell actively promotes its own destruction, as distinct from necrosis. Apoptosis is important in vertebrate development, where tissues and organs are shaped by the death of certain cell lineages.

apoptosis inhibitor protein proteins which block the action of caspases and hence help prevent apoptosis.

apoptotic protease activating factor (Apaf-1) apoptosis pathway protein that promotes caspase activation.

aporepressor protein that binds to an operator region and represses transcription only when it is complexed with another molecule.

aqueous watery.

Arf GTPase that plays a critical role in the formation of coated vesicles.

Argonaute protein protein that associates with processed micro RNA to form an RNA-inducing silencing complex (RISC).

Arp (actin-related protein) an actin nucleation protein. New actin filaments can grow out from an Arp base.

A-site (**aminoacyl site**) site on a ribosome occupied by an incoming tRNA and its linked amino acid.

assay a term for a chemical measurement, for example, one in which the activity of an enzyme reaction is measured.

ATM (ataxia telangiectasia mutated). A component of the checkpoint pathway that arrests the cell cycle in response to DNA damage or incomplete DNA replication. ATM is a protein kinase whose targets include two downstream protein kinases, Cds1 and Chk1.

ATP adenosine triphosphate – adenosine with three phosphates attached to the 5′ carbon of ribose. ATP is a nucleotide, and one of the cell's energy currencies.

ATP synthase (ATP synthetase) a carrier of the inner mitochondrial membrane that is built around a rotary motor. Ten H^+ enter the mitochondrial matrix for every three ATP made.

autoinduction the process of self-induction that occurs when a product of a reaction stimulates the production of more of itself. An example is the activation of the *lux* operon by the small molecule N-acyl-HSL; transcription of the *lux* operon then causes the production of more N-acyl-HSL.

autophagy the controlled digestion of cellular components by the lysosome.

autoradiography a process which detects a radioactive molecule. For example in a Southern blot experiment, the membrane that has been hybridized to a radioactive gene probe is placed in direct contact with a sheet of X-ray film. Radioactive decay activates the silver grains on the emulsion of the X-ray film. When the film is developed, areas that have been in contact with radioactivity will show as black. Most modern techniques avoid the use of radioactivity for safety reasons but the principles are the same.

axon the long process of a neuron, specialized for the rapid conduction of action potentials.

axonal transport movement of material along microtubules within a neuron process; can be outward (anterograde) or inward (retrograde).

9 + 2 axoneme structure of cilium or flagellum; describes the arrangement of nine peripheral microtubules surrounding two central microtubules.

β barrel a domain found in a number of proteins in which a β-pleated sheet is rolled up to form a tube.

β catenin a component of adherens junctions that can also, if free in the cell interior, enter the nucleus to regulate transcription.

β sheet common secondary structure in proteins, in which lengths of fully extended polypeptide run alongside each other, hydrogen bonds forming between the peptide bonds of the adjoining strands.

BAC (**bacterial artificial chromosome**) a cloning vector used to propagate DNAs of about 300 000 bp in bacterial cells.

bacteriophage (sometimes shortened to phage) a virus that infects bacterial cells.

Barr body a highly compacted, inactive X chromosome seen in somatic cells of a female animal.

basal lamina a synonym for the basement membrane, that is, the thin planar layer of extracellular matrix that supports epithelial cells. Some authors reserve the name basal lamina for a very thin sheet of fibers that are directly connected to the plasma membrane and the cytoskeleton by integrins and other integral membrane proteins, distinguishing this from a thicker, but still sheet-like, basement membrane. However, this distinction is rather artificial and is not generally accepted.

base there are two meanings in cell biology:

1. A chemical that will accept an H^+. Many organic bases contain amino groups, which accept H^+ to become $-NH_3^+$.

2. One of a group of ring-containing nitrogenous compounds that combine with a sugar to create a nucleoside; the members include adenine, guanine, hypoxanthine, cytosine, thymine, and uracil.

base excision repair process that repairs DNA double helices that have lost a purine (depurination), or in which a cytosine has been deaminated to uracil (U). The entire damaged monomer is removed and a correct deoxyribonucleotide inserted in its place.

basement membrane the thin planar layer of extracellular matrix that supports epithelial cells. Some authors use the name basal lamina for a very thin sheet of fibers that are directly connected to the plasma membrane and the cytoskeleton by integrins and other integral membrane proteins, distinguishing this from a thicker, but still sheet-like, basement membrane. However, this distinction is rather artificial and is not generally accepted.

base pair the Watson-Crick model of DNA showed that guanine in one DNA strand would fit nicely with cytosine in another strand, while adenine would fit nicely with thymine. The two hydrogen-bonded bases are called a base pair. RNA can also participate in base pairing: instead of thymine it is uracil that now pairs with adenine. The rare base inosine, found in some tRNAs, can base pair with any of uracil, cytosine, or adenine.

BAX a bcl-2 family protein that dimerizes into a channel that allows the release of cytochrome *c* from mitochondria, triggering apoptosis. BAX is phosphorylated and thereby inactivated by Akt.

bcl-2 an anti-apoptotic protein. By binding to BAX it prevents the release of cytochrome *c* from mitochondria.

bcl-2 family a family of related proteins that regulate cytochrome *c* release from mitochondria.

β barrel a domain found in a number of proteins in which a β-pleated sheet is rolled up to form a tube.

β catenin a component of adherens junctions that can also, if free in the cell interior, enter the nucleus to regulate transcription.

β-galactosidase an enzyme that cleaves the disaccharide lactose to produce glucose and galactose, and which also catalyzes the interconversion of lactose and allolactose. β-galactosidase is a product of the *lac* operon.

binary fission the mechanism of cell division in prokaryotes, in which the two copies of the single chromosome are partitioned between the progeny cells without the obvious cell-division machinery seen in eukaryote mitosis.

binding site a region of a protein which specifically binds a ligand. A property of the protein's tertiary structure.

bioluminescence the production of light by a living organism.

bisphosphate of a compound, bearing two independent phosphate groups (as opposed to a diphosphate, which bears a chain of two phosphates in a line). Phosphatidylinositol bisphosphate (PIP_2) is an example.

bivalent the structure formed when homologous chromosomes (one from the mother, one from the father) associate during prophase I of meiosis.

Bloom's syndrome a disease resulting from a deficiency in helicase. Affected individuals cannot repair their DNA and are susceptible to developing skin cancer and other cancers.

blue-green algae the old name for the photosynthetic prokaryotes now known as cyanobacteria.

blunt ends ends of a DNA molecule produced by an enzyme that cuts the two DNA strands at sites directly opposite one another.

bright-field microscopy the most basic form of light microscopy. The specimen appears against a bright background and appears darker than the background because of the light it has absorbed or scattered.

β sheet common secondary structure in proteins, in which lengths of fully extended polypeptide run alongside each other, hydrogen bonds forming between the peptide bonds of the adjoining strands.

bulky lesion distortion of the DNA helix caused by a thymine dimer.

C-terminus (carboxyl terminus) the end of a peptide or polypeptide that has a free α-carboxyl group. This end is made last on the ribosome.

Ca²⁺ ATPase a carrier that uses the energy released by ATP hydrolysis to move calcium ions up their concentration gradient out of the cytosol. Different isoforms of calcium ATPase are located at the plasma membrane and in the membrane of the endoplasmic reticulum.

cadherin a cell-adhesion molecule that helps form adherens junctions.

calcineurin a calcium–calmodulin-activated phosphatase, that is, an enzyme that removes phosphate groups from proteins, opposing the effects of kinases. Calcineurin is inhibited by the immunosuppressant drug cyclosporin.

calcium ATPase a carrier that uses the energy released by ATP hydrolysis to move calcium ions up their concentration gradient out of the cytosol. Different isoforms of calcium ATPase are located at the plasma membrane and in the membrane of the endoplasmic reticulum.

calcium-binding protein any protein that binds calcium. Calmodulin, troponin, and calreticulin are examples found in the cytosol, attached to actin filaments in striated muscle, and in the endoplasmic reticulum respectively.

calcium-calmodulin-dependent protein kinase an important regulatory enzyme, activated when calcium-loaded calmodulin binds, which phosphorylates target proteins on serine and threonine residues.

calcium-induced calcium release a process in which a rise of calcium concentration in the cytosol triggers the release of more calcium from the endoplasmic reticulum. The best understood mechanism of calcium-induced calcium release is via ryanodine receptors.

calcium pump another name for the calcium ATPase, the carrier that moves calcium ions up their electrochemical gradient out of the cytosol into the extracellular medium or into the endoplasmic reticulum.

calmodulin a calcium-binding protein found in many cells. When calmodulin binds calcium it can then activate other proteins such as the enzymes calcineurin and calcium-calmodulin-dependent protein kinase.

CaM-kinase another name for calcium-calmodulin-dependent protein kinase: an important regulatory enzyme, activated when calcium-loaded calmodulin binds, that phosphorylates target proteins on serine and threonine residues.

cAMP (cyclic adenosine monophosphate) a nucleotide produced from ATP by the action of the enzyme adenylate cyclase. cAMP is an intracellular messenger in many cells and exerts many of its actions by activating protein kinase A.

cAMP-dependent protein kinase (protein kinase A; PKA) a serine–threonine protein kinase that is activated by the intracellular messenger cyclic AMP.

cAMP-gated channel a channel found in the plasma membrane of scent-sensitive neurons. The channel opens when cAMP binds to its cytosolic face and allows sodium and potassium ions to pass.

cAMP phosphodiesterase enzyme that hydrolyzes cyclic AMP, producing AMP and hence turning off signaling through the cAMP system.

cap methylated guanine added to the 5′ end of a eukaryotic mRNA.

CAP (catabolite activator protein) a protein found in prokaryotes that binds to cAMP. The CAP–cAMP complex then binds within the promoter region of some bacterial operons and helps RNA polymerase to bind to the promoter.

carbohydrates the strict meaning of a carbohydrate is a chemical whose formula can be written $C_n(H_2O)_m$, as if it was made of carbon and water stuck together. Glucose = $C_6(H_2O)_6$, sucrose = $C_{12}(H_2O)_{11}$, and ribose = $C_5(H_2O)_5$ all fit this rule. But the word is used more generally to mean similar chemicals that do not quite meet this strict rule, for example deoxyribose = $C_5H_{10}O_4$.

carboxyl group the -COOH group. Carboxyl groups give up hydrogen ions to form the deprotonated group -COO⁻, so molecules that bear carboxyl groups are usually acids.

carboxyl terminus (C terminus) the end of a peptide or polypeptide which has a free α-carboxyl group. This end is made last at the ribosome.

cardiac muscle the form of striated muscle that is found in the heart.

carrier there are two meanings used in this book:

1. An integral membrane protein that forms a tube through the membrane that is never open all the way through. Solutes can move into the tube through the open end. When the channel changes shape, so that the end that was closed is open, the solute can leave on the other side of the membrane.

2. A person who has one nonfunctional or mutant copy of a gene, but who shows no effects because the other copy produces sufficient functional protein.

caspase a cysteine-containing protease that cleaves at aspartate residues. Caspases are responsible for the degradative processes that occur during apoptosis.

catabolite activator protein (CAP) a protein that binds to cAMP. The CAP–cAMP complex then binds within the promoter region of some bacterial operons and helps RNA polymerase to bind to the promoter.

catalyst a chemical or substance that reduces the activation energy of a reaction, allowing it to proceed more quickly. Many biological reactions would proceed at an infinitesimal rate without the aid of enzymes, which are protein catalysts.

cation positively charged ion, for example Na⁺, K⁺, and Ca²⁺.

cdc cell-division cycle. The acronym is usually used to denote genes which, when mutated, cause the cell-division cycle to be abnormal.

Cdc25 a protein phosphatase that dephosphorylates and hence activates Cdk1.

Cdk1 (cyclin-dependent kinase 1) a protein kinase involved in the regulation of the G2/M transition of the cell cycle. Associates with cyclins A and B.

Cdk2 (cyclin-dependent kinase 2) a protein kinase involved in the regulation of the G1 phase of the cell cycle. Associates with cyclins A and E.

Cdk4 (cyclin-dependent kinase 4) a protein kinase involved in the regulation of the G1 phase of the cell cycle. Associates with cyclin D.

Cdk6 (cyclin-dependent kinase 6) a protein kinase involved in the regulation of the G1 phase of the cell cycle. Associates with cyclin D.

cDNA (complementary DNA) a DNA copy of an mRNA molecule.

cDNA library collection of bacterial cells each of which contains a different foreign cDNA molecule.

cell the fundamental unit of life. A cell comprises a complex and ordered mass of protein, nucleic acid, and many biochemical species such as protein and nucleic acids separated from the world outside by a limiting membrane.

cell adhesion molecule an integral membrane protein responsible for sticking cells together. The extracellular domain binds a cell adhesion molecule on another cell while the intracellular domain binds to the cytoskeleton, either directly or via a linker protein.

cell center point immediately adjacent to the nucleus of eukaryotes where the centrosome and Golgi apparatus are located.

cell cycle (cell division cycle) ordered sequence of events that must occur for successful eukaryotic cell division; consists of G1, S, G2, and M phases.

cell junctions points of cell–cell interaction in tissues; includes tight junctions, anchoring junctions, and gap junctions.

cell membrane the membrane that surrounds the cell; also known as the plasmalemma or plasma membrane.

cell migration movement of cells within a medium or tissue, typically powered by cytoskeleton dynamics including actin polymerization and depolymerization.

cell surface membrane another name for the plasma membrane.

cell wall a rigid case that encloses plant and fungal cells and many prokaryotes. The cell wall lies outside the plasma membrane. Plant cell walls are composed of cellulose plus other polysaccharide molecules.

central dogma (of molecular biology) "DNA makes RNA makes protein" – the concept that the sequence of bases on DNA defines the sequence of bases on RNA, and the sequence of bases on RNA then defines the sequence of amino acids in protein.

centriole the structure found at the centrosome (= microtubule organizing center) of animal cells; composed of microtubules.

centromere the region of the chromosome at which the kinetochore (where the microtubules of the mitotic or meiotic spindle attach) is formed.

centrosome (microtubule organizing center) structure comprising centrioles and a surrounding cloud of pericentriolar material from which cytoplasmic microtubules arise.

CF *(cystic fibrosis)* inherited disease characterized by failure of the pancreas and by thick sticky mucus in the lungs leading to fatal lung infection unless treated. Cystic fibrosis is caused by failure to make, or properly target, functional plasma membrane chloride channels.

CFTR (cystic fibrosis transmembrane regulator) a chloride channel encoded by the *CTFR* gene, especially important in epithelia.

cGMP (cyclic guanosine monophosphate) a nucleotide produced from GTP by the action of the enzyme guanylate cyclase. cGMP is an intracellular messenger in many cells and exerts many of its actions by activating protein kinase G.

cGMP-dependent protein kinase (protein kinase G) a protein kinase that is activated by the intracellular messenger cyclic GMP. Protein kinase G phosphorylates proteins (for example the calcium ATPase) on serine and threonine residues.

cGMP phosphodiesterase enzyme that hydrolyzes the cyclizing phosphodiester bond in cGMP to generate 5′ GMP.

channel an integral membrane protein that forms a continuous water-filled hole through the membrane.

chaotropic reagents reagents such as urea which cause proteins to lose all their higher levels of structure and adopt random, changing conformations.

chaperone a protein that helps other proteins to remain unfolded for correct protein targeting, or to fold into their correct three-dimensional structure.

charge
1. an excess or deficit of electrons giving a negative or positive charge respectively.
2. a transfer RNA is said to be charged when it has an amino acid attached.

charged tRNA a tRNA attached to an amino acid.

checkpoint a control point in the eukaryote cell cycle, where progression is allowed only if all the necessary processes have been completed.

chiasmata (singular chiasma) structures formed during crossing over between the chromatids of homologous chromosomes during meiosis; the physical manifestation of genetic recombination.

chimera structure formed from two different parts. Chimeric proteins are generated by joining together all or part of the protein-coding sections of two distinct genes. Chimeric organisms are formed by mixing two or more distinct clones of cells.

chimeric protein proteins generated by joining together all or part of the protein-coding sections of two distinct genes, for example GFP and a protein of interest.

chiral structure whose mirror image cannot be superimposed on it. Organic molecules will be chiral if a carbon atom has four different groups attached to it and is therefore asymmetric.

chloroplast the photosynthetic organelle of plant cells.

cholesterol a hydrophobic molecule made up of four fused rings with a short hydrocarbon tail at one end and a single hydroxyl group at the other. Cholesterol is a component of eukaryotic (but not prokaryotic) membranes and helps to keep them flexible. Steroid hormones are synthesized from cholesterol.

chromatid a complete DNA double helix plus accessory proteins subsequent to DNA replication in eukaryotes. At mitosis, the chromosome is seen to be composed of two chromatids; these then separate to form the chromosomes of the two daughter cells.

chromatin a complex of DNA and certain DNA-binding proteins such as histones.

chromosome a single, enormously long molecule of DNA, together with its accessory proteins. Chromosomes are the units of organization of the nuclear chromatin and carry many genes. In eukaryotes chromosomes are linear; in prokaryotes they are circular.

chromosome walking investigating a chromosome bit by overlapping bit, each bit being used to clone the next.

cilium (plural cilia) locomotory appendage of some epithelial cells and protozoa.

circular RNA a form of regulatory RNA in which the free 5′ and 3′ ends are joined. Circular RNAs bind to target micro RNAs and reduce their inhibitory effect.

cis face the side to which material is added. Of the Golgi complex, the surface that receives vesicles from the endoplasmic reticulum.

cisternae flattened membrane-bound sacs, for example, those that make up the Golgi apparatus.

CKI (cyclin dependent kinase inhibitor) a type of cell cycle regulatory protein. Binds to and inactivates CDKs.

class I mutation of **CF** mutations that cause a complete failure to synthesize CFTR, typically a nonsense mutation.

class II mutation of **CF** mutations that generate functional but unstable or misfolded CFTR, most or all of which is therefore destroyed before it can reach the plasma membrane.

class III mutation of **CF** mutations that generate CFTR that fails to gate open.

class IV mutation of **CF** mutations that generate CFTR where the pore, even when gated to the open state, has poor or no chloride permeability.

class V mutation of **CF** mutations that generate normal, functional CFTR that is trafficked to the plasma membrane, but the amounts of CFTR synthesized are too small.

class VI mutation of **CF** mutations that generate functional CFTR that is trafficked to the plasma membrane, but once at the membrane the protein degrades quickly.

clathrin a protein that functions to cause vesicle budding in response to binding of specific ligand.

clathrin adaptor protein protein that binds to specific membrane receptors and which in turn recruits clathrin to form a coated vesicle. The vesicle therefore contains the molecule for which the receptor is specific.

cleavage furrow in animal cells, the structure that constricts the middle of the cell during cytokinesis.

clone a number of genetically identical individuals.

clone library a collection of bacterial clones where each clone contains a different foreign DNA molecule.

cloning strictly, the creation of a number of genetically identical organisms. In molecular genetics, the term is used to mean the multiplication of particular sequences of DNA by an asexual process such as bacterial cell division.

cloning vector a DNA molecule that carries genes of interest, can be inserted into cells, and which will then be replicated inside the cells. Cloning vectors range in size from plasmids to entire artificial chromosomes.

closed promoter complex a structure formed when RNA polymerase binds to a promoter sequence prior to the start of transcription.

coatamer protein complex that encapsulates one class of coated vesicle. Formation of the coatamer coat on a previously flat membrane forces the membrane into a curved shape and therefore drives vesicle formation.

coated vesicle a cytoplasmic vesicle encapsulated by a protein coat. There are two types of coated vesicles, coated by coatamer and clathrin respectively.

coat proteins peripheral membrane proteins that, on associating with a membrane, force it to bend and form a vesicle.

codon a sequence of three bases in an mRNA molecule that specifies a particular amino acid.

cohesin protein complex that holds chromatids together prior to chromosome segregation. Degradation of cohesin allows the chromatids to separate at the start of anaphase in mitosis and meiosis.

colchicine a plant toxin from the autumn crocus, *Colchicum autumnale*; binds to tubulin and causes microtubule disassembly.

collagen the major structural protein of the extracellular matrix.

columnar taller than it is broad. Used as a description of some types of epithelial cells.

competent in molecular genetics this refers to a bacterial culture treated with a solution such as calcium chloride so that uptake of foreign DNA is enhanced.

complementary note the spelling! Two structures are said to be complementary when they fit into or associate with each other. The two strands of the DNA double helix are complementary, as are the anticodon and codon of transfer and messenger RNA.

complementary DNA (cDNA) a DNA copy of an mRNA molecule.

complex (as a noun) an association of molecules that is held together by noncovalent interactions, and often can be readily dissociated.

condenser lens the lens of light and electron microscopes that focusses light (or electrons) onto the specimen.

confocal light microscopy a technique, usually used in fluorescence microscopy, in which only one spot in the specimen is illuminated and the light collection system set up to only collect light from that spot. This means that out of focus and scattered light is rejected.

connective tissue a tissue that contains relatively few cells within a large volume of extracellular matrix.

connexon an integral membrane protein that can open to form a channel about 1.5 nm in diameter when it contacts a second connexon on another cell. This forms a water-filled tube that runs all the way through the plasma membrane of the first cell, across the small gap between the cells, and through the plasma membrane of the second cell, so allowing passage of solute from the cytosol of one cell to the cytosol of the other.

consensus sequence the standard, average form of a sequence of bases in DNA, or amino acids in a protein, that play a particular role. Small deviations from this consensus sequence are often found in enhancer sequences in DNA, or in amino acid sequences in proteins, that are nevertheless functional.

constitutive operating all the time without obvious regulation. Housekeeping genes, expressed all the time, are sometimes called constitutive genes. Proteins that are secreted all the time are said to use the constitutive route.

constitutive exocytosis (constitutive secretion) secretion that continues all the time, without the need for a signal such as an increase of cytosolic calcium concentration.

contact inhibition inhibition of cell division by cell–cell contact. Contact inhibition allows cells to proliferate to fill a gap and then stop dividing.

-COOH carboxyl group. Carboxyl groups give up hydrogen ions to form the deprotonated group -COO⁻, so molecules that bear carboxyl groups are usually acids.

cortex the outer part of any organ or structure. For instance, the tissue that forms the outer region of the brain, and the outer regions of a cell occupied by an actin lattice, are each called cortex.

corticosteroids steroid hormones synthesized in the cortex of the adrenal gland. Corticosteroids regulate glucose metabolism and salt balance and also affect inflammation and the immune response.

covalent bond a strong bond between two atoms in which electrons are shared.

cristae the name given to the folds of the inner membrane of mitochondria.

crossing over process by which maternal and paternal chromosomes exchange portions of DNA during meiosis.

crosstalk two messenger systems show crosstalk when one messenger can produce some or all of the effects of the other.

C-terminus (carboxyl terminus) the end of a peptide or polypeptide that has a free α-carboxyl group. This end is made last on the ribosome.

cyanobacteria the photosynthetic prokaryotes that were formerly known as blue-green algae.

cyclic adenosine monophosphate (cAMP) a nucleotide produced from ATP by the action of the enzyme adenylate cyclase. cAMP is an intracellular messenger in many cells and exerts many of its actions by activating protein kinase A.

cyclic guanosine monophosphate (cGMP) a nucleotide produced from GTP by the action of the enzyme guanylate cyclase. cGMP is an intracellular messenger in many cells and exerts many of its actions by activating protein kinase G.

cyclin one of a family of proteins whose level oscillates (cycles) through the cell division cycle. Cyclins associate with and activate cyclin-dependent kinases and hence allow progression through cell-cycle control points.

cyclin D the cyclin that associates with and activates Cdk4 and Cdk6, promoting cell proliferation by controlling early events in G1.

cyclin-dependent kinase inhibitor (CKI) a type of cell-cycle regulatory protein. Binds to and inactivates CDKs.

cyclin-dependent protein kinase (CDK) one of the family of protein kinases that regulate the cell cycle. Cyclin-dependent kinases are only active when bound to one of the family of cyclin proteins. For example, CDK1 associates with cyclin B and regulates the G2/M transition while CDK2 associates with cyclin D and cyclin E and regulates the G1/S transition.

cystic fibrosis (CF) inherited disease characterized by failure of the pancreas and by thick sticky mucus in the lungs leading to fatal lung infection unless treated. Cystic fibrosis is caused by failure to make, or properly target, functional plasma-membrane chloride channels.

cystic fibrosis transmembrane regulator (CFTR) a chloride channel encoded by the *CTFR* gene, especially important in epithelia.

cystine a double amino acid formed by two cysteine molecules joined by a disulphide bond.

cytochemistry the use of chemical compounds to stain specific cell structures and organelles.

cytochrome proteins with a heme prosthetic group that are able to transfer electrons. Cytochromes form a critical part of the electron transport chain of mitochondria.

cytochrome c a soluble protein of the mitochondrial intermembrane space, often found loosely associated with the inner mitochondrial membrane. Cytochrome *c* transports electrons between components of the electron transport chain. If it is allowed to escape from mitochondria cytochrome *c* activates caspase 9 and hence triggers apoptosis.

cytokine one of a large family of protein transmitters that are especially important in the immune system.

cytokinesis the process by which a cell divides in two; part of the M-phase of the cell division cycle.

cytoplasm the semi-viscous ground substance of the cell. All the volume outside the nucleus and inside the plasma membrane is cytoplasm.

cytoplasmic dynein a motor protein that moves along microtubules in a retrograde direction.

cytosine one of the bases present in DNA and RNA – cytosine is a pyrimidine.

cytoskeleton cytoplasmic filament system consisting of microtubules, microfilaments, and intermediate filaments.

cytosol the viscous, aqueous medium in which the organelles and the cytoskeleton are bathed.

cytostatic factor cytoplasmic activity found only in eggs that causes the cell cycle to arrest at metaphase until fertilization.

DAG (diacylglycerol) two acyl groups (fatty acid chains) joined by ester bonds to a glycerol backbone. Diacylglycerol is produced by the action of phospholipase C on phospholipid and helps to activate protein kinase C.

deamination the removal of an amino group. Deamination of cytosine to form uracil is a form of DNA damage.

death domain a domain found on proteins concerned with regulating apoptosis, such as Fas and the p75 neurotrophin receptor. When activated death domain proteins turn on caspase 8 and hence initiate apoptosis.

default pathway the pathway followed by proteins synthesized on the rough endoplasmic reticulum and having no additional sorting signal; Golgi processing followed by constitutive secretion.

degenerate when a language or code is degenerate, concepts or outputs can be represented in more than one way by the language or code. So, in English ethanol and ethyl alcohol have exactly the same meaning; in the genetic code, AGU and UCA both code for serine.

denature to cause to lose three-dimensional structure by breaking noncovalent intramolecular bonds.

dendrite a branching cell process. The term is commonly used to name those processes of neurons that are too short to be called axons.

deoxyribonucleic acid (DNA) a polymer of deoxyribonucleotides. DNA specifies the inherited instructions of a cell.

deoxyribonucleotide the building block of DNA that is made up of a nitrogenous base and the sugar deoxyribose to which a phosphate group is attached.

deoxyribose a ribose that instead of -OH has only -H on carbon 2. Deoxyribose is the sugar used in the nucleotides that make up DNA.

depolarization any positive shift in the membrane voltage, whatever its size or cause.

depth of focus the distance toward and away from an object over which components of the object remain in clear focus.

depurination the removal of either of the purine bases, adenine and guanine, from a DNA molecule. Depurination is a form of DNA damage.

desmin protein that makes up the intermediate filaments in muscle cells.

desmosome a type of anchoring junction which joins the intermediate filaments of neighboring cells. Desmosomes are common in tissues such as skin.

diacylglycerol (DAG) two acyl groups (fatty acid chains) joined by ester bonds to a glycerol backbone. Diacylglycerol is produced by the action of phospholipase C on phospholipid and helps to activate protein kinase C.

dideoxyribonucleotide a manmade molecule similar to a deoxyribonucleotide but lacking a 3′ hydroxyl group on its sugar residue. Used in DNA sequencing.

differentiation the process whereby a cell becomes specialized for a particular function.

diffusion the movement of a substance that results from the individual small random thermal movements of its molecules.

dimeric formed of two parts. Of a molecule, formed of two parts that are not covalently linked.

diphosphate of a compound, bearing a chain of two phosphates in a line (as opposed to a bisphosphate, which bears two independent phosphate groups). Adenosine diphosphate is an example.

diploid containing two sets of chromosomes, in humans, this means two sets of 23, one from the father, one from the mother. Most of the cells of the body (the somatic cells) are diploid.

dipole a molecule that has a positive and a negative charge (partial or whole) separated by a (usually) small distance.

dishevelled a cytosolic protein that forms a complex with axin and GSK3 on the cytosolic aspect of frizzled receptors when the latter bind Wnt proteins. This disrupts the β-catenin destruction complex, allowing β-catenin concentrations to increase and affect transcription.

distal far from the center.

disulfide bond (bridge) covalent bond between two sulfur atoms. In proteins disulfide bonds form by oxidation of two thiol (-SH) groups of cysteine residues. Found chiefly in extracellular proteins.

DNA (deoxyribonucleic acid) a polymer of deoxyribonucleotides. DNA specifies the inherited instructions of a cell.

DnaA DNA binding protein that causes the two strands of the double helix to separate in the first stages of DNA replication.

DnaB helicase that moves along a DNA strand, breaking hydrogen bonds, and in the process unwinding the helix.

DnaC DNA binding protein that serves to bring DNAb to the DNA strands.

DNA excision process of cutting out damaged DNA prior to repair.

DNA fingerprint the individual pattern of DNA fragments determined by the number and position of specific repeated sequences.

DNA ligase an enzyme that joins two DNA molecules by catalyzing the formation of a phosphodiester bond.

DNA polymerase enzyme that synthesizes DNA by catalyzing the formation of a phosphodiester link. DNA is always synthesized in the 5′ to 3′ direction.

DNA polymerase III the enzyme that copies DNA during chromosome replication in prokaryotes.

DNA repair enzymes enzymes that detect and repair altered DNA.

DNA replication a process in which the two strands of the double helix unwind and each acts as a template for the synthesis of a new strand of DNA.

DNA sequencing determining the order of bases on the DNA strand.

docking protein (signal recognition particle receptor) the receptor on the endoplasmic reticulum to which the signal recognition particle binds during the process of polypeptide chain synthesis and import into the endoplasmic reticulum.

dogma belief. The "central dogma" of molecular biology is that "DNA makes RNA makes protein" – the concept that the sequence of bases on DNA defines the sequence of bases on RNA, and the sequence of bases on RNA then defines the sequence of amino acids on protein.

domain a separately folded segment of the polypeptide chain of a protein.

double helix the structure formed when two filaments wind about each other, most commonly applied to DNA, but also applicable to, for example, F-actin.

downstream a general term meaning the direction in which things move. When applied to the DNA within and adjacent to a gene, it means lying on the side of the transcription start site that is transcribed into RNA. When applied to signaling pathways it means in the direction in which the signal travels, thus MAP kinase is downstream of Ras.

dynamic instability term that describes the behavior of microtubules in which they switch from phases of growth to phases of shortening.

dynamin GTPase that plays a critical role in the formation of clathrin-coated vesicles.

dynein a motor protein. Cytoplasmic dynein moves vesicles along microtubules while dynein arms power ciliary and flagellar beating by generating sliding between adjacent outer doublet microtubules.

E-site (exit site) the site on the ribosome from which the tRNA exits after its amino acid has been transferred to the peptide chain.

E2F transcription factor required for DNA synthesis. In quiescent cells E2F is prevented from activating transcription by being bound to RB, the product of the retinoblastoma gene. E2F is released when RB is phosphorylated by CDK4 or CDK6.

effective stroke the part of the beat cycle of a cilium that pushes on the extracellular medium.

effector caspase a caspase that digests cellular components in the process of apoptosis. Effector caspases are differentiated from caspases 8 and 9 which do not themselves digest cellular components but which activate effector caspases by hydrolyzing particular peptide bonds in them.

eIFs (eukaryotic initiation factors) proteins that assist the small and large ribosomal subunits of eukaryotes to assemble correctly on the mRNA at the start of protein synthesis.

electrically excitable able to produce action potentials.

electrochemical gradient the free energy gradient for an ion in solution. The arithmetical sum of gradients due to concentration and voltage.

electron microscope a microscope in which the image is formed by passage of electrons through, or scattering of electrons by, the object.

electron transport chain the series of electron acceptor/donator molecules found in the inner mitochondrial membrane which uses chemical energy to transport hydrogen ions up their electrochemical gradient out of the mitochondrion.

electrophoresis a method of separating charged molecules by drawing them through a filtering gel material using an electrical field.

electrostatic bond a strong attraction between ions or charged groups of opposite charge.

electrostatic interaction attraction or repulsion between ions or charged groups.

elongation factors proteins that speed up the process of protein synthesis at the ribosome. Elongation factors tu and G are GTPases.

embryonic stem cell cells derived from an early embryo. Embryonic stem cells have the capability to divide indefinitely and to become any cell type in the body.

endocytosis the inward budding of plasma membrane to form vesicles. Endocytosis is the process by which cells retrieve plasma membrane and take up material from their surroundings.

endocytotic pathway the pathway of vesicular traffic that begins with endocytosis, the budding of vesicles from the plasma membrane, and delivers the vesicles and their cargo to the cell interior.

endogenous belonging to the cell or organism; not introduced by a pathogen or human intervention.

endonuclease an enzyme that digests nucleic acids by cleaving phosphodiester bonds away from the ends of the strand, c.f. exonuclease.

endoplasmic reticulum (ER) a network (reticulum) of membrane-delimited tubes and sacs that extends from the outer membrane of the nuclear envelope to the plasma membrane. There are two types of ER, rough endoplasmic reticulum (RER) with a surface coating of ribosomes, and smooth endoplasmic reticulum (SER).

endosome the organelle to which newly formed endocytotic vesicles are translocated and with which they fuse.

endosymbiotic theory the proposal that some of the organelles of eukaryotic cells originated as free-living prokaryotes.

endothelial cells epithelial cells that line blood vessels and other body cavities that do not open to the outside.

endothelium a layer of cells that lines blood vessels and other body cavities that do not open to the outside.

energy currency a source of energy that the cell can use to drive processes that would otherwise not occur because their ΔG is positive. Energy currencies can be coenzymes such as ATP and NADH, which give up energy on conversion to respectively ADP and NAD$^+$, or electrochemical ion gradients such as the hydrogen ion gradient across the mitochondrial membrane and the sodium gradient across the plasma membrane.

enhancer a specific DNA sequence to which a protein binds to increase the rate of transcription of a gene.

envelope a closed sheet enclosing a volume. The term is used to describe, among other things, the double membrane layer enclosing certain organelles, and the outer membrane of certain viruses.

enzyme a biological catalyst. Most are proteins, but some critical (and evolutionarily ancient) enzymes are made of RNA.

epidermis the protective outer cell layer of an organism.

epigenetics chemical modification of DNA that does not destroy the coding sequence but can affect the structure of the double helix or the frequency with which the DNA is transcribed. Methylation of adenines is one example of epigenetics.

epithelial cells the cells that make up an epithelium.

epithelium a sheet of cells.

equilibrium the total balance of opposing forces. A process or object is in equilibrium if the tendency to go in one direction is exactly equal to the tendency to go in the other direction. For an ion this condition is equivalent to saying that the electrochemical gradient for that ion across the membrane is zero. For a chemical reaction, equilibrium occurs when the rate of the forward reaction is equal to the rate of the reverse reaction, e.g. $2H_2O \rightleftarrows OH^- + H_3O^+$.

equilibrium voltage the membrane voltage that will exactly balance the concentration gradient of a particular ion.

ER (endoplasmic reticulum) network (reticulum) of membrane channels that extends from the outer membrane of the nuclear envelope to the plasma membrane; there are two types of ER, rough endoplasmic reticulum (RER) with a surface coating of ribosomes and smooth endoplasmic reticulum (SER).

ES (embryonic stem) cell a cell derived from a very early embryo. ES cells have not yet determined their developmental fate and can therefore, depending on the conditions, generate the entire range of tissue types.

E-site (exit site) the site on the ribosome from which the tRNA exits after its amino acid has been transferred to the peptide chain.

ester bond the bond formed between the hydrogen of an alcohol group and the hydroxyl of a carboxyl group by the elimination of water.

euchromatin that portion of the nuclear chromatin that is not tightly packed. Euchromatin contains genes that code for proteins that are being actively transcribed.

eukaryotic an organism whose cells contain distinct nuclei and other organelles; includes all known organisms except prokaryotes (bacteria and cyanobacteria).

eukaryotic initiation factors (eIFs) proteins that assist the small and large ribosomal subunits of eukaryotes to assemble correctly on the mRNA at the start of protein synthesis.

excitation-contraction coupling the signaling and mechanical control systems that operate in muscle cells to cause contraction when the plasma membrane depolarizes.

excitatory transmitter a transmitter that causes electrical depolarization of the target cell and makes it more likely to fire an action potential.

exit site (E-site) the site on the ribosome from which the tRNA exits after its amino acid has been transferred to the peptide chain.

exocytosis the fusion of a vesicle with the plasma membrane. Exocytosis causes the soluble contents of the vesicle to be released to the extracellular medium, while integral membrane proteins of the vesicle become integral membrane proteins of the plasma membrane.

exocytotic pathway the pathway of vesicular traffic that culminates in the secretion of product from the cell.

exon in a eukaryotic gene, exons are those parts that after RNA processing leave the nucleus. In contrast introns are spliced out before the RNA leaves the nucleus.

exonuclease an enzyme that digests nucleic acids by cleaving phosphodiester links successively from one end of the molecule: c.f. endonuclease.

expression (of a gene) the appearance of the protein for which the gene encodes.

expression vector a cloning vector containing a promoter sequence recognized by the host cell, thus enabling a foreign DNA insert to be transcribed into mRNA.

extracellular matrix the meshwork of filaments and fibers that surrounds and supports mammalian cells. The major protein of the extracellular matrix is collagen.

extracellular medium the aqueous medium outside cells. For a unicellular organism the extracellular medium is the outside world. For a multicellular organism such as a human being the extracellular medium is the interstitial fluid.

extragenic DNA DNA that can neither be identified as coding for protein or RNA nor as being promoters or enhancers regulating transcription.

extrinsic pathway pathway of apoptosis triggered on the receipt of extracellular signals.

Fas a cell-surface receptor that activates caspase 8 and hence triggers apoptosis.

fat a lipid that is solid at room temperature. In contrast, oils are liquid at room temperature.

fat cells (adipocytes) cells that store fats.

fatty acid
a carboxyl group attached to a long chain of carbon atoms with attached hydrogens, that is, a chemical of the general form:

$$C_nH_m - \underset{\underset{O}{\|}}{C} - OH$$

feedback a process in which the result of a process modifies the mechanisms carrying out that process to increase or decrease their rate. In negative feedback a change in some parameter activates a mechanism that reverses the change in that parameter; an example is the effect of tryptophan on expression of the *trp* operon. In positive feedback a change in some parameter activates a mechanism that accelerates the change; an example is the effect of depolarization on the opening of voltage-gated sodium channels.

fertilization fusion of sperm and egg to generate a single-cell embryo.

fibroblast a cell found in connective tissue. Fibroblasts synthesize collagen and other components of the extracellular matrix.

fibroblast growth factor a transmitter that opposes apoptosis and promotes cell division in target cells. Although named for its effect on fibroblasts, FGF triggers proliferation of many tissues and plays critical roles in determining cell fate during development.

fight or flight the set of physiological changes that prepare the body for muscular work: either fighting or running away fast.

fission breakage into two parts. The word is used of prokaryote replication. In addition, it describes the division of a single membrane-bound organelle into two and the process whereby a vesicle breaks away from a membrane.

flagellum a swimming appendage. In eukaryotes flagella are extensions of the cell that use a dynein/microtubule motor system. In prokaryotes flagella are extracellular proteins rotated by a motor at their base.

fluorescence microscope a microscope in which excitation light is shone onto the specimen, and light emitted by fluorescent components within the specimen is collected to form a magnified image.

fluorescent fluorescent molecules and ions emit light of one wavelength when excited with light of a different wavelength. The emitted light is usually of longer wavelength than the excitation light, although some modern techniques, such as two photon microscopy, allow this rule to be evaded.

fmet (formyl methionine) methionine modified by the attachment of a formyl group – fmet is the first amino acid in all newly made bacterial polypeptides.

focal adhesion points at which a locomoting cell makes contact with its substrate.

formyl methionine (fmet) methionine modified by the attachment of a formyl group – fmet is the first amino acid in all newly made bacterial polypeptides.

frameshift mutation a mutation that changes the mRNA reading frame, caused by the insertion or deletion of a nucleotide.

frizzled receptors receptors on the cell surface that bind Wnt proteins and initiate a signaling process that acts through β catenin to affect transcription of particular genes.

G0 describes the quiescent state of cells that have left the cell-division cycle.

G1 (gap 1) period of the cell division cycle that separates mitosis from the following S-phase.

G1/S transition checkpoint the checkpoint controlling entry into the S-phase of the eukaryotic cell cycle. For example, DNA replication is blocked if the DNA is damaged.

G2 (gap2) period of the cell division cycle between the completion of S-phase and the start of cell division or M-phase.

G2/M transition checkpoint the checkpoint controlling entry into mitosis. Cells can proceed into mitosis only when cyclin-dependent kinase 1 is active, which in turn only occurs if the DNA is undamaged and the cell is large enough.

GABA (γ-amino butyric acid) a γ-amino acid that acts to open chloride channels in the plasma membrane of sensitive neurons. Usually called γ-amino butyric acid rather than γ-amino butyrate, even though the latter (NH_3^+-CH_2-CH_2-CH_2-COO^-) is the form in which it is found at neutral pH.

GABA$_A$ receptor an ionotropic receptor for the transmitter GABA. GABA$_A$ receptors allow chloride ions to pass when GABA binds. In contrast GABA$_B$ receptors, not mentioned in this book, are G protein-coupled receptors.

GABAergic using GABA as a transmitter.

G protein-coupled receptor one of a large number of receptors on the cell surface that, upon binding of their transmitter, act as guanine nucleotide exchange factors for a heterotrimeric G protein.

G-actin globular, subunit form of *actin*.

gamete a sperm or egg. Gametes are haploid; that is, they contain just one set of chromosomes (23 in humans).

gamma-amino butyric acid (γ-amino butyric acid, GABA) a γ-amino acid that acts to open chloride channels in the plasma membrane of sensitive neurons. Usually called γ-amino butyric acid rather than γ-amino butyrate, even though the latter (NH_3^+-CH_2-CH_2-CH_2-COO^-) is the form in which it is found at neutral pH.

GAP (GTPase-activating protein) a protein that speeds up the rate at which GTPases hydrolyze GTP and therefore switch from the active to the inactive state.

gap 0 (G0) describes the quiescent state of cells that have left the cell division cycle.

gap 1 (G1) period of the cell division cycle that separates mitosis from the following S-phase.

gap 2 (G2) period of the cell division cycle between the completion of S-phase and the start of cell division or M-phase.

gap junction type of cell junction that allows solute to pass from the cytosol of one cell to the cytosol of its neighbor without passing through the extracellular medium. Gap junctions consist of many paired connexons.

gap junction channel the channel formed by two connexons. Connexons only open when they contact a second connexon on another cell, in this case they open and form a water-filled tube about 1.5 nm in diameter that runs all the way through the plasma membrane of the first cell, across the small gap between the cells, and through the plasma membrane of the second cell, so allowing passage of solute from the cytosol of one cell to the cytosol of the other.

gated, gating a channel is gated if it can switch to a shape in which the tube through the membrane is closed.

gated transport transport of fully folded proteins through intracellular pores that open to allow their passage.

GEF (guanine nucleotide exchange factor) a protein that accelerates the rate at which GDP leaves a GTPase to be replaced by GTP, thus switching the GTPase from its inactive state to its active state.

gene the fundamental unit of heredity. In many cases a gene contains the information needed to code for a single polypeptide.

gene chip a tiny glass wafer to which cloned DNAs are attached. Also known as microarrays.

gene family a group of genes that share sequence similarity and usually code for proteins with a similar function.

gene probe a cDNA or genomic DNA fragment used to detect a specific DNA sequence to which it is complementary in sequence. The probe is tagged in some way to make it easy to detect. The tag could be, for example, a radioactive isotope or a fluorescent dye.

gene therapy a correction or alleviation of a genetic disorder by the introduction of a normal gene copy into an affected individual.

genetically modified (GM) an organism with a genome that has been modified by modern molecular techniques, usually by the addition of novel gene(s) or by swapping in new DNA to replace existing gene(s).

genetic code the relationship between the sequence of the four bases in DNA and the amino acid sequence of proteins.

genetic perturbation an experimental technique in which a specific genetic intervention is applied (for example knocking down the expression of a target gene) and then all the effects of that intervention tabulated, for example by RNA-seq.

genome one complete set of an organism's genes. In humans, the genome resides on 23 chromosomes, and each cell has two sets.

germ cells the cells that give rise to the eggs and sperm.

GFP (green fluorescent protein) a fluorescent protein made by the jellyfish *Aequorea victoria*. Unlike other colored or fluorescent proteins, it contains no prosthetic groups and therefore will fluoresce when expressed by any cell in which the gene is successfully inserted and expressed.

glial cells electrically inexcitable cells found in the nervous system.

glucocorticoid a steroid hormone produced by the adrenal cortex that forms part of the system controlling blood sugar levels.

glucocorticoid receptor the intracellular receptor to which glucocorticoid hormone binds.

glucose the commonest sugar in the blood and the dominant cellular fuel in animals.

glucose carrier a plasma membrane protein that carries glucose into or out of cells. Some cells, such as skeletal muscle cells, will only translocate glucose carriers to their membranes when Akt is active.

glutamatergic using glutamate as a transmitter. A neuron is glutamatergic if it releases glutamate; a synapse is glutamatergic if the transmitter used there is glutamate.

glyceride compound formed by attaching units to a glycerol backbone. The phospholipids in biological membranes are glycerides.

glycerol $CH_2OH\text{-}CHOH\text{-}CH_2OH$. The backbone to which acyl groups (fatty acid chains) are attached to make triacylglycerols (storage fats) and the phospholipids in biological membranes.

glycogen a glucose polymer that can be quickly hydrolyzed to yield glucose.

glycosylation the addition of sugar residues to a molecule. Both proteins and lipids can be glycosylated.

glyoxylate $CHO\text{-}COO^-$; an intermediate in various metabolic pathways.

GM (genetically modified) an organism with a genome that has been modified by modern molecular techniques, usually by the addition of novel gene(s) or by swapping in new DNA to replace existing gene(s).

Golgi apparatus system of flattened cisternae concerned with glycosylation and other modifications of proteins.

G protein (heterotrimeric G protein) a protein that links a class of cell surface receptors with downstream targets. Heterotrimeric G proteins comprise an α subunit that binds and hydrolyzes GTP, and a ßγ subunit that dissociates from the α subunit while the latter is in the GTP-bound state. Important G proteins are G_q, which activates phospholipase Cβ, and G_s, which activates adenylate cyclase.

G protein-coupled receptor one of a large number of receptors on the cell surface that, upon binding of their transmitter, act as guanine nucleotide exchange-factors for a heterotrimeric G protein.

G_q the isoform of trimeric G protein that activates phospholipase Cβ and therefore generates a calcium signal.

gratuitous inducer an inducer of transcription that is not itself metabolized by the resulting enzymes.

Grb2 (growth factor receptor binding protein number 2) a linker protein that has an SH2 domain and is therefore recruited to phosphotyrosine, e.g. on receptor tyrosine kinases. Grb2 in turn recruits SOS, bringing SOS to the plasma membrane where it can act as a guanine nucleotide exchange-protein (GEF) for Ras.

green fluorescent protein (GFP) a fluorescent protein made by the jellyfish *Aequorea victoria*. Unlike other colored or fluorescent proteins, it contains no prosthetic groups and therefore will fluoresce when expressed by any cell in which the gene is successfully inserted and expressed.

growth factor a transmitter that modifies the developmental pathway of the target cell, often by causing cell division.

growth factor receptor binding protein number 2 (Grb2) a linker protein that has an SH2 domain and is therefore recruited to phosphotyrosine, e.g. on receptor tyrosine kinases. Grb2 in turn recruits SOS, bringing SOS to the plasma membrane where it can act as a guanine nucleotide exchange protein (GEF) for Ras.

G_s the isoform of trimeric G protein that activates adenylate cyclase and therefore causes an increase of cAMP concentration.

GTPase an enzyme that hydrolyzes GTP. The name is usually restricted to that family of proteins that bind GTP and adopt a new shape that can then activate target proteins. Once they hydrolyze their bound GTP they switch back to the original form. Examples are EF-tu, Arf, Ran, dynamin, G_q, and G_s.

GTPase-activating protein (GAP) a protein that speeds up the rate at which GTPases hydrolyze GTP and therefore switch from the active to the inactive state.

guanine one of the bases found in DNA and RNA – guanine is a purine.

guanine nucleotide exchange factor (GEF) a protein that accelerates the rate at which GDP leaves a GTPase to be replaced by GTP, thus switching the GTPase from its inactive state to its active state.

guanosine guanine linked to the sugar ribose. Guanosine is a nucleoside.

guanylate cyclase one of a family of enzymes that generates cyclic guanosine monophosphate (cGMP) from GTP. Two isoforms of guanylate cyclase are mentioned in this book. The guanylate cyclase found in photoreceptors is constitutively active, making cGMP all the time. The guanylate cyclase found in smooth muscle cells is activated by nitric oxide.

hairpin loop a loop in which a linear object folds back on itself. Used to describe the loop formed in an RNA molecule due to complementary base pairing.

haploid containing a single copy of each chromosome, in humans, this means 23 chromosomes. Sperm and eggs are haploid while somatic cells contain one set of 23 from each parent and are referred to as being diploid.

head group the hydrophilic group found in phospholipids. The head group is attached to the glycerol backbone by a phosphodiester link. Examples are choline and inositol.

heat-shock protein a class of chaperone proteins that help denatured proteins refold.

helicase an enzyme that helps unwind the DNA double helix during replication.

α helix a common secondary structure in proteins, in which the polypeptide chain is coiled, each turn of the helix taking 3.6 amino acid residues. The nitrogen atom in each peptide bond forms a hydrogen bond with the oxygen four residues ahead of it in the polypeptide chain.

helix-turn-helix a motif in protein structures that consists of a length of α helix separated from another section of α helix by a turn. The motif is found in many DNA-binding proteins.

hemidesmosome a strong connection between intracellular intermediate filaments and the extracellular matrix.

hemoglobin the oxygen-carrying, iron-containing protein of the blood.

heterochromatin that portion of the nuclear chromatin that is tightly packed. Much of the heterochromatin is repetitive DNA with no coding function.

heterotrimeric G protein protein with three subunits, α, β, and γ, where the α subunit is a GTPase that dissociates from the βγ units when it is in its GTP-bound state. Both the α subunit, in its GTP-bound state, and the now independent βγ subunit, can activate target proteins. Examples are G_s that activates adenylate cyclase and G_q that activates phospholipase Cβ.

histone a positively charged protein that binds to negatively charged DNA and helps to fold DNA into chromatin.

homologous, homology objects that are similar because they have a common ancestor.

homologous chromosomes chromosomes carrying the same set of genes. One of a pair of homologous chromosomes is inherited from the mother, the other from the father.

homologous proteins proteins that are similar because they have a common evolutionary origin. For example, small GTPases such as Ras, Ran, Arf, and Rab are thought to have a common ancestor encoded by a gene that was then duplicated and mutated.

homologous recombination process in which a length of DNA with ends that are homologous to a section of chromosome swaps in, replacing the existing length of DNA in the chromosome. Homologous recombination occurs naturally at chiasmata during crossing over during meiosis. It can also occur in some cells when they transfected with the appropriate exogenous DNA, and in those cases is thought to use some of the same enzymes.

hormone long-lived transmitter that is released into the blood and travels around the body before being broken down.

hormone response element (HRE) a specific DNA sequence to which a steroid hormone receptor binds.

housekeeping gene a gene that is transcribed into mRNA in nearly all the cells of a eukaryotic organism; in bacterial cells a housekeeping gene is one that is always being transcribed.

HRE (hormone response element) a specific DNA sequence to which a steroid hormone receptor binds.

hybridization the association of unlike things. In molecular genetics, the association of two nucleic acid strands (either RNA or DNA) by complementary base pairing.

hydrocarbon a chemical compound formed of carbon and hydrogen only.

hydrocarbon tail the long chain of carbon atoms with attached hydrogens found in phospholipids and triacylglycerols (storage fats). The tail represents all of a molecule of fatty acid except the carboxyl group.

hydrogen bond a relatively weak bond formed between a hydrogen atom and two electron-grabbing atoms (such as nitrogen or oxygen) where the hydrogen is shared between the other atoms.

hydrolysis breakage of a covalent bond by the addition of water. -H is added to one side, -OH to the other.

hydropathy plot a running average of side chain hydrophobicity along a polypeptide chain. From the hydropathy plot one can predict, for example, membrane-spanning domains in integral membrane proteins.

hydrophilic an ion, molecule, or part of a molecule that can interact with water.

hydrophobic a molecule or part of a molecule that will associate with other hydrophobic molecules in preference to water.

hydrophobic effect the tendency of hydrophobic molecules or parts of molecules to cluster together away from water, such as hydrophobic amino acid residues in the center of a protein, or the hydrocarbon tails of phospholipids in lipid bilayers.

hydroxyl group the -OH group. The term is specifically *not* used of an -OH that forms part of a carboxyl (-COOH) group.

hydroxylation the addition of an -OH group to a molecule. Hydroxylation of proteins is a form of post-translational modification. Hydroxylation of collagen is critical to its function as an extracellular matrix protein.

hypoxia low oxygen concentration.

hypoxia-inducible factor a transcription factor whose concentration increases when oxygen concentrations are low.

hypoxia response element enhancer regions of DNA that bind hypoxia-inducible factor, enhancing the transcription of certain genes including the hormone erythropoietin.

imino acid an organic molecule containing both carboxyl (-COOH) and imino (-NH-) groups. Proline is an imino acid although it is usually called an amino acid.

immunofluorescence the use of fluorescently labeled antibodies to reveal the location of specific chemicals, e.g. in fluorescence microscopy or in western blotting.

inactivation (of voltage-gated channels) blockage of the open channel with a plug that is attached to the cytosolic face of the protein.

inducible operon an operon that is transcribed only when a specific substance is present.

inhibitory transmitter a transmitter that makes the target cell less likely to fire an action potential.

initiation factor (IF) a protein that assists the small and large ribosomal subunits of eukaryotes to assemble correctly on the mRNA at the start of protein synthesis.

inositol a cyclic polyalcohol $(CHOH)_6$, that forms the head group of the phospholipid phosphatidylinositol. Phosphorylation of inositol yields inositol trisphosphate.

inositol trisphosphate (IP$_3$, InsP$_3$) a small (Mr = 420) phosphorylated cyclic polyalcohol that is released into the cytosol by the action of phospholipase C on the membrane lipid phosphatidylinositol bisphosphate and which acts to cause release of calcium ions from the endoplasmic reticulum.

inositol trisphosphate-gated calcium channel a channel found in the endoplasmic reticulum of many cells. The channel opens when inositol trisphosphate binds to its cytoplasmic aspect. It allows only calcium ions to pass.

in situ *hybridization* the binding of a particular labeled sequence of RNA (or DNA) to its matching sequence in the genome as a way of searching for the location of that sequence in a tissue sample.

insulin a hormone produced by endocrine cells in the pancreas. It activates its own receptor tyrosine kinase, which in turn acts mainly through activation of PI-3-kinase and hence Akt.

insulin receptor a receptor tyrosine kinase specific for insulin, which acts mainly through activation of PI-3-kinase and hence Akt.

insulin receptor substrate number 1 (IRS-1) a protein phosphorylated on tyrosine by the insulin receptor. Once phosphorylated it recruits PI-3-kinase, which can then be phosphorylated and hence activated.

integral protein (of a membrane) class of protein that is tightly associated with a membrane, usually because its polypeptide chain crosses the membrane at least once. Integral membrane proteins can only be isolated by destroying the membrane, e.g. with detergent. In contrast, peripheral membrane proteins are more loosely associated.

integrin dimeric proteins with an extracellular domain that binds to extracellular matrix proteins and an intracellular domain that attaches to actin microfilaments.

intercellular junction structures formed when cells form close appositions. There are three types: tight, gap, and anchoring junctions.

intermediate filament one of the filaments that makes up the cytoskeleton; composed of various subunit proteins.

intermembrane space in organelles such as mitochondria, chloroplasts, and nuclei that are bound by two membranes the intermembrane space is the aqueous space between the inner and outer membranes. The intermembrane space of nuclei is continuous with the lumen of the ER. The intermembrane space of mitochondria has the ionic composition of cytosol because VDAC in the outer mitochondrial membrane allows solutes of $M_r < 10\,000$ to pass.

interphase period of synthesis and growth that separates one cell division from the next; consists of the G1, S, and G2 phases of the cell division cycle.

interstitial fluid the extracellular medium that lies between the cells of a multicellular organism.

intracellular junction (membrane contact site) structures formed when intracellular organelles of different types form close appositions, allowing ions and molecules to be passed between the different organelles.

intracellular messenger a cytosolic solute that changes in concentration in response to external stimuli or internal events, and which acts on intracellular targets to change their behavior. Calcium ions, cyclic AMP, and cyclic GMP are the three common intracellular messengers.

intracellular receptors receptors that are not on the plasma membrane but which lie within the cell and bind transmitters that diffuse through the plasma membrane.

intrinsic pathway pathway of apoptosis arising from mitochondrial stress.

intron in a eukaryotic gene, introns are those parts that are spliced out before the RNA leaves the nucleus. In contrast exons are those parts that after RNA processing leave the nucleus.

ion a charged chemical species. A single atom that has more or less electrons than are required to exactly neutralize the charge on the nucleus is an ion (e.g. Na^+, Cl^-). A molecule with one or more charged regions is also an ion (e.g. phosphate, HPO_4^{2-}, and leucine, NH_3^+-CH(CH_2·CH[CH_3]$_2$)-COO$^-$.

ionotropic glutamate receptor a channel which opens when glutamate binds to its extracellular aspect, and which allows sodium and potassium ions to pass; some isoforms also pass calcium.

ionotropic receptors channels that open when a specific chemical binds to the extracellular face of the channel protein.

IP_3 (inositol trisphosphate, InsP$_3$) a small (Mr = 420) phosphorylated cyclic polyalcohol that is released into the cytosol by the action of phospholipase C on the membrane lipid phosphatidylinositol bisphosphate and which acts to cause release of calcium ions from the endoplasmic reticulum.

IP_3 receptor a calcium channel in the membrane of the endoplasmic reticulum that opens when inositol trisphosphate binds to its cytosolic aspect.

IRS-1 a protein phosphorylated on tyrosine by the insulin receptor. Once phosphorylated it recruits PI-3-kinase, which can then be phosphorylated and hence activated.

isoforms related proteins that are the products of different genes or differential splicing of mRNAs from one gene.

isomers different compounds with the same molecular formula. For example leucine and isoleucine are both $C_6H_{13}NO_2$ and are therefore isomers.

JAK a cytoplasmic tyrosine kinase that associates with type 1 cytokine receptors and phosphorylates targets, including the STAT transcription factors, when the cytokine ligand binds to the receptor.

K^+/Na^+ ATPase (potassium/sodium ATPase) a plasma membrane carrier. For every ATP hydrolyzed three Na^+ ions are moved out of the cytosol and two K^+ ions are moved in. Usually called the sodium/potassium ATPase.

keratin a protein that makes up the intermediate filaments in epithelial cells.

kinase an enzyme that phosphorylates a molecule by transferring a phosphate group from ATP to the molecule.

kinesin the molecular motor protein responsible for movement along microtubules in the anterograde direction.

kinetochore the point of attachment of the chromosome to the spindle. The kinetochore is a multi-protein structure that forms on the centromere.

knockdown artificial reduction of expression of a particular target gene.

knockout a knocked-out gene is one whose function has been disrupted artificially and which can no longer code for a functional protein. A knockout animal is one in which this procedure has been performed.

lac (lactose) operon the cluster of three bacterial genes which encode enzymes involved in metabolism of lactose.

lactose a sugar comprising galactose linked to glucose.

lactose (lac) operon the cluster of three bacterial genes which encode enzymes involved in metabolism of lactose.

lagging strand the strand of DNA that grows discontinuously during replication, in contrast to the leading strand which is synthesized continuously.

lamellipodium a projection extended by an amoeba or other crawling cell in the direction of movement.

lamins intermediate filament proteins that make up the nuclear lamina.

leading strand a strand of DNA that grows continuously in the 5′ to 3′ direction by the addition of deoxyribonucleotides.

ligand when two molecules bind together, one (often the smaller one) is called the ligand, and the other (often the bigger one) is called the receptor.

linkage, linked (of genes) the physical association of genes on the same chromosome. Linked genes tend to be inherited together.

linker DNA a stretch of DNA that separates two nucleosomes.

lipid any cellular component that is soluble in an organic solvent such as octane. The term includes triacylglycerols, cholesterol, steroid hormones, and many other materials.

lipid bilayer two layers of lipid molecules that form a membrane. In bacteria and eukaryotes (but not in archaea) the majority of the lipids in the bilayer are phospholipids.

localization sequence (targeting sequence) a stretch of polypeptide that determines the cellular compartment to which a synthesized protein is sent.

long noncoding RNA a form of regulatory RNA that does not code for protein. Long noncoding RNA modifies the expression of target genes.

lumen the inside of a closed structure or tube.

lysosome a membrane-bound organelle containing digestive enzymes.

MCS (multiple cloning site) a stretch of DNA within an artificial plasmid that contains several restriction endonuclease recognition sites and which therefore allows an experimenter a choice of restriction enzymes to cut the plasmid and insert DNA of interest.

M-phase the period of the cell division cycle during which the cell divides; M-phase consists of mitosis and cytokinesis.

macromolecule a large molecule.

macrophage a phagocytic housekeeping cell that engulfs and digests bacteria and dead cells.

major groove (of DNA) the wider of the two grooves along the surface of the DNA double helix.

major histocompatability complex protein an integral protein of the plasma membrane that presents short lengths of peptide to patrolling T cells.

mannose a sugar.

mannose-6-phosphate mannose that is phosphorylated on carbon number 6. Mannose-6-phosphate is the sorting signal that identifies lysosomal proteins.

MAP kinase (MAPK, mitogen-associated protein kinase) an enzyme that phosphorylates numerous targets, including transcription factors that trigger transcription of the cyclin D gene.

MAPKK (MAPK kinase) the enzyme that phosphorylates and activates MAP kinase, it itself being phosphorylated and hence activated by MAPKK kinase.

MAPKKK (MAPKK kinase) the enzyme that phosphorylates and activates MAPK kinase. The most important isoform of MAPKKK is also called Raf. Raf is activated by the G protein Ras.

matrix a vague term meaning a more-or-less closed location, often but not exclusively one that is a solid basis on which things can grow or attach. The term is used of the extracellular matrix in animals, formed of collagen and other fibers, and of the aqueous volume inside various organelles.

MCC (mitotic checkpoint complex) protein complex that is generated by the spindle assembly checkpoint and serves as an inhibitor of the anaphase-promoting complex, preventing anaphase until all chromosomes are correctly attached to the spindle.

meiosis the form of cell division that produces gametes, each with half the genetic material of the cells that produce them.

meiosis I and II the first and second meiotic divisions.

meiotic spindle a bipolar, microtubule-based structure on which chromosome segregation occurs during meiosis I and II.

membrane a planar sheet. Biological membranes comprise a lipid bilayer plus protein.

membrane contact site (intracellular junction) structures formed when intracellular organelles of different types form close appositions, allowing ions and molecules to be passed between the different organelles.

membrane fission the breaking of what was a continuous sheet of membrane to allow budding off of a vesicle.

membrane fusion the integration of the membrane of a vesicle into a sheet of membrane, delivering the contents of the vesicle to the lumen of an intracellular organelle, or to the extracellular medium.

membrane voltage the voltage difference between one side of a membrane and the other. It is usually stated as the voltage inside with respect to outside.

mesenchymal cell migration a type of cell locomotion that uses contractile actin stress fibers to pull the cell forward.

messenger RNA (mRNA) the RNA molecule that carries the genetic code. The order of bases on mRNA specifies the amino acid sequence of a polypeptide chain. In eukaryotes, the mRNA leaves the nucleus and is translated into protein in the cytoplasm.

metabolism all of the reactions going on inside a cell.

metaphase the period of mitosis or meiosis at which the chromosomes align prior to separation at anaphase.

metaphase plate the equator of the mitotic or meiotic spindle; the point at which the chromosomes congregate at metaphase of mitosis or meiosis.

7-methyl guanosine cap modified guanosine found at the 5′ terminus of eukaryotic mRNA. A guanosine is attached to the mRNA by a 5′ – 5′ phosphodiester bond and is subsequently methylated on atom number 7 of the guanine.

MHC proteins a family of integral plasma membrane proteins expressed by all somatic cells whose function is to present short lengths of peptide, generated by proteolysis of cytosolic or endocytosed proteins, to patrolling T cells.

microarray a tiny glass wafer to which cloned DNAs are attached. Also known as gene-chips or DNA chips.

microfilament one of the major filaments of the cytoskeleton; also known as actin filament or F-actin; synonymous with the thin filament of striated muscle.

microorganism any single-celled organism, whether prokaryote or eukaryote. Bacteria like *E. coli*, paramecium, and yeast are microorganisms.

micro RNA (miRNA) a form of regulatory RNA that does not code for protein. Micro RNAs form part of an RNA-inducing silencing complex that binds to and inhibits translation of target mRNAs.

microsatellite DNA repetitious DNA of unknown function comprising many repeats of a unit of four or fewer base pairs.

microtubule tubular cytoplasmic filament composed of tubulin. Microtubules are the structural element in 9 + 2 cilia and flagella and in the mitotic and meiotic spindle.

microtubule motor protein protein that moves cargo along microtubules; examples of microtubule-based motors are cytoplasmic dynein and kinesin.

microtubule organizing center (abbr. ***MTOC***) structure from which cytoplasmic microtubules arise; synonymous with the centrosome.

microvilli (sing. ***microvillus***) projections from the surface of epithelial cells that increase the absorptive surface; contain actin filaments.

midbody microtubule-rich structure that forms following anaphase to direct cytokinesis.

migration (of cells) movement of cells within a medium or tissue, typically powered by cytoskeleton dynamics including actin polymerization and depolymerization.

minisatellite DNA repetitious DNA of unknown function comprising up to 20000 repeats of a unit of about 25 base pairs.

minor groove (of DNA) the narrower of the two grooves along the surface of the DNA double helix.

miRNA (micro RNA) a form of regulatory RNA that does not code for protein. Micro RNAs form part of an RNA-inducing silencing complex that binds to and inhibits translation of target mRNAs.

mismatch repair a cell marks its DNA strands by methylation. This allows the cell to distinguish the template strand from a newly synthesized DNA. If an incorrect base, the mismatch, is introduced into the new strand, the repair enzymes will remove the incorrect nucleotide and insert the correct nucleotide in place using the methylated strand as the template.

missense mutation a base change in a DNA molecule that changes a codon so that it now specifies a different amino acid.

mitochondrial inner membrane the inner membrane of mitochondria that is elaborated into cristae. The electron transport chain, and ATP synthase, are integral membrane proteins of the mitochondrial inner membrane.

mitochondrial matrix the aqueous space inside the mitochondrial inner membrane, where the enzymes of the Krebs cycle are located.

mitochondrial outer membrane the outer membrane of mitochondria, permeable to solutes of $M_r < 10000$ because of the presence of the channel VDAC.

mitochondrial pathway another name for the intrinsic pathway, the pathway of apoptosis arising from mitochondrial stress.

mitochondrion the organelle where most of the cell's ATP is generated by aerobic (using oxygen) respiration.

mitogen anything that promotes cell division. FGF and PDGF (fibroblast growth factor and platelet-derived growth factor) are potent mitogens for endothelial and smooth muscle cells.

mitogen-activated protein kinase (MAP kinase, MAPK) an enzyme that phosphorylates numerous targets, including transcription factors that trigger transcription of the cyclin D gene.

mitosis the type of cell division found in somatic cells, in which each daughter cell receives the full complement of genetic material present in the original cell.

mitotic checkpoint (spindle assembly checkpoint) the checkpoint controlling entry into anaphase. Cells can only proceed if the anaphase-promoting complex is active, which in turn only occurs when all the chromatids are correctly lined up on the metaphase plate.

mitotic checkpoint complex (MCC) protein complex that is generated by the spindle assembly checkpoint and serves as an inhibitor of the anaphase-promoting complex, preventing anaphase until all chromosomes are correctly attached to the spindle.

mitotic spindle microtubule-based structure upon which chromosomes are arranged and translocated during mitosis.

mole an amount of substance comprising 6.023×10^{23} (Avogadro's number) molecules. One mole has a mass equal to the value of the relative molecular mass expressed in grams.

monomer a single unit, usually used to refer to a single building block of a larger molecule. Thus, DNA is formed of nucleotide monomers. By analogy, the word is sometimes used to describe proteins that act as a single unit to distinguish them from related proteins that act as larger units, so myoglobin is said to be monomeric by comparison with hemoglobin, which has four subunits. GTPases such as Ran, Arf, and Ras are sometimes called "monomeric G proteins" to distinguish them from trimeric G proteins such as G_q and G_s.

motif a recognizable conserved sequence of bases (in DNA) or amino acids (in a polypeptide). A motif in DNA may bind one transcription factor, for example 5′AGAACA3′ binds the glucocorticoid receptor. A motif in a polypeptide may, for example, fold in a particular way (for example a helix-turn-helix motif), or bind to a target (e.g. EFFDAXE which binds to VAPs).

motor neuron (motoneuron, motoneurone) the neuron that carries action potentials from the spine to the muscles. It releases the transmitter acetylcholine onto the muscle cells, causing them to depolarize and hence contract.

M-phase period of the cell division cycle during which the cell divides; consists of mitosis and cytokinesis.

mRNA (messenger RNA) the RNA molecule that carries the genetic code. The order of bases on mRNA specifies the amino acid sequence of a polypeptide chain.

MTOC (microtubule organizing center) structure from which cytoplasmic microtubules arise; synonymous with the centrosome.

multiple cloning site (MCS) a stretch of DNA within an artificial plasmid that contains several restriction endonuclease recognition sites and which therefore allows an experimenter a choice of restriction enzymes to cut the plasmid and insert DNA of interest.

muscle tissue specialized for generating contractile force.

mutation a change in the structure of a gene or chromosome that is inherited by daughter cells or, in the case of gametes, by children.

myelin a fatty substance that is wrapped around neuronal axons by glial cells.

MyoD a transcription factor of skeletal muscle that acts to increase the expression of muscle-specific proteins such as myosin.

myosin type of motor protein that moves along, or pulls on, actin filaments. The thick filaments in skeletal muscle are formed of the myosin II isoform.

myosin II an isoform of myosin that forms the large filaments seen in striated muscle but which is also found in other cell types.

myosin V an isoform of myosin that carries cargo along actin filaments.

N terminus (amino terminus) the end of a peptide or polypeptide that has a free α-amino group. This end is made first on the ribosome.

Na+/K+ ATPase a plasma membrane carrier. For every ATP hydrolyzed three Na^+ ions are moved out of the cytosol and two K^+ ions are moved in.

necrosis cell death that is due to damage so severe that the cell cannot maintain the level of its energy currencies and therefore falls apart – distinct from apoptosis.

negative feedback a control system in which a change in some parameter activates a mechanism that reverses the change in that parameter; an example is the effect of tryptophan on expression of the *trp* operon.

negative regulation (of transcription) the inhibition of transcription due to the presence of a particular substance which is often the end-product of a metabolic pathway.

Nernst equation
the equation that allows the equilibrium voltage of an ion across a membrane to be calculated. The general form is

$$V_{eq} = \frac{RT}{zF} \log_e \left(\frac{[I_{outside}]}{[I_{inside}]} \right) \text{volts}$$

where R is the gas constant (8.3 J per degree per mole), T is the absolute temperature, z is the charge on the ion in elementary units, and F is 96 500 coulombs in a mole of monovalent ions. The value of RT/F is 0.025 at a room temperature of 22 °C, and 0.027 at human body temperature.

nerve cell an electrically excitable cell with a long axon specialized for transmission of (usually sodium) action potentials. Also called neuron or neurone.

nervous tissue a tissue formed of neurons and glial cells that carries out electrical data processing.

neurofilament a type of intermediate filament found in neurons.

neuron, neurone an electrically excitable cell with a long axon specialized for transmission of (usually sodium) action potentials. Also called a nerve cell.

neurotrophin one of a family of transmitters that act upon neurons to prevent apoptosis and trigger differentiation. Neurotrophin number one used to be called nerve growth factor or NGF.

next generation sequencing one of a number of high throughput DNA sequencing techniques.

NFAT a transcription factor that only translocates to the nucleus when dephosphorylated by the calcium: calmodulin-activated phosphatase calcineurin.

niche a location and environment where stem cells reside. The niche is not only a physical location where the stem cells are found, but also an environment that keeps the cells in the stem cell state and regulates the rate at which they divide.

nicotinic acetylcholine receptor a channel which opens when acetylcholine binds to its extracellular aspect, and which allows sodium and potassium ions to pass; some isoforms also pass calcium.

nine plus two (9 + 2) axoneme) structure of cilium or flagellum; describes the arrangement of nine peripheral microtubules surrounding two central microtubules.

Nitric oxide (NO) a transmitter that acts on intracellular receptors, the most important of which is guanylate cyclase.

nitric oxide synthase cytosolic enzyme that synthesizes nitric oxide, NO. Some isoforms of NO synthase are activated by calcium.

node (of a myelinated axon) the gap between adjacent lengths of myelin sheath, where the neuronal plasma membrane directly contacts the extracellular medium. Also called node of Ranvier, after its discoverer, Louis Ranvier.

noncoding RNA RNA that does not code for protein, that is, all RNA other than mRNA. Ribosomal RNA and transfer RNA are noncoding, but a large and increasing repertoire of other noncoding RNAs that control and modify gene expression is currently being discovered.

nonsense mutation a change in the base sequence that generates a stop codon.

noradrenaline a transmitter released by some neurons. Noradrenaline and adrenaline are closely related but adrenaline has an additional methyl group. To a first approximation, noradrenaline acts mainly on α-adrenergic receptors and adrenaline acts mainly on β-adrenergic receptors.

northern blotting a blotting technique in which RNAs, separated by size, are probed using a single-stranded cDNA probe or an antisense RNA probe.

NO synthase cytosolic enzyme that synthesizes nitric oxide, NO. Some isoforms of NO synthase are activated by calcium.

N-terminus (amino terminus) the end of a peptide or polypeptide that has a free α-amino group. This end is made first on the ribosome.

nuclear envelope double membrane system enclosing the nucleus; it contains nuclear pores and is continuous with the endoplasmic reticulum.

nuclear lamina meshwork of intermediate filaments lining the inner face of the nuclear envelope, formed of the monomer lamin.

nuclear pores holes running through the nuclear envelope that regulate traffic of proteins and nucleic acids between the nucleus and the cytoplasm.

nuclear pore complex multiprotein complex that lies within and around the nuclear pore and which regulates import into and export from the nucleus in a process called gated transport.

nuclease an enzyme that degrades nucleic acids.

nucleic acid a polymer of nucleotides joined together by phosphodiester links – DNA and RNA are nucleic acids.

nucleoid the region of a bacterial cell which contains the chromosome.

nucleolar organizer regions region of one or more chromosomes at which the nucleolus is formed.

nucleolus, nucleoli the region(s) of the nucleus concerned with the production of ribosomes.

nucleoside a purine or a pyrimidine attached to either ribose or deoxyribose.

nucleosome the bead-like structure formed by a stretch of DNA wrapped around a histone octamer.

nucleotide building block of nucleic acids; a nucleoside that is phosphorylated on its 5′ carbon atom.

nucleotide-binding domain a domain of a protein that binds nucleotides, typically ATP. Binding of ATP by the nucleotide-binding domains of CFTR is necessary but not sufficient for the channel to open.

nucleotide excision repair process that removes a thymine dimer, together with some 30 surrounding nucleotides, from DNA. The gap is then repaired by the actions of DNA polymerase I and DNA ligase.

nucleus the cell organelle housing the chromosomes; enclosed within a nuclear envelope.

objective lens the lens of a light or electron microscope that forms a magnified image of a specimen.

Okazaki fragment a series of short fragments that are joined together to form the lagging strand during DNA replication.

oligonucleotide a short fragment of DNA or RNA.

oligosaccharide a chain of up to a 100 or so sugar molecules linked by glycosidic bonds.

oocyte a cell that undergoes meiosis to give rise to an egg.

open complex a complex between DNA and protein which causes the two strands of DNA to separate, e.g. during DNA replication and transcription.

open promoter complex a structure formed when the two strands of the double helix separate so that transcription can commence.

operator the DNA sequence to which a repressor protein binds to prevent transcription from an adjacent promoter.

operon a cluster of genes that encode proteins involved in the same metabolic pathway and which are transcribed as one length of mRNA under the control of one promoter. As far as we know, operons are only found in prokaryotes.

optical isomers two molecules that differ only in that one is the mirror image of the other.

organelle a membrane-bound, intracellular structure such as a mitochondrion, chloroplast, lysosome, etc.

organic in chemistry, an organic compound is one that contains carbon atoms. When applied to farming and food, organic means the avoidance of high-intensity agricultural techniques including genetic engineering.

organism a cell or clone of cells that functions as a discrete and integrated whole to maintain and replicate itself.

origin of replication the site on a chromosome at which DNA replication can commence.

osmolarity the overall strength of a solution; the more solute that is dissolved in a solution, the higher the osmolarity.

osmosis the movement of water down its concentration gradient.

outer doublet microtubules the paired microtubules that make the "9" of the 9 + 2 axoneme in cilia and flagella.

oxidation the removal of electrons from a molecule, e.g. by adding oxygen atoms, which tend to take more than their fair share of electrons in any bonds they make. Removal of hydrogen atoms from a molecule oxidizes it.

P site (peptidyl site) the site on a ribosome that is occupied by the growing polypeptide chain.

p53 a transcription factor that stimulates cell repair but also apoptosis. Cancer cells are frequently found to have mutated, nonfunctional *P53* genes.

p75 neurotrophin receptor a death domain receptor for neurotrophins. Upon binding neurotrophins p75 activates caspase 8 and hence initiates apoptosis unless a countermanding signal to survive is present.

PAC (P1 artificial chromosome) a cloning vector, derived from the bacteriophage P1, that is used to propagate, in *E. coli*, DNAs of about 150 000 bp.

pain receptor a neuron whose distal axon terminal is depolarized by potentially damaging events such as heat or stretch.

parallel β sheet β sheet in which all the parallel polypeptide chains run in the same direction.

patch clamp a technique in which a glass micropipette is sealed to the surface of a cell to allow electrical recording of cell properties.

PCR (polymerase chain reaction) a method for making many copies of a DNA sequence where the sequence, at least at each end, is known.

PDGF (platelet-derived growth factor) a transmitter that opposes apoptosis and promotes cell division in target cells such as endothelial cells and smooth muscle.

peptide short linear polymer of amino acids.

peptide bond the bond between amino acids. The bond is formed between the carboxyl group of one amino acid and the amino group of the next.

peptidyl site (P-site) the site on a ribosome that is occupied by the growing polypeptide chain.

peptidyl transferase enzyme that catalyzes the formation of a peptide bond between two amino acids. *E. coli* peptidyl transferase is an example of a ribozyme.

pericentriolar material electron-dense material that surrounds the centrioles to make up the centrosome.

peripheral membrane protein class of protein that is easily detached from a cell membrane, unlike integral membrane proteins, which can only be isolated by destroying the membrane, e.g. with detergent.

peroxisome class of cell organelles of diverse function. Peroxisomes frequently contain the enzyme catalase, which breaks down hydrogen peroxide into oxygen and water.

pH measure of the acidity of a solution, equal to minus the logarithm to base ten of the hydrogen ion concentration in moles liter^{-1}. The smaller the pH value, the more acid the solution. A neutral solution has a pH of 7, that is, the H$^+$ concentration is 10^{-7} mol liter^{-1} or 100 nmoles liter^{-1}.

PH domain a protein domain that binds phosphorylated inositols. The PH domain on Akt is recruited to the membrane by the intensely charged lipid phosphatidylinositol trisphosphate.

phage short for bacteriophage, that is, a virus that infects bacterial cells.

phase contrast microscopy the type of light microscopy in which differences in the refractive index of a specimen are converted into differences in contrast.

PH domain a protein domain that binds phosphorylated inositols. The PH domain on Akt is recruited to the membrane by the intensely charged lipid phosphatidylinositol trisphosphate.

phosphatase an enzyme that removes phosphate groups from substrates.

phosphate properly, a name for the ions $H_2PO_4^-$, HPO_4^{2-} and PO_4^{3-}. The word is also very commonly used to mean the group:

$$-O-\overset{\overset{\displaystyle O^-}{|}}{\underset{\underset{\displaystyle O^-}{|}}{P}}= O$$

and we have used this convention in this book. If one wants to specifically describe the group, *not* the ions, one can refer to a phosphoryl group (a term that does not include the left-most oxygen in the diagram above).

phosphoinositide 3-kinase (PI-3-kinase) an enzyme that phosphorylates phosphatidylinositol bisphosphate on the 3 position of the inositol ring to generate the intensely charged lipid phosphatidylinositol trisphosphate (PIP$_3$). PIP$_3$ can then recruit proteins containing PH domains to the plasma membrane.

phosphatidylinositol bisphosphate **(*PIP$_2$*)** a membrane lipid. PIP$_2$ releases inositol trisphosphate into the cytosol upon hydrolysis by phospholipase C. Alternatively, PIP$_2$ can be further phosphorylated to yield phosphatidylinositol trisphosphate.

phosphatidylinositol trisphosphate (PIP$_3$) a membrane lipid with a highly charged head group comprising inositol with three phosphate groups. PIP$_3$ recruits proteins containing PH domains to the plasma membrane.

phosphodiester bond a link between two parts of a molecule in which a phosphate is attached through an oxygen atom to each of the two parts. In phospholipids, the head group is attached to glycerol by a phosphodiester bond. In DNA and RNA, successive nucleotides are joined together by phosphodiester bonds.

phospholipase C an enzyme that hydrolyzes the membrane lipid phosphatidylinositol bisphosphate (PIP$_2$) to release inositol trisphosphate into the cytosol. The β isoform is activated by the trimeric G protein G$_q$, while the γ isoform is activated by tyrosine phosphorylation.

phospholipid the building blocks of biological membranes. Phospholipids comprise a glycerol backbone to which are attached two fatty acids with long hydrophobic tails, and a hydrophilic, usually electrically charged, head group.

phosphorylated having had a phosphate group added. The phosphate groups are usually substituted into a hydroxyl group to form a phosphoester bond but are sometimes substituted into an acid to form a phosphoanhydride or attached to a nitrogen atom to form a phosphoimide.

phosphorylation addition of a phosphate group to a molecule. Usually substituted into a hydroxyl group to form a phosphoester bond. Sometimes substituted into an acid to form a phosphoanhydride, or attached to a nitrogen atom to form a phosphoimide.

PI-3-kinase (phosphoinositide 3-kinase) an enzyme that phosphorylates phosphatidylinositol bisphosphate on the

3 position of the inositol ring to generate the intensely charged lipid phosphatidylinositol trisphosphate (PIP$_3$). PIP$_3$ can then recruit proteins containing PH domains to the plasma membrane.

PIP$_2$ (phosphatidylinositol bisphosphate) a membrane lipid. PIP$_2$ releases inositol trisphosphate into the cytosol upon hydrolysis by phospholipase C. Alternatively, PIP$_2$ can be further phosphorylated to yield phosphatidylinositol trisphosphate.

PIP$_3$ (phosphatidylinositol trisphosphate) a membrane lipid with a highly charged head group comprising inositol with three phosphate groups. PIP$_3$ recruits proteins containing PH domains to the plasma membrane.

pKa a parameter equal to $-\log_{10} K_a$ and representing the pH at half the molecules have H$^+$ bound and the other half do not.

PKA (protein kinase A; cAMP-dependent protein kinase) a protein kinase that is activated by the intracellular messenger cyclic AMP. PKA phosphorylates proteins (for example, CFTR) on serine and threonine residues.

PKB (protein kinase B) another name for Akt, a protein kinase that is activated when it is itself phosphorylated; this in turn only occurs when it is recruited to the plasma membrane by phosphatidylinositol trisphosphate (PIP$_3$). Akt phosphorylates proteins on serine and threonine residues (for example, the bcl-2 family protein BAX).

PKC (protein kinase C) a protein kinase that is activated by a rise of cytosolic calcium concentration; it is also activated by diacylglycerol. PKC phosphorylates proteins on serine and threonine residues.

PKG (protein kinase G; cGMP-dependent protein kinase) a protein kinase that is activated by the intracellular messenger cyclic GMP. PKG phosphorylates proteins (for example, calcium ATPase) on serine and threonine residues.

plasmalemma the membrane that surrounds the cell. Also called the plasma membrane or the cell membrane.

plasma membrane the membrane that surrounds the cell. Also called the plasmalemma or the cell membrane.

plasmid a circular DNA molecule that is replicated independently of the host chromosome in bacterial cells.

platelet small fragment of cell that contains no nucleus, but which has a plasma membrane and some endoplasmic reticulum. Platelets are critical in the process of blood clotting, and also release platelet-derived growth factor.

platelet-derived growth factor (PDGF) a transmitter that opposes apoptosis and promotes cell division in target cells such as endothelial cells and smooth muscle.

PLC (phospholipase C) an enzyme that hydrolyzes the membrane lipid phosphatidylinositol bisphosphate (PIP$_2$) to release inositol trisphosphate into the cytosol. The β isoform is activated by the trimeric G protein G_q, while the γ isoform is activated by tyrosine phosphorylation.

pluripotent able to differentiate into a wide range of cell types. Embryonic stem cells and induced pluripotent stem (iPS) cells are pluripotent.

polar body small cell that forms during female meiosis that receives half of the chromosomes, but very little cytoplasm. This allows the egg to retain the majority of the oocyte cytoplasm, but only half of the chromosomes.

polarized divided into two parts with disparate properties. More specific uses in chemistry and biology include: *polarized light* whose electromagnetic oscillations occur in defined planes; *polarized molecules* where the electrons are not shared equally so that one end of the molecule has a +ve charge and another end has a –ve charge; *electrically polarized* membranes that separate volumes at different voltages; and *polarized cells* that exhibit different morphologies, and express different proteins, at different ends.

poly A tail (poly adenosine tail) a string of adenine residues added to the 3′ end of a eukaryotic mRNA.

polyadenylation the process whereby a poly A tail is added to the 3′ end of a eukaryotic mRNA.

polycistronic mRNA an mRNA that, when translated, yields more than one polypeptide.

polymer a chemical composed of a long chain of identical or similar subunits.

polymerase an enzyme that makes polymers, that is, long chains of identical or very similar subunits. DNA and RNA polymerase respectively are involved in making DNA and RNA.

polymerase chain reaction (PCR) a method for making many copies of a DNA sequence when the sequence, at least at each end, is known.

polypeptide a polymer of more than 50 amino acids joined by peptide bonds.

polyribosome (polysome) a chain of ribosomes attached to an mRNA molecule.

polysome (polyribosome) a chain of ribosomes attached to an mRNA molecule.

positive feedback a process in which the consequences of a change act to increase the magnitude of that change, so that a small initial change tends to get bigger and bigger.

positive regulation (of transcription) a process whereby transcription is activated in the presence of a particular substance.

postsynaptic cell the cell upon which a neuron releases its transmitter at a synapse.

potassium channel a channel in the plasma membrane of many cells that allows potassium ions to pass.

potassium/sodium ATPase (K⁺/Na⁺ ATPase) a plasma membrane carrier. For every ATP hydrolyzed three Na^+ ions are moved out of the cytosol and two K^+ ions are moved in. Usually called the sodium/potassium ATPase.

preinitiation complex complex formed between transcription factors and RNA polymerase at the promoter of a eukaryotic gene.

presynaptic terminal the region of an axon terminal that is specialized for exocytosis of transmitter.

primary antibody the first antibody applied to the preparation. In many cases the primary antibody is not itself labeled, but must be revealed by applying a labeled secondary antibody.

primary immunofluorescence technique in which the location or presence of a chemical is revealed by treating the sample with a dye-labeled antibody.

primary structure (of a protein) the sequence of amino acids held together by peptide bonds making up a polypeptide.

primase the enzyme that synthesizes the RNA primers needed for the initiation of synthesis of the leading and lagging DNA strands.

primer a short sequence of nucleic acid (RNA or DNA) which acts as the start point at which a polymerase can initiate synthesis of a longer nucleic acid chain.

programmed cell death (apoptosis) a process in which a cell actively promotes its own destruction, as distinct from necrosis. Apoptosis is important in vertebrate development, where tissues and organs are shaped by the death of certain cell lineages.

projector lens lens of light or electron microscope that carries the image to the eye; more commonly known as the "eyepiece."

prokaryotic a type of cellular organization found in bacteria in which the cells lack a distinct nucleus and other organelles.

prometaphase the period of mitosis or meiosis that sees the breakdown of the nuclear envelope and the attachment of the chromosomes to the mitotic spindle.

promoter the region of DNA to which RNA polymerase binds to initiate transcription.

prophase the period of mitosis or meiosis in which the chromosomes condense.

prophase I the first prophase of meiosis.

prophase II the second prophase of meiosis.

prosthetic group a nonprotein molecule necessary for the activity of a protein. The heme group of hemoglobin is a prosthetic group.

proteasome a barrel-shaped proteolytic machine characteristic of eukaryotic cells that degrades proteins that are no longer needed. Proteasomes are found in bacteria but in most prokaryotes the components are not so obviously organized into a discrete barrel-shaped structure.

protein a polypeptide (a polymer of α-amino acids) that has a preferred way of folding.

protein engineering designing a novel protein and then causing it to be built by altering the sequence of DNA on a vector or in the chromosome.

protein kinase an enzyme that phosphorylates a protein by transferring a phosphate group from ATP to the molecule.

protein kinase A a protein kinase that is activated by the intracellular messenger cAMP. Protein kinase A phosphorylates proteins (for example, CFTR) on serine and threonine residues.

protein kinase B another name for the protein kinase Akt. Akt is activated when it is itself phosphorylated; this in turn only occurs when it is recruited to the plasma membrane by phosphatidylinositol trisphosphate (PIP$_3$). Akt phosphorylates proteins on serine and threonine residues (for example, the bcl-2 family protein BAX).

protein kinase C (PKC) a protein kinase that is activated by a rise of cytosolic calcium concentration; it is also activated by diacylglycerol. PKC phosphorylates proteins on serine and threonine residues.

protein kinase G (cGMP-dependent protein kinase) a protein kinase that is activated by the intracellular messenger cyclic GMP. Protein kinase G phosphorylates proteins (for example, calcium ATPase) on serine and threonine residues.

protein phosphatase an enzyme that removes phosphate groups from proteins.

protein phosphorylation the addition of a phosphate group to a protein. The addition of the charge on the phosphate group can markedly alter the tertiary structure and therefore function of a protein.

protein translocator channel-like protein of the endoplasmic reticulum that allows polypeptide chains to cross the membrane as they are synthesized.

proteolysis the hydrolysis of peptide bonds within a protein to release separate fragments.

proteome the complete protein content of the cell.

proteomics the study of the proteome. For example, one might compare the protein profiles of cells in two tissues.

protofilaments the chains of subunits that make up the wall of a microtubule.

proximal close to the center.

pseudogene a gene that has mutated such that it no longer codes for a functional protein.

P-site (peptidyl site) the site on the ribosome occupied by the growing polypeptide chain.

purine a nitrogenous base found in nucleotides and nucleosides – adenine, guanine, and hypoxanthine are purines.

pyrimidine a nitrogenous base found in nucleotides and nucleosides – cytosine, thymine, and uracil are pyrimidines.

quaternary structure the structure in which subunits of a protein, each of which has a tertiary structure, associate to form a more complex molecule. The subunits associate tightly, but not covalently, and may be the same or different.

quiescent at rest. Nondividing cells are often said to be quiescent.

Rab family family of GTPases that mediate fusion of vesicles with other membranes.

Ran a GTPase that gives direction to transport through the nuclear pore.

Ras a GTPase that activates the MAP kinase pathway and hence promotes cell division. Ras is activated by SOS, its GTP exchange factor (GEF), which is in turn recruited to the plasma membrane via receptor tyrosine kinases and the adaptor protein Grb2.

Rb a protein that binds to the transcription factor E2F and therefore prevents transcription of proteins required for DNA synthesis. Phosphorylation of Rb by CDK4 causes it to release E2F, allowing entry to S-phase. Mutations in *Rb* lead to the formation of the eye cancer called retinoblastoma.

reading frame a reading of the genetic code in blocks of three bases – there are three possible reading frames for each mRNA, only one of which will produce the correct protein.

receptor a protein that specifically binds a particular solute. Receptors can be integral membrane proteins or cytosolic proteins. Particular receptor proteins perform additional functions (ion channel, enzyme, activator of endocytosis, transcription factor, etc.) that are activated by the binding of the solute.

receptor tyrosine kinase an integral membrane protein with a binding site for transmitter on the extracellular aspect and tyrosine kinase catalytic ability on the cytosolic aspect. Binding of transmitter causes self-phosphorylation on tyrosine and hence recruitment of proteins with SH2 domains such as Grb2, phospholipase Cγ, and phosphoinositide 3-kinase, which may then themselves be phosphorylated.

The PDGF receptor, the FGF receptor, the Trk family of receptors, and the insulin receptor are all receptor tyrosine kinases.

recessive a gene is recessive if its effects are hidden when the organism possesses a second, dominant, version. Recessive genes usually code for nonfunctional proteins, so that if the individual can make the protein using the other, working gene, no effects are seen.

recombinant DNA an artificial DNA molecule formed of DNA from two or more different sources.

recombinant plasmid a plasmid into which a foreign DNA sequence has been inserted.

recombinant protein a protein expressed from the foreign DNA inserted into a recombinant plasmid or other cloning vector. Recombinant proteins are often expressed in bacteria, yeast, insect or mammalian cells.

recombination a "cut and paste" of DNA. Recombination occurs at chiasmata during meiosis, and also in embryonic stem cells, allowing insertional mutagenesis.

recovery stroke part of the beat cycle of a cilium in which the cilium is moved back into a position where it can push again; used in contrast to the effective stroke which generates the force.

refractive index a measure of the capacity of any material to slow down the passage of light.

regulated exocytosis (regulated secretion) exocytosis that only occurs in response to a signal, such as a rise in the cytosolic concentration of calcium ions.

release factors proteins that occupy the A site in the ribosome when a stop codon is encountered, and which act to trigger termination of polypeptide synthesis.

repetitive DNA a DNA sequence that is repeated many times within the genome.

replication (of DNA) the process whereby two DNA molecules are made from one.

replication fork Y-shaped structure formed when the two strands of the double helix separate during replication.

repolarization a movement of the membrane voltage back toward the normal resting voltage following a depolarization.

repressible operon operon whose transcription is repressed in the presence of a particular substance – often the final product of the metabolic pathway.

RER (rough endoplasmic reticulum) portion of the endoplasmic reticulum associated with ribosomes and concerned with the synthesis of secreted proteins. Proteins destined to remain within the majority of single-membrane-bound organelles (Golgi, lysosomes, etc.) are also made on the rough

ER, as are integral proteins of these organelles and of the plasma membrane.

resting voltage the voltage across the plasma membrane of an unstimulated cell, typically −70 to −90 mV.

restriction endonuclease an enzyme that cleaves phosphodiester bonds within a specific sequence in a DNA molecule.

retention signal a sorting signal on a polypeptide that tells the cell that the polypeptide should remain in a particular location and not be trafficked further.

retinoblastoma a cancer of the eye, usually caused by a mutation in the *Rb* gene.

retrograde backward movement; when applied to axonal transport it means toward the cell body.

retrovirus a virus whose genetic information is stored in RNA.

reverse genetics research that begins with a gene of known sequence but unknown effect and works toward deducing its effects and therefore function.

reverse transcriptase an enzyme of some viruses that copies RNA into DNA.

reverse transcription the process whereby RNA is copied into DNA.

ribonuclease an enzyme that cleaves phosphodiester links in RNA.

ribonuclease H an enzyme that cleaves phosphodiester links in an RNA molecule that is joined to a DNA molecule by complementary base pairing.

ribonucleic acid (RNA) a polymer of ribonucleoside monophosphates. The best-known types of RNA are messenger RNA, ribosomal RNA, and transfer RNA, but we now know of many other RNAs with control functions.

ribonucleic acid polymerase enzyme that synthesizes RNA.

ribonucleic acid primer a short length of RNA, complementary in sequence to a DNA strand, that allows DNA polymerase III to attach and begin DNA synthesis.

ribonucleoprotein a complex of protein and RNA. Ribosomes are ribonucleoprotein complexes.

ribonucleoside monophosphate a nitrogenous base attached to the sugar ribose which has one phosphate group on its 5′ carbon atom – also known as a ribonucleotide.

ribose a sugar used to make the nucleotides that form RNA.

ribosome the ribosome has two subunits, the large and the small, each of which is made up of several different proteins and rRNA molecules. The two subunits come together to provide the platform on which tRNAs can bind to mRNA so that protein synthesis can take place.

ribosome recycling factor (RRF) a protein that assists the ribosomal subunits to dissociate after protein synthesis is complete. The dissociated subunits can then be reused in another round of protein synthesis.

ribosomal RNA (rRNA) the RNA component of ribosomes. rRNA forms a major part of the ribosome and participates fully in the translation process.

ribozyme an RNA molecule with enzyme-like catalytic activity.

RISC (RNA-inducing silencing complex) a complex of protein and RNA that binds to and inhibits the translation of target mRNAs.

RNA (ribonucleic acid) a polymer of ribonucleoside monophosphates. The best-known types of RNA are messenger RNA, ribosomal RNA, and transfer RNA, but we now know of many other RNAs with control functions.

RNA-inducing silencing complex (RISC) a complex of protein and RNA that binds to and inhibits the translation of target mRNAs.

RNA interference (RNAi) the inhibition of translation of an mRNA by a second RNA. RNA interference, for example by micro RNAs, is a normal part of the regulation of gene expression, however artificial RNAs can be engineered to inhibit transcription of target mRNAs.

RNA polymerase enzyme that synthesizes RNA.

RNA polymerase I the enzyme that transcribes most rRNA genes in eukaryotes.

RNA polymerase II the enzyme that transcribes mRNA genes in eukaryotes.

RNA polymerase III the enzyme that transcribes tRNA and some other small RNAs in eukaryotes.

RNA primer a short length of RNA, complementary in sequence to a DNA strand, that allows DNA polymerase III to attach and begin DNA synthesis.

RNA-seq the sequencing of cDNA molecules using next generation sequencing. RNA-seq allows the identification of all the mRNAs in a single cell or tissue.

RNA splicing the removal of introns from an RNA molecule and the joining together of exons to form the final RNA product.

rough endoplasmic reticulum (RER) portion of the endoplasmic reticulum associated with ribosomes and concerned with the synthesis of secreted proteins. Proteins destined to remain within the majority of single-membrane-bound organelles (Golgi, lysosomes . . .) are also made on the rough ER, as are integral proteins of these organelles and of the plasma membrane.

RRF (ribosome recycling factor) a protein that assists the ribosomal subunits to dissociate after protein synthesis is complete. The dissociated subunits can then be reused in another round of protein synthesis.

rRNA (ribosomal RNA) the RNA component of ribosomes. rRNA forms a major part of the ribosome and participates fully in the translation process.

ryanodine a plant toxin that binds to the ryanodine receptor, with complex effects on gating of the channel.

ryanodine receptor a calcium channel found in the membrane of the endoplasmic reticulum. In most cells it opens in response to a rise of calcium concentration in the cytosol. In skeletal muscle ryanodine receptors are directly linked to voltage-gated calcium channels in the plasma membrane and open when the latter open.

σ factor (sigma factor) subunit of bacterial RNA polymerase that recognizes the promoter sequence.

S-phase phase of the cell cycle during which DNA replication occurs. S stands for "synthesis."

S value (Svedberg unit) a value that describes how fast macromolecules and organelles sediment in a centrifuge.

saltatory conduction the jumping of an action potential from node to node down a myelinated axon.

salt bridge (in protein structure) an interaction between a positively charged amino acid residue (such as arginine) and a negatively charged residue (such as aspartate).

sarcomere contractile unit of striated muscle.

sarcoplasmic reticulum type of smooth endoplasmic reticulum found in striated muscle; concerned with the regulation of the concentration of calcium ions.

satellite DNA a DNA sequence that is tandemly repeated many times.

scanning electron microscope (SEM) a type of electron microscope in which the image is formed from electrons that are reflected back from the surface of a specimen as the electron beam scans rapidly back and forth over it. The scanning electron microscope is particularly useful for providing topographical information about the surfaces of cells or tissues.

Schwann cell glial cell of the peripheral (outside the brain and spinal cord) nervous system.

SDS-PAGE (sodium dodecyl sulphate-polyacrylamide gel electrophoresis) technique for separating proteins by their relative molecular mass.

secondary antibody in immunofluorescence, a labeled antibody that is used to reveal the location of an unlabeled primary antibody. Secondary antibodies are produced by one species of animal in response to the injection of another animal's antibodies, thus a "goat anti-rabbit" secondary will bind to any primary antibody that has been generated by a rabbit. Labeled secondary antibodies can be used to study many different questions because the specificity in the final image or western blot is provided by the primary antibody, not by the secondary.

secondary immunofluorescence a technique in which a fluorescent secondary antibody is used to label a pre-applied primary antibody specific for a particular protein or subcellular structure. Also called indirect immunofluorescence.

secondary structure regular, repeated folding of the backbone of a polypeptide. The side chains of the amino acids have an influence but are not directly involved. There are two common types of secondary structure: the α helix and the β sheet.

secretion the synthesis and release from the cell of a chemical.

secretory vesicles vesicles derived from the Golgi apparatus which transport secreted proteins to the cell membrane with which they fuse.

second messenger a subset of intracellular messengers (the "first messenger" being the extracellular transmitter chemical). A cytosolic solute that changes in concentration in response to external stimuli or internal events, and which acts on intracellular targets to change their behavior. Cyclic AMP and cyclic GMP are often called second messengers. We prefer the name intracellular messenger, which allows one to refer to all soluble signals without worrying about counting the steps from the receptor to the signal.

semi-conservative replication the mode of replication used for DNA – both strands of the double helix serve as templates for the synthesis of new daughter strands.

serine–threonine kinase an enzyme that phosphorylates proteins on serine or threonine residues, by transferring the γ phosphate of ATP to the amino acid side chain. With a few exceptions, protein kinases are either serine–threonine kinases (which can phosphorylate on serine or on threonine) or tyrosine kinases (which only phosphorylate on tyrosine).

seven-methyl guanosine cap modified guanosine found at the 5′ terminus of eukaryotic mRNA. A guanosine is attached to the mRNA by a 5′ – 5′ -phosphodiester link and is subsequently methylated on atom number 7 of the guanine.

sex steroids steroid hormones synthesized in the gonads (ovaries and testes). Estrogens, progesterone, and testosterone are sex steroids.

SH2 domain a domain found in a number of proteins that binds to phosphorylated tyrosine. Many proteins with SH2 domains are recruited to receptor tyrosine kinases when the latter phosphorylate themselves in response to ligand binding. Important proteins with SH2 domains are Grb2, PI 3-kinase, and phospholipase Cγ.

β sheet common secondary structure in proteins, in which lengths of fully extended polypeptide run alongside each other, hydrogen bonds forming between the peptide bonds of the adjoining strands.

Shine-Dalgarno sequence a sequence on a bacterial mRNA molecule to which the ribosome binds.

shRNA (small hairpin RNA) a small artificial RNA molecule that assumes a hairpin structure by intramolecular base pairing and which binds to and inhibits translation of a target mRNA.

side chain (of amino acids) the group attached to the α carbon of an amino acid.

sigma factor (σ factor) subunit of bacterial RNA polymerase that recognizes the promoter sequence.

signal peptidase enzyme that cleaves the signal sequence from a polypeptide as it enters the lumen of the endoplasmic reticulum.

signal recognition particle ribonucleoprotein particle that recognizes and binds to the signal sequence at the N terminus of a polypeptide.

signal recognition particle receptor receptor on the endoplasmic reticulum to which the signal-recognition particle binds during the process of polypeptide chain synthesis and import into the endoplasmic reticulum. Also called the "docking protein."

signal sequence short stretch of amino acids found at the N terminus of polypeptides which targets them to the endoplasmic reticulum.

simple diffusion not a specific term: we often say "simple diffusion" to emphasize that a solute movement is passive, down a concentration gradient, and does not require a carrier or channel.

single base substitution mutation that replaces a single base in DNA with another. For example the G542X mutation that causes type I cystic fibrosis is caused by one guanine mutating to thymine (uracil in the mRNA).

single-stranded binding protein the protein that binds to the separated DNA strands to keep them in an extended form during replication, thus preventing the double helix from reforming.

siRNA (small interfering RNA) a small RNA molecule that binds to and inhibits translation of a target mRNA. Small interfering RNAs are a normal part of the regulation of gene expression, however siRNAs can be engineered to inhibit transcription of target mRNAs.

site-directed mutagenesis a technique used to change the base sequence of a DNA molecule at a specific site.

skeletal muscle cells large multinucleate muscle cells that are attached to bone. Most cuts of meat are mainly skeletal muscle.

small hairpin RNA (shRNA) a small artificial RNA molecule that assumes a hairpin structure by intramolecular base pairing and which binds to and inhibits translation of a target mRNA.

small interfering RNA (siRNA) a small RNA molecule that binds to and inhibits translation of a target mRNA. Small interfering RNAs are a normal part of the regulation of gene expression, however siRNAs can be engineered to inhibit transcription of target mRNAs.

small nuclear RNAs (snRNAs) small RNA molecules found in the nucleus that play a role in RNA splicing.

smooth endoplasmic reticulum (SER) the portion of the endoplasmic reticulum without attached ribosomes, among its functions are the synthesis of lipids and the storage and stimulated release of calcium ions.

smooth muscle nonstriated muscle, found in many places in the body, including blood vessels and the intestine.

smooth muscle cells small muscle cells that lack the characteristic striations seen in skeletal and cardiac muscle.

SNARES proteins that mediate fusion of vesicles with other membranes.

snRNAs (small nuclear RNAs) small RNA molecules found in the nucleus that play a role in RNA splicing.

sodium action potential an action potential driven by the opening of voltage-gated sodium channels and the resulting sodium influx.

sodium/calcium exchanger a carrier in the plasma membrane. Three sodium ions move into the cell down their electrochemical gradient and one calcium is moved out up its concentration gradient.

sodium/potassium ATPase a plasma membrane carrier. For every ATP hydrolyzed three Na$^+$ ions are moved out of the cytosol and two K$^+$ ions are moved in.

sodium pump (sodium/potassium ATPase) a plasma membrane carrier. For every ATP hydrolyzed three Na$^+$ ions are moved out of the cytosol and two K$^+$ ions are moved in.

solute a substance that is dissolved in a liquid.

somatic cells cells that make up all the normal tissues of the human body; distinct from the germ cells that form the gametes (eggs and sperm).

somatic mutation a mutation that occurs in the DNA of a somatic cell. Somatic mutations are not passed on to children. Rather they lead to a falloff in the performance of tissues with age, and to cancers.

sorting signal a section of a protein that causes the cell to direct the protein to a specific cell compartment such as the nucleus or mitochondrion. Sorting signals can be lengths of peptide (targeting sequences), such as the signal sequence that targets a protein to the endoplasmic reticulum, or can be the result of post-translational modification, for example mannose-6-phosphate, which targets a protein to the lysosome.

SOS guanine nucleotide exchange factor for a GTPase called Ras. SOS is recruited to the plasma membrane by binding to growth factor receptor binding protein number 2 which in turn binds activated receptor tyrosine kinases.

Southern blotting a blotting technique in which DNAs, separated by size, are probed using a single-stranded cDNA probe.

spatial summation the phenomenon whereby although one presynaptic action potential does not generate a depolarization large enough to elicit an action potential in a postsynaptic cell, action potentials that occur more-or-less simultaneously in a number of presynaptic cells do depolarize the postsynaptic cell to threshold.

spermatozoon motile male gamete.

S-phase phase of the cell cycle during which DNA replication occurs. S stands for "synthesis."

spliceosome a complex of proteins and small RNA molecules involved in RNA splicing.

spindle microtubule-based structure by which the chromosomes are organized then segregated during cell division.

spindle assembly checkpoint the checkpoint controlling entry into anaphase. Cells can only proceed if the anaphase-promoting complex is active, which in turn only occurs when all the chromatids are correctly lined up on the metaphase plate.

squamous flat. A term used of epithelial cells.

start signal the start signal for protein synthesis is the codon AUG specifying the incorporation of methionine.

STAT one of a family of transcription factors that dimerize and translocate to the nucleus when phosphorylated on tyrosine. STATs are phosphorylated by JAKs associated with activated type 1 cytokine receptors.

steady state a state whose parameters do not change with time. A steady state may be very far from equilibrium and be maintained by the constant expenditure of energy. Examples include the constant concentration of ATP in cells and the constant 37° temperature of the human body.

stem cell an undifferentiated cell that retains the ability to divide and generate more stem cells as well as cells committed to differentiate. Some stem cells are committed to one differentiated fate, others are more-or-less pluripotent and are capable of creating a range of cell types.

steroid hormones these transmitters act on intracellular receptors to activate transcription of particular genes. Glucocorticoids are one type of steroid hormone; the sex hormones (oestradiol, testosterone, progesterone) are another.

steroid hormone receptor transcription factors that, upon binding their appropriate steroid hormone, move from the cytosol to the nucleus and activate transcription of particular genes.

sticky ends (of DNA) the short single-stranded ends produced by cleavage of the two strands of a DNA molecule at sites which are not opposite to one another.

stop codon the codons UAA, UAG, and UGA are codons that signal protein synthesis to stop. Also known as termination codons.

stop signal the signal, to stop protein synthesis, given to the ribosome by a stop codon on mRNA.

stratified epithelium type of epithelium consisting of several layers, such as the skin.

stress fiber a strong, cable-like assembly of actin microtubules.

stretch-activated channel a channel found in the plasma membrane of endothelial and other cells that opens when the membrane is stretched. It is permeable to sodium, potassium, and calcium ions. Although well characterized it has not to date been genetically identified.

striated muscle striped muscle; includes skeletal and cardiac muscle.

substrate in normal English, a substrate is a solid base. In biology, the term is used to mean: (1) a reactant in a reaction catalyzed by an enzyme, e.g. "lactose is a substrate for β-galactosidase"; (2) a base that cells grow and move on, e.g. "collagen is a good substrate for cell attachment."

sucrose a sugar comprising glucose linked to fructose.

summation (at synapses) the additive effects of more than one presynaptic action potential upon the postsynaptic voltage.

supercoiling the organization of a linear structure into coils at more than one spatial scale.

super-resolution microscopy advanced fluorescence microscopy techniques that allow the resolution of objects considerably smaller than the wavelength of light.

S value (Svedberg unit) a value that describes how fast macromolecules and organelles sediment in a centrifuge.

synapse the structure formed from the axon terminal of a neuron and the adjacent region of the postsynaptic cell.

Transmitter released by the axon terminal diffuses across the synapse gap and acts upon the postsynaptic cell.

T killer cell a cell of the immune system that kills somatic cells that present novel peptides.

tandem repeats many copies of the same DNA sequence that lie side by side on the chromosome.

targeting sequence a stretch of polypeptide that determines the cellular compartment to which a synthesized protein is sent.

TATA box a sequence found about 20 bases upstream of the beginning of many eukaryotic genes that forms part of the promoter sequence and is involved in positioning RNA polymerase for correct initiation of transcription.

taxol compound obtained from the bark of the Pacific yew, *Taxus brevifolia*; binds to tubulin. Taxol is a powerful anti-cancer drug.

telomeres specialized regions at the ends of eukaryotic chromosomes. Telomeres are rich in minisatellite DNA.

telophase final period of mitosis or meiosis in which the chromosomes decondense and the nuclear envelope reforms.

telophase I telophase of the first meiotic division (meiosis I).

telophase II telophase of the second meiotic division (meiosis II).

temporal summation the phenomenon whereby although one presynaptic action potential does not generate a depolarization large enough to elicit an action potential in a postsynaptic cell, a series of action potentials at rapid succession in one presynaptic cell does depolarize the postsynaptic cell to threshold.

terminally differentiated term that describes a cell that cannot return to the cell-division cycle. Neurons are terminally differentiated; glial cells are not.

termination codon the codons UAA, UAG, and UGA are codons that signal protein synthesis to stop. Also known as stop codons.

terminator (of transcription) a DNA sequence that, when transcribed into mRNA, causes transcription to terminate.

tertiary structure the three-dimensional folding of a polypeptide chain into a biologically active protein molecule. It usually includes regions of secondary structure. Interactions of the amino acid side chains are central in its formation.

thick filament one of the two filaments that form the cytoskeleton of striated muscle; composed of the motor protein myosin II.

thin filament one of the two filaments that form the cytoskeleton of striated muscle; composed of actin.

thiol group the -S-H group.

threshold (voltage) the membrane voltage at which enough voltage-gated sodium (or sometimes calcium) channels open to initiate an action potential.

thymine one of the four bases found in DNA – thymine is a pyrimidine. It is replaced in RNA by uracil.

tight junctions type of cell junction in which a tight seal is formed between adjacent cells occluding the extracellular space.

tissue group of cells having a common function.

topoisomerases enzymes that cut and rejoin DNA strands. Topoisomerase I relieves simple torsional stress by cutting one strand, allowing rotation about the phosphodiester link of the other strand, then rejoining the cut strand. Topoisomerase II cuts both strands of a double helix and passes another complete double helix through the gap, keeping hold of the ends and rejoining them when the other strand has passed through. Topoisomerases are essential during DNA replication.

totipotent able to differentiate into (or divide to create cells that then differentiate into) any cell type of the organism. While all plant cells are totipotent, in animals only embryonic stem cells are naturally totipotent.

toxin poison.

transcribed, transcription synthesis of an RNA molecule from a DNA template.

transcription bubble structure formed when two strands of DNA separate and one acts as the template for synthesis of an RNA molecule.

transcription factor a protein (other than RNA polymerase) that is required for gene transcription.

transcriptome the entire complement of RNAs expressed in a cell or tissue.

trans face the side from which material is removed. Of the Golgi apparatus, the surface from which vesicles bud to pass to the plasma membrane and to lysosomes.

transfected, transfection a cell, prokaryotic or eukaryotic, that has been infected by a foreign DNA molecule(s) is said to be transfected. The process of DNA infection is called transfection.

transfer RNA (tRNA) the RNA molecule that carries an amino acid to an mRNA template.

transfer vesicle vesicles that carry cargo from one membrane-bound compartment to another.

transform, transformation in addition to its common English meaning, this is used in molecular genetics to mean introduction of foreign DNA into a cell.

transgenic animal an animal carrying a gene from another organism – the foreign gene is usually injected into the nucleus of a fertilized egg.

trans Golgi network the complex network of tubes and sheets that comprise the trans face of the Golgi apparatus. It is in the trans-Golgi network that proteins made on the rough endoplasmic reticulum are sorted as to their final destination.

translation the synthesis of a protein molecule from an mRNA template.

translocation active movement. Used generally of, for example, the movement of a newly synthesized protein to its final destination. When used of the ribosome, it means the movement, three nucleotides at a time, of the ribosome on the mRNA molecule.

transmembrane helix an α helix that crosses a biological membrane. The majority of the amino acid residues are hydrophobic and therefore sit comfortably in the interior of the membrane.

transmembrane proteins class of proteins that span the plasma membrane.

transmembrane translocation form of protein transport in which unfolded polypeptide chains are threaded across one or more membranes as a simple polypeptide chain and then (re)folded at their final destination.

transmission electron microscope type of microscope in which the image is formed by electrons that are transmitted through the specimen.

transmission light microscope a microscope in which visible light passes through the specimen and then forms a magnified image.

transmitter a chemical that is released by one cell and which changes the behavior of another cell.

trimeric formed of three subunits.

trimeric G protein protein with three subunits, α, β, and γ, where the α subunit is a GTPase that dissociates from the $\beta\gamma$ units when it is in its GTP-bound state. Both the α subunit, in its GTP-bound state, and the now independent $\beta\gamma$ subunit, can activate target proteins. Examples are G_s that activates adenylate cyclase and G_q that activates phospholipase Cβ.

triphosphate having a chain of three phosphate groups attached. Nucleoside triphosphates such as ATP are the most familiar examples.

trisphosphate having three phosphate groups attached at three different points on the molecule. Inositol trisphosphate and phosphatidylinositol trisphosphate are examples.

Trk a family of receptor tyrosine kinases that bind growth factors of the neurotrophin class.

tRNA (transfer RNA) the RNA molecule that carries an amino acid to an mRNA template.

troponin a calcium-binding protein found in muscle cells. When calcium binds to troponin, myosin is able to pull on the actin microfilament.

tryptophan (trp) operon a cluster of five bacterial genes involved in the synthesis of the amino acid tryptophan.

tubulin subunit protein of microtubules; exists as α-, ß-, and γ- isoforms.

tumor proliferative cell mass associated with many cancers.

tyrosine kinase an enzyme that phosphorylates tyrosine residues in proteins by transferring a phosphate group from ATP. With a few exceptions, protein kinases are either tyrosine kinases (which only phosphorylate on tyrosine) or serine–threonine kinases (which can phosphorylate on serine or on threonine).

ubiquitin, ubiquitinated a small protein of only 76 amino acids that acts as a flag, generally marking for destruction any protein to which it is attached.

ubiquitin ligase an enzyme that attaches ubiquitin to a second protein, usually marking that protein for destruction in the proteasome.

ultrastructure fine structure of the cell and its organelles revealed by electron microscopy.

unambiguous when a language or code is unambiguous, there is only ever one meaning. English is ambiguous: for example "carrier" can mean a protein that moves solute across a membrane, a person with one copy of a recessive gene, or indeed many other concepts we don't use in this book. In contrast the genetic code is unambiguous: each triplet corresponds to one amino acid or to STOP.

untranslated sequence sequence of bases in a mRNA molecule that does not code for protein. Untranslated regions are found at the 5′ and 3′ ends of an mRNA.

upstream a general term meaning the direction from which things have come. When applied to the DNA within and adjacent to a gene, it means lying on the side of the transcription start site that is not transcribed into RNA. When applied to signaling pathways it means opposite to the direction in which the signal travels, thus the insulin receptor is upstream of Akt.

uracil one of the four bases found in RNA – uracil is a pyrimidine.

uracil-DNA glycosidase a DNA repair enzyme that recognizes and removes uracil from DNA molecules.

urea $H_2N\text{-}(CO)\text{-}NH_2$, compound made in the liver that contains lots of nitrogen but is less toxic than ammonia. Urea is

chaotropic: at high concentration it reversibly denatures proteins.

Valium an anti-anxiety drug that increases the chance that the GABA receptor channel will open, allowing chloride ions to pass.

Van der Waals force weak close-range attraction between atoms.

vascular tissue blood vessels. The term is also used to describe the water transporting and support tissue of plants.

VDAC a channel found in the mitochondrial outer membrane. It is always open and allows all solutes of $M_r < 10\,000$ to pass. It is the eukaryote homolog of the bacterial protein porin. The name, VDAC, stands for voltage-dependent anion channel, a rather unfortunate name – it has a slight preference for anions over cations but basically allows anything to pass.

vector something that carries something else. The term is often used to describe a plasmid or bacteriophage that carries a foreign DNA molecule and is capable of independent replication within a bacterial cell.

vesicle a small, closed bag made of membrane.

vesicular trafficking the precisely controlled movement of vesicles between different organelles and/or the plasma membrane.

villin type of actin-binding protein that cross-links actin filaments.

villus (plural villi) fingerlike extension of an epithelial surface that increases the surface area.

vimentin protein that makes up the intermediate filaments in cells of mesenchymal origin such as fibroblasts.

virus a packaged fragment of DNA or RNA that uses the synthetic machinery of a host cell to replicate its component parts.

VNTRs (variable number tandem repeats) DNA sequences that occur many times within the human genome. Each person carries a different number of these repeats.

voltage clamp a technique in which the experimenter passes current to one side of a membrane to artificially set the value of the membrane voltage to a desired level.

voltage-gated calcium channel a channel that is selective for calcium ions and which opens upon depolarization. Found in the plasma membrane of many cells.

voltage-gated sodium channel a channel that is selective for sodium ions and which opens upon depolarization. Found in the plasma membrane of neurons and muscle cells.

von Hippel–Lindau tumor suppressor a ubiquitin ligase that ubiquitinylates hydroxylated hypoxia-inducible factor α, marking the latter for destruction in the proteasome.

Wee1 protein kinase that phosphorylates and hence inactivates CDK1.

western blotting a technique in which specific proteins on a gel are revealed by using antibodies.

Wnt Proteins protein transmitters that play key roles during development and in the adult body. Wnts often promote cell division but have other effects as well.

wobble (in tRNA binding) flexibility in the base pairing between the 5' position of the anticodon and the 3' position of the codon.

Xeroderma pigmentosum an inherited human disease caused by defective DNA repair enzymes – affected individuals are sensitive to ultraviolet light and contract skin cancer when exposed to sunlight.

YAC (yeast artificial chromosome) a cloning vector used to propagate DNAs, of about 500 000 bp, in yeast cells.

Z disc disc that is set within and at right angles to the actin microfilaments in striated muscle, holding them in a regularly spaced array.

zinc fingers a structural motif in some families of DNA-binding proteins in which a zinc ion coordinated by cysteines and/or histidines stabilizes protruding regions which touch the edges of the base pairs exposed in the major groove of DNA.

INDEX

Cell Biology: A Short Course, Fourth Edition. Stephen Bolsover, Andrea Townsend-Nicholson,
Greg FitzHarris, Elizabeth Shephard, Jeremy Hyams and Sandip Patel.
© 2022 John Wiley & Sons Ltd. Published 2022 by John Wiley & Sons Ltd.
Companion website: www.wiley.com/go/bolsover/cellbiology4